# Modern Masonry
## Brick, Block, Stone

by

**Clois E. Kicklighter**

Dean, School of Technology
and Professor of Construction Technology
Indiana State University
Terre Haute, Indiana

Publisher
**THE GOODHEART-WILLCOX COMPANY, INC.**
Tinley Park, Illinois

2

Library of Congress Catalog Card Number 96-22363
International Standard Book Number 1-56637-342-5

3 4 5 6 7 8 9 10  97  01 00 99

**Library of Congress Cataloging in Publication Data**

Kicklighter, Clois E.
Modern masonry: brick, block, stone / by Clois E. Kicklighter
   p. cm.
  Includes index.
  ISBN 1-56637-342-5
  1. Masonry. I. Title.
TH1199.K53 1996
693' .1—dc20
                                      96-22363
                                          CIP

Materials used for the cover courtesy of **Kimpling's Ace
Hardware**, Washington, Illinois; and **Builders Square**,
Homewood, Illinois.

# Introduction

***Modern Masonry—Brick, Block, Stone*** provides a thorough grounding in the methods of laying brick, block, and stone. The book presents a broad understanding of materials and their properties. Students in community colleges, trade schools, vocational and technical schools, or any individual training for the masonry trades, will find this title a source book of near-encyclopedic dimensions. It is equally suitable for apprenticeship training.

Simply and clearly written, this book covers all the following important aspects of the masonry trade:
- ❖ The makeup, properties, uses, and sizes of every type of masonry unit
- ❖ Uses, descriptions, and illustrations of tools and equipment
- ❖ Descriptions, illustrations, and uses of anchors, ties, and reinforcement
- ❖ Types of courses and bonds
- ❖ Properties and technical details of entire masonry systems such as foundations, floors, roofs, and walls
- ❖ Accepted techniques for laying all kinds of masonry units in all kinds of bonds
- ❖ Construction details for masonry walls, foundations, pavement, steps, garden walls, and masonry arches
- ❖ Concrete materials; procedures for placing and finishing concrete
- ❖ Concrete reinforcement and concrete design data
- ❖ Types of forms for concrete construction, their uses, and how to build them
- ❖ Math for masonry trades in US Customary and SI metric
- ❖ Print reading
- ❖ Methods of cleaning completed masonry structures
- ❖ Over 75 procedures for laying brick, block, and stone

The book includes the newest materials and building techniques used in the industry. The number of step-by-step procedures for laying brick and block walls of all types and setting stone in a number of applications has been greatly increased in this edition. New products such as insulated concrete wall forms, insulated blocks, and special masonry units have been included. The latest ASTM, ACI, OSHA, ASCE, NCMA, SBCCI, and BOCA standards related to concrete and masonry construction were consulted in the preparation of this revision. Reliance on the latest research from the ***Brick Institute of America***, ***Portland Cement Association***, and the ***Wire Reinforcement Institute*** adds to the validity of the content of this text

## Flexibility in Design

The text is organized into functional sections more closely related to the needs of an apprenticeship training program, community college program, or other related instruction. The coverage of topics is comprehensive and presented in a well-illustrated and interesting fashion.

Many new illustrations have been added to enhance the communication and clarity. Many of the hundreds of illustrations are color and show patterns and bonds used in various types of masonry construction. There are numerous individual color samples of brick, block, stone, terra cotta, and manufactured products in a variety of bonds. Line drawings and photographs illustrate methods of bricklaying and provide technical details of masonry structures.

## Enhancing the text

To help develop masonry skills, Goodheart-Willcox offers the **Modern Masonry Job Practice Manual,** which is designed to be used with the text, **Modern Masonry—Brick, Block, Stone**. The jobs and review questions are designed to help you become a professional brick mason, block mason, stone mason, or cement finisher. Every job is related to the skills and knowledge that professional masons must have to perform their duties. Most of the jobs are supported by additional information in the text in addition to the step-by-step procedures. A comprehensive list of the jobs covered in the manual is included in its contents section.

The jobs are action-oriented laboratory and field experiences that are designed to build skill and knowledge of accepted practices in the trade. They cover the areas specified in apprenticeship training programs. Each job is a standalone activity that can be undertaken in any preferred sequence, but they become more complex as the novice mason progresses through the jobs in the recommended sequence. Also, more detailed information is given in the beginning jobs than in later ones. Each job adds to the experience and lays the foundation for a more advanced experience in a related area. The **Modern Masonry Job Practice Manual** can be ordered directly from Goodheart-Willcox.

## Notices to users

When using the text you will see two notices throughout. These notices include Trade Tips and Warnings that help you through your development of masonry skills.

 **Trade Tips.** These ideas and suggestions are aimed at increasing your productivity and enhancing your use of masonry skills, techniques, and knowledge.

 **Warning!** These warnings alert you to potential problems. These warnings are either related to personal safety topics or to a situation where damage to equipment could be a result. If you are in doubt after reading a warning, always consult your instructor or supervisor.

## Reviewing the text

Each chapter includes a set of review questions at the end of the chapter. Questions require you to give the proper answers as they apply to the chapter material just studied. These questions are either true/false, multiple choice, fill-in-the-blank, or short answer. Do not write in this text. Write all answers on a separate sheet of paper.

## Additional resources

A reference section of more than 30 charts and drawings is located at the back of the text. This section includes useful information on such things as sizes of masonry units, mortar types and classes, metric conversions, an outline of a bricklayer's apprenticeship course, abbreviations used on prints, spans for concrete garden walls, and many other useful trade-related items.

The glossary entries are numerous and provide you with an easy access to a large resource. A complete index is included so you can easily locate important topics used throughout the text.

## In addition...

A new Chapter 18, *Job Performance, Leadership, Ethics, and Entrepreneurship,* has been added to encourage professionalism in the field. There are many factors that can affect progress on the job. Special care must be taken to address these so that a position will not be in jeopardy. These factors include job performance, ethics, teamwork and leadership, and entrepreneurship.

# Contents

## SECTION II
# Materials

## Chapter 5

Building Codes for Masonry Structures ❖ History and Development ❖ Structural Clay Products ❖ Manufacture of Brick ❖ Soft Mud Process ❖ Stiff Mud Process ❖ Dry-Press Process ❖ Drying and Burning ❖ Classification of Brick ❖ Building Brick ❖ Facing Brick ❖ Hollow Brick ❖ Paving Brick ❖ Ceramic Glazed Brick ❖ Thin Brick ❖ Sewer and Manhole Brick ❖ Brick Sizes and Nomenclature ❖ Why Modular Bricks Are Popular ❖ Color and Finishes ❖ Properties of Brick ❖ Solid Masonry Units ❖ Hollow Masonry Units ❖ Bonds and Patterns ❖ Structural Bonds ❖ Pattern Bonds ❖ Mortar Joints ❖ Special Shapes ❖ Hollow Masonry Units (Tile) ❖ Structural Clay Tile ❖ Structural Clay Facing Tile ❖ Architectural Terra Cotta ❖ Classification of Architectural Terra Cotta

## Chapter 6

How Concrete Masonry Units Are Made ❖ Aggregates ❖ Classification of Concrete Masonry Units ❖ Concrete Brick ❖ Slump Brick ❖ Concrete Block ❖ Calcium Silicate Face Brick (Sand-Lime Brick) ❖ Glass Block ❖ Metric Concrete Masonry Units

## Chapter 7

Classification of Stone ❖ Igneous Stone ❖ Sedimentary Stone ❖ Metamorphic Stone ❖ Grades of Stone ❖ Classification of Indiana Limestone ❖ Stone Surface Finishes ❖ Stone Wall Patterns ❖ Rubble ❖ Roughly Squared Stone ❖ Dimensioned or Ashlar Stone ❖ Stone Applications ❖ Stone Characteristics Summary ❖ Manufactured Stone

## Chapter 8

Mortar Materials ❖ Cementitious Materials ❖ Portland Cement ❖ Masonry Cements ❖ Hydrated Lime ❖ Sand (Aggregate) ❖ Water ❖ Measuring Mortar Materials ❖ Mixing Mortar ❖ Machine Mixing ❖ Hand Mixing ❖ Retempering Mortar ❖ Mortar Properties ❖ Properties of Plastic Mortar ❖ Properties of Hardened Mortar ❖ Mortar Proportions and Uses ❖ Type M Mortar ❖ Type S Mortar ❖ Type N Mortar ❖ Type O Mortar ❖ Type K Mortar ❖ Admixtures ❖ Specific Mortar Uses ❖ Coloring Mortar ❖ Cold Weather Mortar ❖ Recommended Practices ❖ Estimating Mortar Quantities ❖ Estimating Mortar for Concrete Block Wall ❖ Estimating Mortar for a Single Wythe Brick Wall ❖ Grout ❖ Codes, Specifications, and Requirements ❖ Admixtures ❖ Fluid Consistency ❖ Strength ❖ Mixing ❖ Placing Grout ❖ Curing Grout ❖ Sampling and Testing

## Chapter 9

Metal Ties or Wires ❖ Types of Ties ❖ Reinforcing Rods or Bars ❖ Anchors ❖ Joint Reinforcement

*SECTION* **III**
**Techniques** ▬▬▬▬▬▬▬▬▬▬

## Chapter 10

Basic Operations ❖ Spreading Mortar ❖ Holding a Brick ❖ Using Both Hands at Once ❖ Forming a Head Joint ❖ Cutting Brick ❖ Using a Mason's Line ❖ Related Tasks ❖ Laying Common Brick Walls ❖ Laying a Four Course, Single Wythe, Running Bond Lead ❖ Laying a 4" Running Bond Wall with Leads ❖ Laying an 8" Common Bond, Double Wythe Wall with Leads ❖ Laying An 8", Two Wythe Intersecting Brick Wall ❖ Laying a 12" Common Bond Solid Wall With Leads ❖ Corner Layout in Various Bonds ❖ Constructing a 10" Brick Masonry Cavity Wall ❖ Laying a 10" Brick Cavity Wall with Metal Ties ❖ Constructing a Single Wythe Brick Bearing Wall ❖ Advanced and Specialized Brickwork ❖ Corbelling a 12" Wall ❖ Hollow Brick Pier ❖ Cleaning New Masonry ❖ Cleaning Brick ❖ Acid Solutions ❖ Sandblasting

## Chapter 11

Types of Concrete Masonry Walls ❖ Solid Masonry Walls ❖ Hollow Masonry Walls ❖ Cavity Walls ❖ Composite Walls ❖ Veneered Walls ❖ Reinforced Concrete Masonry Walls ❖ Grouted Masonry Walls ❖ Basic Operations ❖ Spreading Mortar ❖ Handling Concrete Blocks ❖ Applying the Head Joint ❖ Cutting Block ❖ Using a Mason's Line ❖ Laying Concrete Block Walls ❖ Laying an 8" Running Bond Concrete Block Wall ❖ Control Joints in a Concrete Block Wall ❖ Wall Intersections ❖ Anchorage to Masonry Walls ❖ Laying a 10" Concrete Block Cavity Wall ❖ Laying an 8" Composite Wall with Concrete Block Backup ❖ Cleaning Concrete Block Masonry

## Chapter 12

Basic Operations ❖ Spreading Mortar ❖ Handling Stone ❖ Forming Joints ❖ Splitting, Shaping, and Cutting Stone ❖ Types of Stone Masonry Construction ❖ Rubble Masonry ❖ Ashlar Veneer ❖ Trimmings ❖ Setting a Rubble Stone Veneer Wall ❖ Building a Solid Stone Wall ❖ Limestone Panels ❖ Design and Construction Criteria ❖ Panel Sizes ❖ Anchors, Supports, and Embeds ❖ Mortars ❖ Pointing ❖ Cold Weather Setting ❖ Sealant Systems ❖ Joint Movement ❖ Expansion Joints ❖ Typical Cornice Detail ❖ Cleaning New Masonry ❖ Cleaning Stonework

## Chapter 13

Concrete and Masonry Foundation Systems ❖ Spread Foundations ❖ Foundation Walls ❖ Masonry Wall Systems ❖ Solid Masonry Walls ❖ Four-Inch RBM Curtain and Panel Walls ❖ Hollow Masonry Walls ❖ Anchored Veneered Walls ❖ Ties ❖ Composite Walls ❖ Resisting Moisture Condensation ❖ Reinforced Masonry Walls ❖ Grouted Masonry Walls ❖ Thin Brick Veneer ❖ Wall Openings ❖ Arches ❖ Window and Door Details ❖ Sills ❖ Movement Joints ❖ Brick Masonry Soffits ❖ Floors, Pavements, and Steps ❖ Fireplace and Chimney Construction ❖ Design and Construction ❖ Prefabricated Steel Heat Circulating Fireplaces ❖ Stone Quoins ❖ Garden Walls ❖ Straight Walls ❖ Pier and Panel Walls ❖ Serpentine Walls ❖ Caps and Copings ❖ Corbels and Racking ❖ Mortarless Retaining Walls ❖ Typical Guidelines for Installation

## SECTION IV
## Concrete

## SECTION V
## Succeeding on the Job

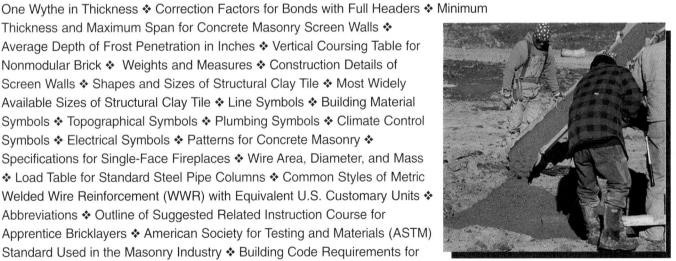

# Chapter 1
# Tools and Equipment

Tools and equipment and their safe use are a very important part of all trades. Every tool is designed for a specific purpose. Beginning masons must learn how to use each tool skillfully if they are to be successful at their trade.

## Masonry Tools and Equipment

This chapter describes the tools and equipment commonly used in the masonry trade and concrete work. It will show their intended use and how to use them safely.

### Trowels

The mason's trowel is the most used tool by a mason. The trowel has a blade and a handle. The blade is made from a flat piece of forged steel ground to the proper balance, taper, and shape. The narrow end of the blade is the point and the wide end nearest the handle is the heel. The blade is connected to the handle by the shank. The handle is either made of wood or plastic. Wooden handles usually have a metal band, called a *ferrule*, around the shank end. This adds strength and prevents the handle from splitting.

Before purchasing a trowel, you will need to consider the following factors: weight, size, materials, construction, and angle of the handle to the blade. A fine trowel should have a flexible blade of high-grade steel that will withstand long and hard use. A trowel should be light in weight and well balanced.

Trowels are available in lengths from about 2" to 9" and in widths from about 4 1/2" to 7". Mason's trowels are produced with wide, sharp heels or narrow, rounded heels. See Figure 1-1.

Special purpose trowels allow the mason to perform some specialized jobs better. These trowels are shown in Figure 1-2.

A trowel must be clean to perform properly. Old mortar should be removed from the blade and shank. Form the

**Figure 1-1.** Mason trowels are manufactured in several sizes and shapes. The handle may be wood or plastic.

habit of cleaning your tools at the end of each day. They will then be in good condition the next time you use them.

### Jointers

Jointers are used to finish the surface of mortar joints. See Figure 1-3. They are also called *joint tools* or *finishing tools*. Several types and sizes are commonly used. They are usually either cast or forged rods or stamped sheet metal. The cross-sectional shape of the jointer gives the mortar joint its shape.

Long horizontal joints are easily finished with a sled runner or joint runner. This tool has a handle and is shown in Figure 1-3.

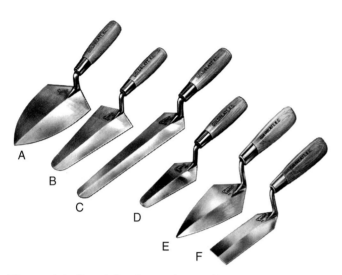

**Figure 1-2.** Specialized trowels used by masons. A—Buttering. B—Gauging. C—Duck bill. D—Cross joint. E—Margin. F—Pointing. (Stanley Goldblatt)

**Figure 1-3.** Jointers used by masons to finish mortar joints. Above—Sled runner jointers. Center—Forged heavy duty jointer. Below—Stamped jointer.

## Joint rakers

A *joint raker* is a tool used to remove a portion of the mortar from the joint just before the mortar hardens. The rake joint requires the use of this tool. Two styles of joint rakers are shown in Figure 1-4.

**Figure 1-4.** Two tools used to produce the rake joint. Above—Plain joint raker. Below—Skate wheel joint raker. (Stanley Goldblatt)

## Brick hammers

The **brick hammer** is a hammer that is frequently used to drive nails, strike chisels, and break or chip masonry materials. The head is flat on one side so that it may be used as a conventional hammer. The other side is drawn out to form a chisel for dressing up cuts. Figure 1-5 shows two styles of brick hammers. One has a wooden handle and the other is one-piece forged metal. Most brick hammers weigh from 12 ounces to 24 ounces.

**Figure 1-5.** Mason's brick hammer is designed to drive nails, strike chisels, and chip masonry units. Above—Wooden handle. Below—Metal handle with plastic grip.

## Rules

The mason usually has two kinds of rules—the folding rule and the retractable steel tape. See Figure 1-6.

1. A 6' folding rule sometimes with a 6" sliding scale on the first section for inside measurements.
2. A 10' retractable steel tape.

The 6' folding rule is a standard tool used by the mason. It usually has special markings on the backside representing course heights for various unit sizes and joint thicknesses. A well constructed rule of quality materials will have a long life and give dependable service.

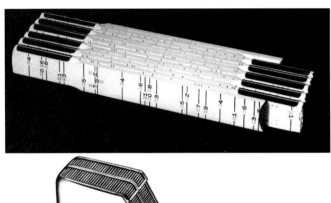

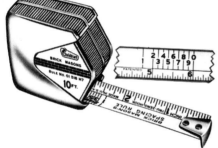

**Figure 1-6.** The 6" folding rule and retractable steel tape are standard measuring tools used by the mason. Numbers from 1 to 0 on inset represent height of various brick and mortar joints.

## Levels

The **mason's level,** or **plumb rule,** is a level that is constantly used to check the wall to be sure that it is built absolutely vertical and level. It is the most delicate piece of equipment the mason uses. See Figure 1-7.

Several spirit bubble vials are built into the level to permit plumb (vertical) and level (horizontal) use. The vials are generally embedded in a plaster of Paris setting. They are very accurately adjusted.

Masonry levels are usually 42" to 48" in length. A good level is lightweight and absolutely straight. It may be made from wood or metal or a combination of the two.

## Chisels

Different chisels are used by masons for different kinds of materials. A **brick set** is used to cut brick to exact dimensions. See Figure 1-8. A **blocking chisel** is used to cut concrete block. See Figure 1-9. They are made in a number of sizes, shapes, and weights.

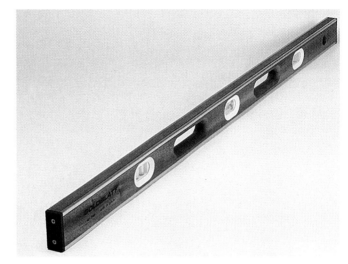

**Figure 1-7.** Common type of level used in masonry work. (Stanley Goldblatt)

**Figure 1-8.** Brick set is a type of chisel used to cut brick to specified length. Cutting edge is very blunt.

**Figure 1-9.** Blocking chisel is used to cut concrete block.

There are a variety of stone mason's chisel types. See Figure 1-10. They are used for scoring or splitting stone.

All chisels must be kept sharp and straight if they are to work properly. Burrs must be removed from the top end to prevent injury during use. Quality chisels are made from tempered tool steel. They are very hard and must be sharpened by grinding.

## Line and holders

The **mason's line** is a strong nylon or Dacron cord that is used to keep each course level and the wall true and out-of-wind (no bulges or hollows). See Figure 1-11. The mason's line is secured at either end with line holders or line pins.

Figure 1-12 shows a pair of adjustable line holders and braided nylon line. Line is available in strengths from 100 lb. to 350 lb. test and in white, yellow, and green. It is produced in lengths from 100' to 1000'.

**Figure 1-10.** Stone mason's chisels are used to split stone. (Stanley Goldblatt)

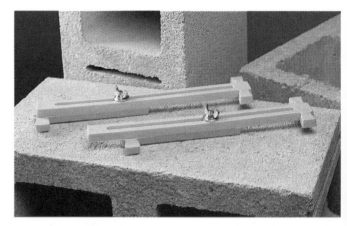

**Figure 1-11.** Nylon masons line on plastic winder in three highly visible colors. (Stanley Goldblatt)

**Figure 1-12.** Line and line holders are used to keep each course level and wall true and out-of-wind. (Stanley Goldblatt)

## Brushes

Brushes are produced in a variety of shapes and textures. See Figure 1-13. They are used for the three following purposes:

1. To remove mortar from the masonry units after the wall has been constructed.
2. To wash brick surfaces with muriatic acid.
3. For general cleaning.

## Brick tongs

**Brick tongs** are used to carry brick. See Figure 1-14. They are designed in such a way that they do not chip the brick. They are adjustable for various size units.

**Figure 1-13.** General purpose masonry brushes have fairly stiff bristles.

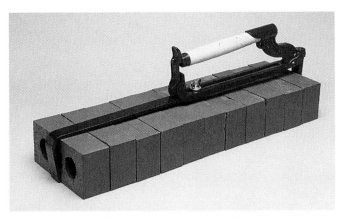

**Figure 6-14.** Brick tongs make carrying brick easier. (Stanley Goldblatt)

## Scaffolding

Few masonry projects are at a height where they can be reached while standing on the ground. For this reason, much of a mason's work is done on scaffolding. Scaffolding makes masons' work easier by providing a place for their materials and tools at a convenient height.

Several types of scaffolding are used by masons. They are the tubular, tower, and swing stage scaffolding. Figure 1-15 shows the tubular type most frequently used. An example of swing stage scaffolding is shown in Figure 1-16.

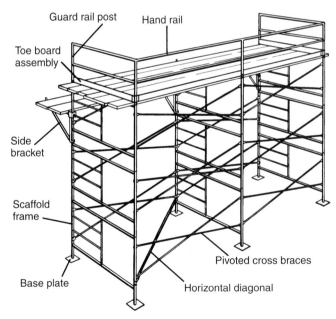

**Figure 1-15.** Tubular scaffolding is frequently used by masons to reach high places. Side brackets hold planking to provide deck for workers. Materials are never placed here but on upper deck.

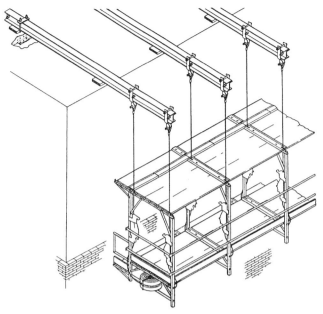

**Figure 1-16.** Swing stage scaffolding provides flexibility in height adjustment, a good smooth work surface, easy assembly, and top or side access for material staging. (Safway Steel Products)

## Mortar mixers

Mortar is usually prepared in a mechanical mixer similar to the one shown in Figure 1-17. Mixers may be electric or gasoline powered, depending on job conditions or preference. Mixers are produced in several sizes, but a typical size mixes about 4 cubic feet (cu. ft.) of mortar at a time. For best results, the dry materials should be mixed first. Then add water.

**Figure 1-17.** Mechanical mortar mixer is driven by a gasoline engine.

Mortar should not be allowed to harden in the mixer. This will prevent proper mixing of ingredients because the metal surfaces become mortar clogged.

# Cement Masonry Tools and Equipment

Most tools used in cement masonry are designed for use in finishing horizontal concrete surfaces. These tools are known as *flatwork finishing tools*.

## Screeds

A *screed* is a straightedge or strike-off rod made of any straight piece of wood or metal that has sufficient rigidity (stiffness). It is the first finishing tool used by the cement mason after the concrete is placed. It is used to strike off or screed the concrete surface to the proper level. Figure 1-18 shows a power screed which is used for large jobs.

**Figure 1-18.** Power screed is used to strike off concrete on larger jobs.

## Tampers

**Hand tampers** are produced in several basic designs and are used to compact the concrete into a dense mass. They are used on flatwork construction with low-slump concrete which is usually stiff and hard to work. Two styles of tampers are shown in Figure 1-19.

## Darbies

A *darby* is a long, flat, rectangular piece of wood, aluminum, or magnesium used to remove any high or low spots left by the screed. It is usually 30" to 80" long and

**Figure 1-19.** Two styles of hand tampers. (Marshalltown Trowel Co.)

from 3" to 4" wide with a handle on top as shown in Figure 1-20. It is used to float the surface of the concrete slab immediately after it has been screeded. It also helps to embed the course aggregate for later floating and troweling.

**Figure 1-20.** Darbies are used to float the surface of concrete immediately after it has been screeded. (Stanley Goldblatt)

## Bull floats

A *bull float* is a large, flat, rectangular piece of wood, aluminum, or magnesium, that does essentially the same job as the darby, but it enables the mason to float a much larger area. See Figure 1-21. It is usually 8" wide and 42" to 60" long with a long handle. It is particularly suited for outdoor use since the long handle (up to 16' long) is difficult to use inside.

## Edgers

*Edgers* are tools used to produce a radius on the edge of a slab. They are produced in several sizes and styles. See Figure 1-22. A typical length is 6" and widths vary from 1 1/2" to 4". Edgers have radii from 1/8" to 1 1/2". The radius improves the appearance of the slab and reduces the risk of damage to the edge. The curved-end edger is very popular.

**Figure 1-21.** Lightweight magnesium bull float. Such floats are also made from wood and aluminum. (Stanley Goldblatt)

## Jointers or groovers

*Jointers,* or *groovers,* are used to cut a joint partly through fresh concrete to control the location of any possible cracks. They are usually about 6" long and from 2" to 4 1/2" wide. Figure 1-23 shows two styles. The cutting edge or bits are available in depths from 3/16" to 3/4".

## Power joint cutters

Another method of cutting joints in concrete slabs is to use a power joint cutter. See Figure 1-24. This machine may be either electric or gasoline powered. It has a shatterproof abrasive or diamond blade which produces a narrow joint. The joint is usually cut when the concrete has hardened 4 to 12 hours.

**Figure 1-22.** Assortment of edgers used to produce a radius on edges of concrete slabs. (Stanley Goldblatt)

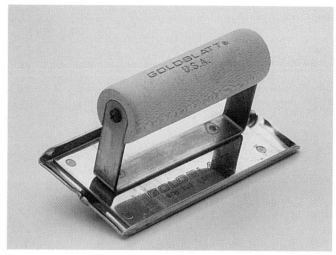

**Figure 1-23.** Two popular styles of jointers or groovers. Jointer is used to cut a joint partly through fresh concrete to control location of any possible cracks. (Stanley Goldblatt)

**Figure 1-25.** Hand floats are made from wood, aluminum, and cork as well as other materials.

**Figure 1-24.** Power joint cutter for large jobs must be handled carefully to avoid injury from flying particles.

## Hand and power floats

**Hand floats** are used to prepare the concrete surface for troweling. They are made from wood, aluminum, magnesium, cork, or molded rubber. See Figure 1-25. They range in size from 10" to 18" long and 3 1/2" to 4 1/2" wide.

Some floats are powered by electricity or gasoline engines. They have a rotating disk about 2' in diameter that performs the same task as the hand float.

## Hand and power trowels

A cement mason's steel hand trowel is the last tool used in the finishing process of a slab of concrete. It is available in many different sizes ranging from 10" to 20" long and 3" to 4 3/4" wide. See Figure 1-26. The first troweling of a slab is generally performed with a wide trowel 16" to 20" long. The last few troweling operations are usually done with a "fanning" trowel, which is 14" to 16" long and 3" to 4" wide.

A power trowel has three or four rotating steel trowel blades. See Figure 1-27. They may be powered by electricity or a gasoline engine. The purpose of troweling is to give the surface a smooth, dense finish.

## Masonry saws

Cutting masonry units with a power masonry saw is faster and more accurate than using the blocking chisel

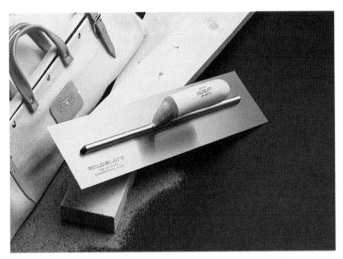

Figure 1-26. Cement mason's steel hand trowel. (Stanley Goldblatt)

Figure 1-28. Cutting masonry units with a power saw is faster and more accurate than using the blocking chisel.

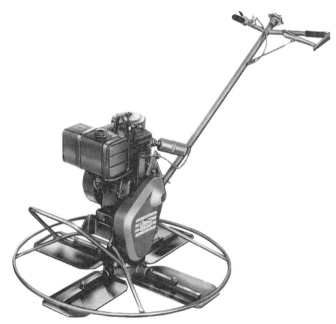

Figure 1-27. Power trowel used for large jobs has rotating steel blades.

and hammer. A typical power masonry saw is shown in Figure 1-28. The blade on a masonry saw is usually 6" or 7" in diameter and about 1/8" thick. It is made of very hard material such as silicon carbide or industrial diamonds.

 **Warning!** This is a dangerous piece of equipment to operate. Goggles should always be worn to protect the eyes from flying chips.

# REVIEW QUESTIONS
# CHAPTER 1

Write all answers on a separate sheet of paper. Do not write in this book.

1. The tool most frequently used by the mason is the _____.
2. Name five factors which should be considered when purchasing a mason's trowel.
3. Name three special purpose trowels.
4. A(n) _____ is used by masons to finish the surface of mortar joints.
5. A joint raker is used to make which of the following joints?
   A. Concave joint
   B. Weathered joint
   C. Struck joint
   D. Raked joint
6. The measuring tool used most frequently by the mason is the _____.
7. A mason uses a(n) _____ to check if a wall is plumb or level.
8. A mason laying concrete block would use a(n) _____ chisel to cut a concrete block.
9. Quality chisels are made from _____, which has been tempered.
10. A mason's _____ is used to keep each course level and the wall true and out-of-wind.
11. _____ helps make a mason's work easier by providing a place for materials and tools at a convenient height.
12. Most tools used in concrete masonry are designed for use in finishing horizontal concrete surfaces. True or False?
13. What is a screed?
14. What is the purpose of a tamper?

15. The tool used to float the surface of a concrete slab immediately after it has been screeded is a(n) _____.

16. A(n) _____ is used to produce a radius on the edge of a slab.

17. A(n) _____ is used to cut a joint partly through fresh concrete to control the location of any possible cracks.

18. The tool that is used for the final finishing process on a concrete slab is usually the _____.

19. What device is used for cutting masonry units more accurately than using the blocking chisel and hammer?

20. What piece of safety equipment should be worn when operating a power masonry saw?

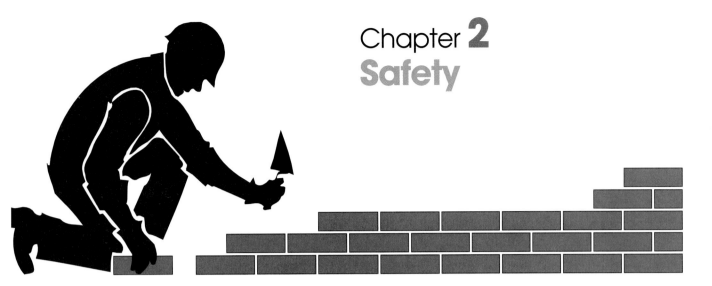

# Chapter 2
## Safety

Most occupations have some hazards. It is important to recognize them and have protection against them. A worker's personal safety and the safety of other workers is everyone's concern.

Accidents are quite frequent in the building industry. These accidents often result in lost time on the job, partial or total disability, or even loss of life. Accidents can be reduced if each person works safely and uses the precautions that their work requires.

## Job Site Safety

Construction is a dangerous business! In fact, construction is cited by the **Occupational Safety and Health Administration (OSHA)** as the most dangerous occupation in the United States. When all occupations are considered, an employee sustains an on-the-job injury every 18 seconds and a worker is killed every 47 minutes in the U.S. Job safety must, by necessity, be a primary concern for every employer and every employee.

Statistics show that 5% of all accidents are caused by unsafe conditions; the other 95% are caused by unsafe actions. Unsafe workers will find a way to injure themselves regardless of safety gear, safety rails, red warning signs, safety nets, guards, etc. The solution is to train workers to work safely. Working safely involves a thorough understanding of the tools, equipment, and materials they are working with. It also involves thinking safety.

Some of the techniques that can be employed to sharpen concern for safety on the job site include: testing the knowledge of workers to be sure they understand the proper use of their tools and machines, discussing safety procedures every few weeks, demonstrating safe work practices, developing a set of company safety rules, and enforcing safe work practices.

A safety program isn't something you can just write up and distribute; it is an ongoing effort between a contractor, his crew leaders, and the workers on the site. When an accident does happen, everyone should participate in a discussion about the particulars—how it happened and what could have prevented it from happening.

Some residential construction workers like to work in shorts and tennis shoes, but this is not safe. See Figure 2-1. In a recent survey of small contractors by the *Journal of Light Construction*, several unsafe practices on the job site were detailed. No less than 25% of those surveyed used no respiration protection at all when doing general demolition. When removing asbestos, 6% used nothing. When installing fiberglass insulation, 13% went without respiration protection. Only 5% wore goggles or safety glasses all the time they were on the job site. When questioned about using saw guards, 13% did not use guards on portable circular saws, 37% did not use guards on power miter boxes, 57% did not use guards on radial arm saws, and 72% did not use guards on table saws. 52% of the contractors reported that they used bounce-nailing when installing sheathing—a dangerous procedure. About 33% surveyed never used ear protection on the job, and 48%

**Figure 2-1.** Several safety practices are being violated by these workers—no hard hats, short pants, and a cluttered site are most obvious.

percent said they weren't sure if their scaffolding would pass OSHA inspection.

Finally, exactly 50% of the contractors reported that they never hold safety meetings or provide first aid instruction. But, 42% reported workers had experienced cuts and punctures, 18% reported an object in the eye, 18% reported sprains and muscle pulls, and 7% reported a back injury in the last three years. Clearly, safety is a problem in the construction industry.

# Safe Work Practices

Everyone on the job must know where accidents are likely to occur and how to prevent them. Above all, there must be a keen sense of responsibility for other workers. Learning safe work practices is just as much a part of learning a trade as using the tools.

Always follow these general safety precautions:

1. Be alert and spot accidents before they happen.
2. Follow safe work practices. Do not take short-cuts or expose yourself to danger unnecessarily.
3. Use the proper safeguards for a given job. Protective clothing, blade guard, or goggles could save a hand, an eye, or a foot.
4. Keep the work area clean. Many accidents are caused by litter underfoot.
5. Wear proper clothing and keep it in good repair. Shoes with steel toe protection are a must. See Figure 2-2. Never wear shoes with a loose sole. They can hang on a scaffold board and cause a fall. Trouser legs should not be too long because they may get caught on a projecting edge and cause an accident. A hard hat could save your life if a brick were to fall from above and strike your head.

6. Never engage in horseplay on the job.
7. Pick up your tools and store them properly. This should become a habit.
8. Lift heavy objects with your legs rather than your back. Keep the arms and back as straight as possible when lifting. If the object is too heavy, get help.
9. Never place articles on a ledge, ladder, or scaffold where they may fall and injure someone. Check scaffolds and ladders for loose articles before moving them.
10. Get first aid for all injuries immediately. Do not chance getting infection or aggravating a wound.
11. Tell your supervisor about unsafe conditions and violations of safety rules.
12. Do not use faulty tools and equipment. Repair or replace broken tools.
13. Do not count on "luck." Learn to work safely.

The concern for job safety prompted the Creation of OSHA, but they only set minimum standards. Some state requirements are more strict and should be reviewed regularly for the benefit of everyone on the job.

## *Proper dress*

Proper dress is an important part of safety on the job. The following general guidelines should be helpful in selecting and wearing proper dress for a given job.

1. Keep clothing in good repair.
2. Roll up or button shirtsleeves.
3. Tuck shirttails in the pants.
4. Turn pant cuffs down or wear pants without cuffs.
5. Do not wear short pants.
6. Protect the feet with steel-toed, thick soled shoes.
7. Wear a hard hat (required on all construction sites). See Figure 2-3.

**Figure 2-2.** These workers are wearing the proper footwear on the construction site.

**Figure 2-3.** These workers should be wearing hard hats to prevent injury from falling debris.

8. Wear eye protection—either goggles, safety glasses, or face shield when hazardous materials are present.
9. Wear gloves to protect the hands from abrasive materials and cold weather.
10. Long, loose hair is a hazard. Cut it or tuck it under your hard hat.

## Tool and equipment safety

Keeping tools in good repair and using them properly is probably the best way to avoid accidents with them. However, when working near others, remember that a trowel or chisel is sharp. Care must be taken not to injure anyone.

Electrical and mechanical equipment is widely used in the masonry trades. This equipment presents a special hazard because of moving parts. Safety is important when using any power equipment—mixer, masonry saw, or sandblaster. Follow these guidelines when using a power mixer:

1. Wear eye protection. See Figure 2-4.
2. Never allow the shovel to extend past the grate when adding ingredients.
3. Never reach inside the mixer for any reason when it is running.
4. Do not exceed the capacity of the machine.
5. Be sure the machine is in a stable position before operating.
6. Always clean the machine properly when you are finished.

**Figure 2-4.** Proper eye protection is necessary when using tools on the construction site. A hard hat is also required.

Using a power masonry saw decreases work time and produces a nice finish, but it can be a dangerous machine if it is not used properly. Observe the follow safety rules:

1. Wear eye protection.
2. Properly secure the masonry unit before cutting.
3. Maintain the blade and don't force the cut.
4. Never have your hands in the path of the blade.
5. Never operate the machine when someone is standing in the path of an object that might be thrown from the machine.
6. Follow the manufacturer's instructions for safe operation.

Sandblasting is sometimes used to clean new walls, even though other methods are preferred. When using a sandblasting machine, observe the following rules:

1. Wear eye protection.
2. Protect exposed skin, such as face and hands.
3. Never point the stream at someone else.
4. Be sure the pressure is properly adjusted.
5. Do not use a higher pressure than necessary to get the job done.
6. Use the proper abrasive for the particular job.
7. Be sure hoses are clear of walkways and work areas.
8. Properly secure the machine when you are finished with it.

Equipment should be serviced properly and checked for safe operation. Electricity is the power source for much of the mechanical equipment used on construction jobs. Electricity can be deadly if it is used improperly. Great care should be taken not to touch any bare wire.

 **Warning!** Be sure that all electrical equipment is grounded. Keep electrical wires off the ground and never operate electrical equipment in wet locations without proper grounding and safeguards.

If an electrical accident occurs, never touch a person that may have live current flowing through them. You may be severely injured or killed. If possible, shut off the power. If this cannot be done, use a dry piece of wood to break the contact. Before using any portable equipment, be sure to check for worn or defective insulation, loose or broken connections, and a bad ground wire connection.

## Handling materials

Your own safety, and the safety of others, depend upon proper handling of materials. Observe the following suggestions for handling materials:

1. When lifting heavy loads, lift with your legs, not your back. The proper procedure is to squat down and pick up the load by straightening your legs. Keep the back straight. If the load is too heavy, get help.

2.  Store materials such as brick, concrete blocks, and bags of cement or mortar on a paved surface, on boards, or on shipping paper.
3.  Be careful not to overload a wheelbarrow or power buggy. Place the load over the wheel, not toward the handles. Be sure the wheelbarrow or buggy is in proper working condition. Do not use equipment that has a cracked, loose, or broken handle; or an improperly inflated tire.
4.  Do not exceed the maximum height of 7'-0" for open, unsupported stacks of brick. A setback of 1" for every 1'-0" of height above 4'-0" is the recommended practice.
5.  Concrete block and brick should always be stacked in tiers that rest on a solid foundation. Tall stacks over 4'-0" high should be stepped back, braced, and propped.
6.  Store cut stone for sills and trim on level boards. Proper support of stone is necessary to prevent breakage. A cracked piece of stone could break off and become a hazard.
7.  When stacking bags of cement or mortar, the bags should be positioned with tops facing inward. Lay every other layer crosswise and do not exceed 10'-0" high. Be sure the stack is stable at all times.
8.  Keep materials clear of walkways, doors, and hoists. Also, keep materials at least 10'-0" away from the edge when they are above the first floor of a building.
9.  Protect masonry units from rain by covering with plastic or tarp.

## Safe use of scaffolds

Much of a mason's work is performed on some type of scaffolding. See Figure 2-5. The three main sources of injury associated with scaffolding are:
1.  Falling from the scaffold
2.  Being struck by tools or materials falling from the scaffold
3.  Faulty scaffolding

Scaffolding must be erected properly and designed to support the load it is expected to carry. The **National Safety Council** advises that scaffolding should be able to support four times the anticipated load of workers and their materials.

The **Scaffolding and Shoring Institute** has developed several safety rules that are directly related to safe use of scaffolding. See Figure 2-6.

**Figure 2-5.** This makeshift scaffold is unsafe because it is not properly constructed and has no guardrail.

## Safety nets

The safety net is a device that is used to catch a worker if they would happen to fall from a high place. Nets made from synthetic fibers are commonly used. They are placed in elevator shafts or between floors of high-rise construction. Be sure safety nets are properly positioned before working high above the ground.

## Enclosure safety

In many places, year-round construction is made possible by enclosing the structure. This is done so that work can continue during adverse weather conditions. A tubular scaffold enclosure can provide some protection from the weather. See Figure 2-7.

# SCAFFOLDING SAFETY RULES
*as Recommended by*
## SCAFFOLDING AND SHORING INSTITUTE
(SEE SEPARATE SHORING SAFETY RULES)

Following are some common sense rules designed to promote safety in the use of steel scaffolding. These rules are illustrative and suggestive only, and are intended to deal only with some of the many practices and conditions encountered in the use of scaffolding. The rules do not purport to be all-inclusive or to supplant or replace other additional safety and precautionary measures to cover usual or unusual conditions. They are not intended to conflict with, or supersede, any state, local, or federal statute or regulation; reference to such specific provisions should be made by the user. (See Rule II.)

I. **POST THESE SCAFFOLDING SAFETY RULES** in a conspicuous place and be sure that all persons who erect, dismantle or use scaffolding are aware of them.

II. **FOLLOW ALL STATE, LOCAL AND FEDERAL CODES, ORDINANCES AND REGULATIONS** pertaining to scaffolding.

III. **INSPECT ALL EQUIPMENT BEFORE USING**—Never use any equipment that is damaged or deteriorated in any way.

IV. **KEEP ALL EQUIPMENT IN GOOD REPAIR.** Avoid using rusted equipment—the strength of rusted equipment is not known.

V. **INSPECT ERECTED SCAFFOLDS REGULARLY** to be sure that they are maintained in safe condition.

VI. **CONSULT YOUR SCAFFOLDING SUPPLIER WHEN IN DOUBT**—scaffolding is his business, **NEVER TAKE CHANCES.**

---

A. **PROVIDE ADEQUATE SILLS** for scaffold posts and use base plates.

B. **USE ADJUSTING SCREWS** instead of blocking to adjust to uneven grade conditions.

C. **PLUMB AND LEVEL ALL SCAFFOLDS** as the erection proceeds. Do not force braces to fit—level the scaffold until the proper fit can be made easily.

D. **FASTEN ALL BRACES SECURELY.**

E. **DO NOT CLIMB CROSS BRACES.** An access (climbing) ladder, access steps, frame designed to be climbed or equivalent safe access to the scaffold shall be used.

F. **ON WALL SCAFFOLDS PLACE AND MAINTAIN ANCHORS** securely between structure and scaffold at least every 30' of length and 25' of height.

G. **WHEN SCAFFOLDS ARE TO BE PARTIALLY OR FULLY ENCLOSED,** specific precautions must be taken to assure frequency and adequacy of ties attaching the scaffolding to the building due to increased load conditions resulting from effects of wind and weather. The scaffolding components to which the ties are attached must also be checked for additional loads.

H. **FREE STANDING SCAFFOLD TOWERS MUST BE RESTRAINED FROM TIPPING** by guying or other means.

I. **EQUIP ALL PLANKED OR STAGED AREAS** with proper guardrails, midrails and toeboards along all open sides and ends of scaffold platforms.

J. **POWER LINES NEAR SCAFFOLDS** are dangerous—use caution and consult the power service company for advice.

K. **DO NOT USE** ladders or makeshift devices on top of scaffolds to increase the height.

L. **DO NOT OVERLOAD SCAFFOLDS.**

M. **PLANKING:**
1. Use only lumber that is properly inspected and graded as scaffold plank.
2. Planking shall have at least 12" of overlap and extend 6" beyond center of support, or be cleated at both ends to prevent sliding off supports.
3. Fabricated scaffold planks and platforms unless cleated or restrained by hooks shall extend over their end supports not less than 6" nor more than 12".
4. Secure plank to scaffold when necessary.

N. **FOR ROLLING SCAFFOLD THE FOLLOWING ADDITIONAL RULES APPLY:**
1. **DO NOT RIDE ROLLING SCAFFOLDS.**
2. **SECURE OR REMOVE ALL MATERIAL AND EQUIPMENT** from platform before moving scaffold.
3. **CASTER BRAKES MUST BE APPLIED** at all times when scaffolds are not being moved.
4. **CASTERS WITH PLAIN STEMS** shall be attached to the panel or adjustment screw by pins or other suitable means.
5. **DO NOT ATTEMPT TO MOVE A ROLLING SCAFFOLD WITHOUT SUFFICIENT HELP**—watch out for holes in floor and overhead obstructions.
6. **DO NOT EXTEND ADJUSTIG SCREWS ON ROLLING SCAFFOLDS MORE THAN 12".**
7. **USE HORIZONTAL DIAGONAL BRACING** near the bottom and at 20' intervals measured from the rolling surface.
8. **DO NOT USE BRACKETS ON ROLLING SCAFFOLDS** without consideration of overturning effect.
9. **THE WORKING PLATFORM HEIGHT OF A ROLLING SCAFFOLD** must not exceed four times the smallest base dimension unless guyed ot otherwise stabilized.

O. For **"PUTLOGS"** and **"TRUSSES"** the following additional rules apply.
1. **DO NOT CANTILEVER OR EXTEND PUTLOGS/TRUSSES** as side brackets without thorough consideration for loads to be applied.
2. **PUTLOGS/TRUSSES SHOULD EXTEND AT LEAST** 6" beyond point of support.
3. **PLACE PROPER BRACING BETWEEN PUTLOGS/TRUSSES** when the span of putlogs/truss is more than 12".

P. **ALL BRACKETS** shall be seated correctly with side brackets parallel to the frames and end brackets at 90° to the frames. Brackets shall not be bent or twisted form normal position. Brackets (except mobile brackets designed to carry materials) are to be used as work platforms only and shall not be used for storage of material or equipment.

Q. **ALL SCAFFOLDING ACCESSORIES** shall be used and installed in accordance with the manufacturers recommended procedure. Accessories shall not be altered in the field. Scaffolds, frames and their components, manufactured by different companies shall not be intermixed.

**Figure 2-6.** Scaffolding and shoring safety rules. (Scaffolding and Shoring Institute)

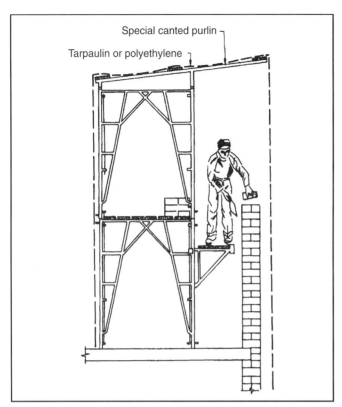

Special canted purlin

Tarpaulin or polyethylene

**Figure 2-7.** This type of tubular scaffold enclosure may be used to provide some protection during cold or rainy weather. (Brick Institute of America)

Enclosures can create a potentially dangerous condition for workers. The enclosures are generally heated with temporary heaters and may not be properly vented. Salamanders or gas fired heaters give off deadly carbon monoxide gas. This gas is odorless and lethal.

 **Warning!** Be sure that all heating devices are properly installed and vented. If the heating device is electric, make sure it is properly grounded.

## Ladder safety

Masons use ladders frequently in their work. Accidents are greatly reduced if a few common sense safety rules are used. Observe the following safety rules when using a ladder:

1. Be sure the ladder is in good repair. Cracked or broken rungs are very dangerous.
2. Report unsafe equipment and do not use it.
3. Keep ladders out of walkways and traffic lanes.
4. Place the ladder on a firm and level surface.
5. Avoid placing a ladder on slippery surfaces.
6. Be sure the ladder extends at least 3'-0" above the point where you plan to step off it.

7. The ladder should be placed against a strong support at a sufficient angle to prevent tipping backward.
8. The area around the bottom of the ladder must be clear of clutter.
9. Avoid aluminum ladders if possible, because they slip more easily than heavier wood ladders. Aluminum ladders also conduct electricity.
10. When climbing a ladder, transport tools in an over-the-shoulder bag.
11. Never stand on the highest rung of a ladder.
12. If there is a chance of the ladder slipping, have someone hold the bottom of the ladder.

## Safe use of chemicals

The first step when using any chemical is to read and follow the manufacturer's directions. More chemicals are being used by masons and some can cause serious burns or loss of sight. If you are not familiar with the use of a chemical, ask someone for proper handling instructions.

Wear the proper protective clothing when using chemicals. Rubber gloves and safety glasses are standard items. Have a bucket of water available to wash off any chemical that comes in contact with the skin. Know what to expect from the chemical that you use. Know the proper safety and clean-up procedures in case of a spill or other accident.

# First Aid

Treat and report all injuries immediately. If chemicals such as lime, cement, or cleaning agents get into your eyes, flush immediately with lots of clean water. Then see a doctor immediately. Burn first aid may be used for chemical burns before going to the doctor. Refer to a first aid manual for all first aid procedures.

Contact with wet (plastic) concrete, cement, mortar, grout, or cement mixtures can cause skin irritation, severe chemical burns, or serious eye damage. Wear waterproof gloves, a long-sleeved shirt, full-length trousers, and proper eye protection when working with these materials. If you must stand in wet concrete, wear high top waterproof boots. Wash wet concrete, mortar, grout, cement, or cement mixtures from your skin immediately. Seek medical attention if you experience a reaction when coming in contact with these materials.

Three good first aid rules to follow are:

1. Seek first aid immediately.
2. See a doctor for serious injuries or anything that deals with the eyes.
3. Report all accidents to the proper supervisor.

# REVIEW QUESTIONS
# CHAPTER 2

Write all answers on a separate sheet of paper. Do not write in this book.

1.  Learning safe work practices is just as much a part of learning a trade as using the tools. True or False?
2.  What is the most dangerous occupation in the U.S.?
3.  What percent of construction accidents are caused by unsafe actions?
4.  List three techniques that can be employed to sharpen concern for safety on the job site.
5.  Some residential construction workers like to work in shorts and tennis shoes, but this is not safe. True or False?
6.  Why is horseplay out of place on the job?
7.  A hard hat is required on all construction sites. True or False?
8.  There is no need to worry about a frayed electrical cord. True or False?
9.  Which of the following illustrates a safe work practice when using any power equipment?
    A.  Keep your eye protection handy.
    B.  Do not exceed the capacity of the machine.
    C.  Some guards may be removed because they are a nuisance.
    D.  Whether a machine is clean or dirty does not affect its performance.
10. List five common sense rules when sandblasting.
11. To prevent electrical shock, be sure that all electrical equipment is properly _____.
12. Describe the proper procedure for lifting a heavy load.
13. Where should the load in a wheelbarrow be placed?
14. What is the maximum height allowed for an open, unsupported stack of brick?
15. Scaffolding should be able to support _____ times the anticipated load of workers and materials.
16. What is the purpose of a safety net?
17. Gas-fired heaters give off deadly _____ gases.
18. To use a ladder safely, be sure that it extends at least _____ feet above the point when you plan to step off it.
19. What should be done before using chemicals?
20. What should you do if you get chemicals such as lime, cement, or cleaning agents in your eyes?

This house is a good example of stone work.

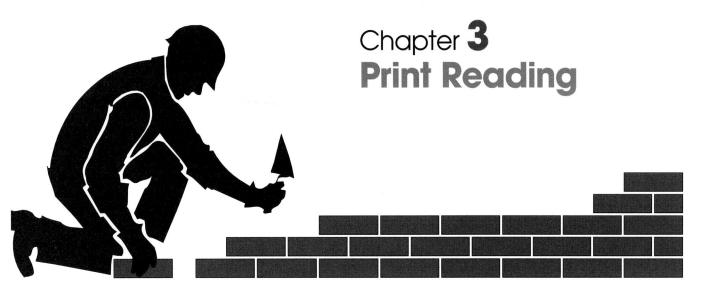

# Chapter 3
# Print Reading

*Construction drawings* describe the size, shape, location, and specifications of the elements of a structure. They are frequently called working drawings, prints, blueprints, or plans. A skilled worker must be able to read these drawings and understand the information contained in them. Otherwise, they could not build the structure as the designer intended. The drawings use standard symbols and notes that are recognized and understood by all the professionals associated with the construction field.

If you are to understand the drawings, you must visualize what the architect has drawn. You must know the meaning of the lines, symbols, abbreviations, and notes shown. Together, these items make up the language of the design and construction industry. Learning to read drawings is essentially learning a new language.

## Line Symbols

Lines make up a large part of the symbols used on construction drawings. The different types of lines are called the *Alphabet of Lines*. See Figure 3-1. A specific line symbol is used to communicate more precisely than just a plain line. It is important to learn the use of each line.

### Border lines

*Border lines* are very heavy, and are used to form a boundary for the drawing. They assure the reader that no part of the drawing has been removed. What is more, they give a "finished" appearance to the drawing.

### Object lines

*Object lines* are heavy lines that show the outline of the visible features of an object. They should be easily seen since they represent important elements. Such things as walls, windows, patios, and roof lines are represented by object lines.

### Hidden lines

*Hidden lines* are lines that represent an edge or intersection of two surfaces that are not visible in a given view. For example, the foundation wall and footings are represented by hidden lines on an elevation because they are below grade and, therefore, not visible. Also, hidden lines may be used to indicate features above the cutting plane, such as wall cabinets in a kitchen or an archway. Hidden lines are usually not as thick as object lines and are considered to be medium weight lines.

### Center lines

*Center lines* are thin lines that indicate the center of symmetrical objects. For example, a window or door may have a center line through it on the floor plan. Center lines simplify dimensioning and are used for location of features.

### Extension lines

*Extension lines* are thin lines used to show where a dimension line ends. They extend from a portion of the object past the dimension line about 1/16". They extend the object for dimensioning purposes.

### Dimension lines

*Dimension lines* are thin lines used to show size or location of a feature of the structure. They may be placed outside or, if there is sufficient space, inside the object. All dimension lines have a dimension figure (number or letter) about halfway between the ends. Each end has some type of termination (ending) symbol. See Figure 3-2.

### Long break lines

*Long break lines* are thin lines used to show that all of the part is not shown. They extend past the object about 1/16" on either side and have an *S* shape symbol in the center.

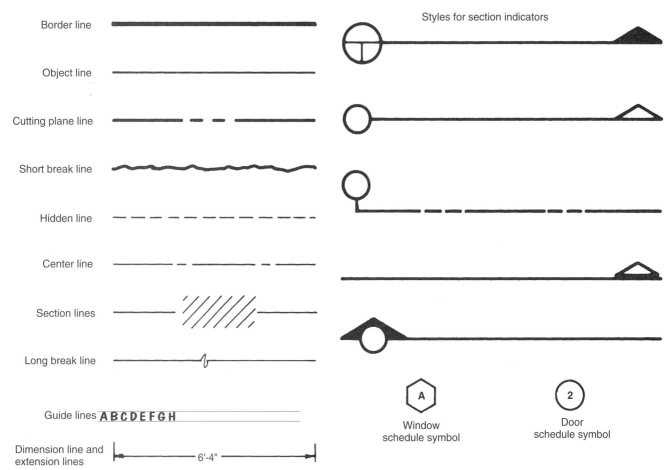

Border line

Object line

Cutting plane line

Short break line

Hidden line

Center line

Section lines

Long break line

Guide lines **ABCDEFGH**

Dimension line and extension lines — 6'-4" —

Styles for section indicators

A
Window schedule symbol

2
Door schedule symbol

**Figure 3-1.** Alphabet of Lines used on construction drawings.

**Figure 3-2.** Shown are the various methods of terminating a dimension line.

## Short break lines

**Short break lines** are heavy lines used when part of the object is shown broken away to reveal a hidden feature. They are drawn freehand.

## Cutting plane lines

**Cutting plane lines** are heavy lines used to indicate where the object has been sectioned to show internal features. They are generally labeled so that the proper section drawing will be identified for a specific cutting plane.

## Section lines

**Section lines,** or **crosshatch lines,** are very thin lines used to show that the feature has been sectioned. Section lines may represent a specific material or may be a general symbol. General section lines are usually drawn at 45° and about 1/16" to 1/8" apart.

## Guidelines

**Guidelines** are very light, thin lines used in lettering. They are for the drafter's use and help produce a neat clear drawing which is easy to read.

## Construction lines

**Construction lines** are also very light, thin lines which are drawn by the drafter in the process of making the drawing. They usually are not visible on a construction drawing.

The various line symbols are identified on a simple floor plan in Figure 3-3. Study these lines. They are the basic symbols used on all construction drawings.

# Symbols and Abbreviations

The purpose of symbols and abbreviations is to conserve space. Some symbols look very much like the actual feature they represent and many others do not. Many of these symbols are found on each set of construction drawings. Figure 3-4 shows:

1. Building material symbols
2. Topographical symbols
3. Plumbing symbols
4. Climate control symbols
5. Electrical symbols

Symbols may represent an elevation (front, side, or rear) or a section view of the material or feature. Elevation symbols are easiest to recognize because they are picture-like. Section symbols are not pictorial representations. They must be memorized or otherwise identified.

 **Trade Tip.** Some symbols are used to represent several different materials or conditions. Care must be exercised in reading them. For example, the face brick section symbol in Figure 3-4 is the same as the general material symbol. The sand symbol (a series of dots) is the same in elevation and section. It is exactly like the section of cut stone, plaster, and the elevation of cast concrete. Study the location of the symbol and type of drawing before deciding what the symbol represents.

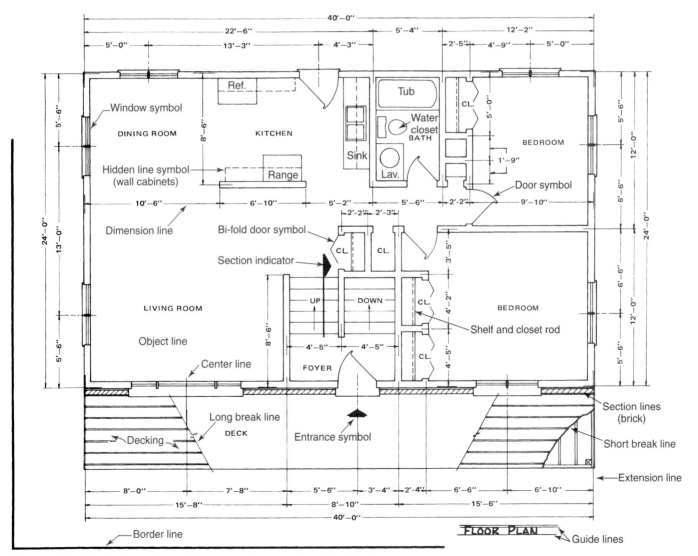

**Figure 3-3.** Simple floor plan with various line symbols identified.

Building Material Symbols

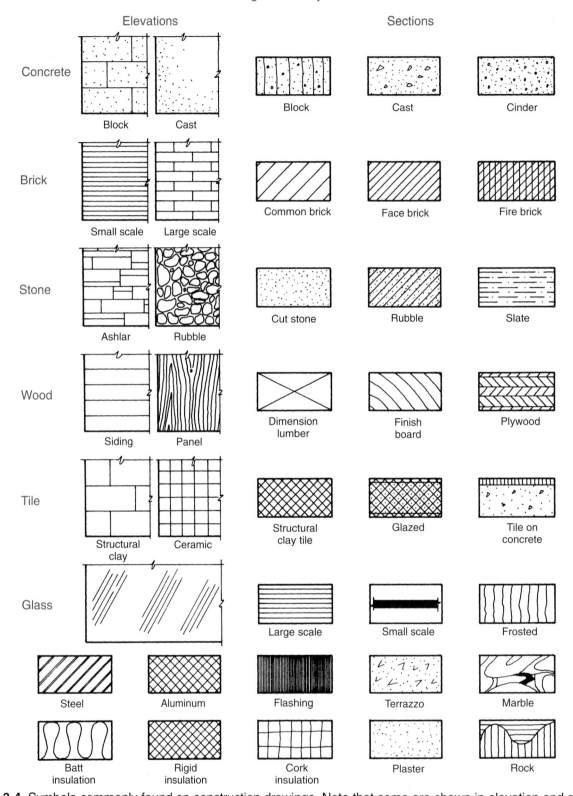

**Figure 3-4.** Symbols commonly found on construction drawings. Note that some are shown in elevation and section.

## Topographical Symbols

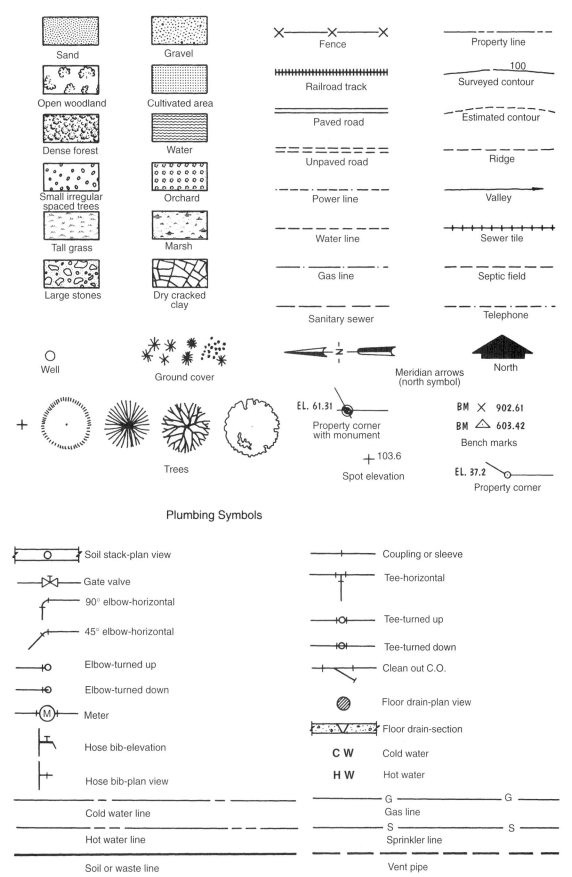

## Plumbing Symbols

**Figure 3-4 (Continued).** Symbols commonly found on construction drawings.

Climate Control Symbols

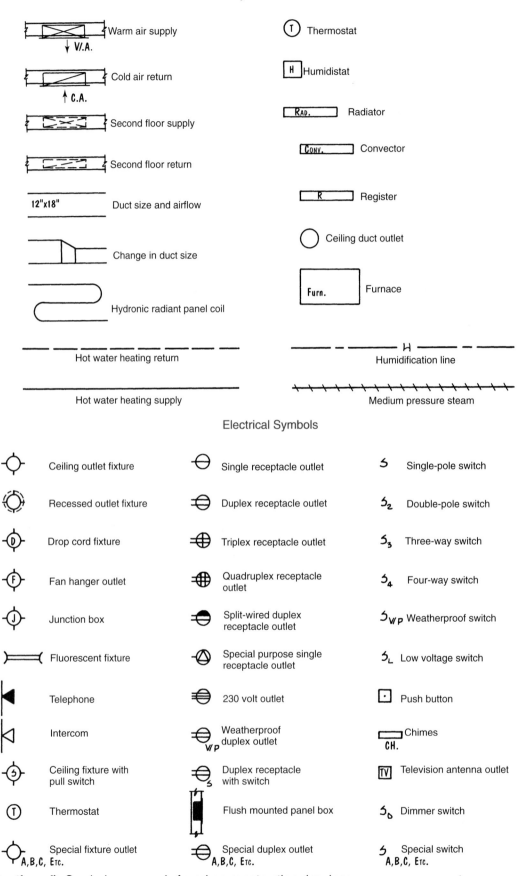

Electrical Symbols

**Figure 3-4 (Continued).** Symbols commonly found on construction drawings.

Abbreviations also save space and are widely used on construction drawings. A tradesperson must be able to read and interpret all the abbreviations that have anything to do with their specific function. The *Reference Section* of this book lists many frequently-used abbreviations.

## Scale and Dimensions

Most drawings are drawn at a particular scale. Residential floor plans, elevations, foundation plans, etc., are generally drawn at 1/4" = 1'-0" scale. This means that every 1/4" on the drawing is equal to 1'- 0" on the house. Details are almost always drawn at a larger scale such as 1/2" = 1'-0" or 1" = 1'-0". Do not be confused by the terms "size" and "scale." *Size* means inches to inches (1/4" to 1" or quarter size) and *scale* means inches to feet (1/4" to 1'-0" or quarter scale).

For example, if the side of a building were 40' long and drawn at 1/4 scale, it would be 10" on the drawing. However, if it were drawn at 1/4 size it would be 10' long.

Drawings of large commercial buildings are many times drawn at scales other than 1/4"= 1'-0". A scale of 1/8" = 1'-0" is common. Be sure to look for the scale on the drawing so that you can visualize the relative size of the building elements.

**Trade Tip.** As a rule, never measure a drawing to determine a length.

Look for the dimension. If the dimension is not given, then try to add or subtract other dimensions to determine the length you need. If you must measure the drawing, check it several places for accuracy.

*Dimensions* indicate size and location of building elements. They are very important and must be followed accurately. Dimensions are usually shown in feet and inches but they may be shown in inches if the length is less than 1' or, the length is some standard distance that is more easily recognized in inches.

Interior frame walls may be dimensioned several ways. See Figure 3-5. They may be dimensioned to the center of the wall, to the outside of the studs, or to the outside of the finished wall. The technique used may be easily determined by remembering that a 2 × 4 stud is actually 1 1/2" × 3 1/2". The thickness of the finished wall material can vary from 1/4" to 1" or more. Dimensioning to the center of the wall is the most commonly used method.

Interior masonry or concrete walls are dimensioned to the outside of the walls. See Figure 3-6. They are never dimensioned to the center.

Exterior walls, if they are frame, are dimensioned to the outside of the stud wall. This usually includes the sheathing but not the siding. See Figure 3-7. Again,

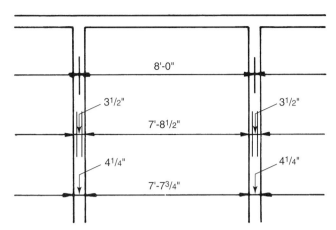

**Figure 3-5.** Accepted methods of dimensioning interior frame walls.

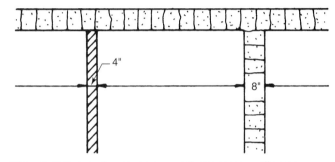

**Figure 3-6.** Interior masonry and concrete walls are dimensioned to the outside of the walls and wall thickness is also shown.

masonry walls are dimensioned to the outside of the wall. Brick veneer is dimensioned to the outside of the stud wall.

Notes are frequently required to present information that cannot be represented by a conventional dimension or symbol. Be sure to read all notes. They contain information that will be needed to build the structure.

**Trade Tip.** Study the entire set of plans before starting work on a building. If there are any questions or problems, these should be cleared up before work begins.

## Metric System of Dimensioning

Each year the metric system of measurement is used more world wide. It is anticipated that the metric system will eventually be the standard system of measurement. Even though no standards have been agreed upon in the lumber and building industry, a simple floor plan has been included to demonstrate the use of metrics in construction. See Figure 3-8.

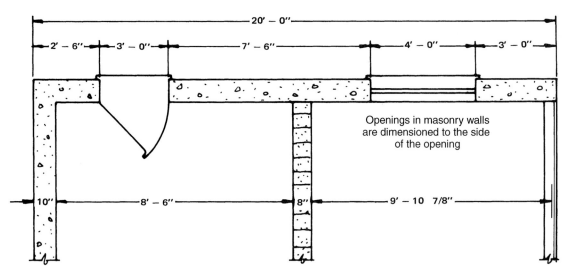

**Figure 3-7.** Exterior frame walls are dimensioned to outside of stud wall. Sheathing is usually included because it is attached before walls are lifted into place. Exterior masonry or concrete walls are dimensioned to the outside with thickness shown.

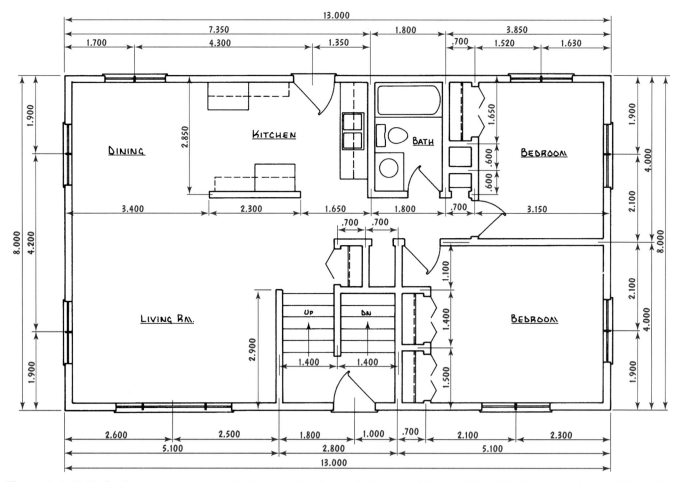

**Figure 3-8.** Units for linear measurement in the construction industry should be restricted to the meter (m) and the millimeter (mm). Thus on drawings, whole number dimensions will always indicate millimeters, and decimal numbers (to three places of decimals) will always indicate meters.

In construction, the basic metric unit of measure is the meter. However, before universal adoption of the system can be made by the building industry, new lumber sizes and standards must be approved. Clay products and concrete masonry products have made a "soft conversion." This means the actual size of the products have not changed, but metric sizes have been identified. Units sized to metric dimensions will most likely be adopted in the future. Presently, reinforcing steel has adopted a soft conversion, but work is progressing toward a hard conversion.

# Working Drawings

*Working drawings* are one group of drawings that are required to construct a building. *Preliminary drawings* are drawings that are often prepared during the promotional stage of the building's design. Corrections and revisions are recorded on the preliminary drawings. These drawings form the basis for the working drawings.

Presentation drawings are generally pictorial (picture-like) drawings. A pictorial drawing shows several faces of the structure to be built. It gives a better idea of how the structure will actually look.

There are several types of pictorial drawings. The one used most often by architects is called a *perspective drawing*. It is more natural in appearance—like a photograph. In a perspective you will notice that as parallel lines move away from the foreground they move closer together. If you study a photograph of a building the same thing happens.

There are many other types of working drawings needed to construct a building. Shop drawings are usually prepared by the trades participating in the project. They will use them to complete their part of the work.

Working drawings are generally prepared by the architect or designer. They are in graphic form showing the size, shape, location, quantity, and relationship of the parts of the building. They are produced in sufficient quantity so that each tradesperson has access to a set.

A set of working drawings for a typical structure might include the following:

1. Site plan
2. Foundation plan
3. Floor plan
4. Elevations
5. Electrical plan
6. Mechanical plan
7. Construction details

The plans for a small School Concessions building illustrate the type of information commonly shown on the various drawings. See Figures 3-9 through 3-19. Figure 3-9 shows a pictorial drawing of the School Concessions building. The following section briefly describes each drawing:

The *site plan* shown in Figure 3-10 shows the location of the building on the site. It also may show utilities, topographical features, site dimensions, other buildings on the property, landscaping, walks, drives, and retaining walls.

The *foundation plan* shown in Figure 3-11 shows the foundation size and materials. It may also give information about the excavation, waterproofing, and supporting structures such as footings and/or piles. Accuracy of this drawing is very important. Upper parts of the building will depend on the layout and construction of the foundation.

The *floor plan* shown in Figure 3-12 shows all exterior and interior walls, doors, windows, patios, walks, decks, fireplaces, built-in cabinets, and appliances. A separate plan view is drawn for each floor. A floor plan is a section drawing usually cut about 4' above the floor. Small buildings are usually drawn at 1/4" = 1'-0" scale, however, larger buildings may be drawn at 1/8" = 1'-0" or some other scale.

CONCESSION BUILDING

**Figure 3-9.** A pictorial view of a School Concessions building. Drawings on the following pages will be used to construct this building. (The InterDesign Group, Indianapolis, Indiana)

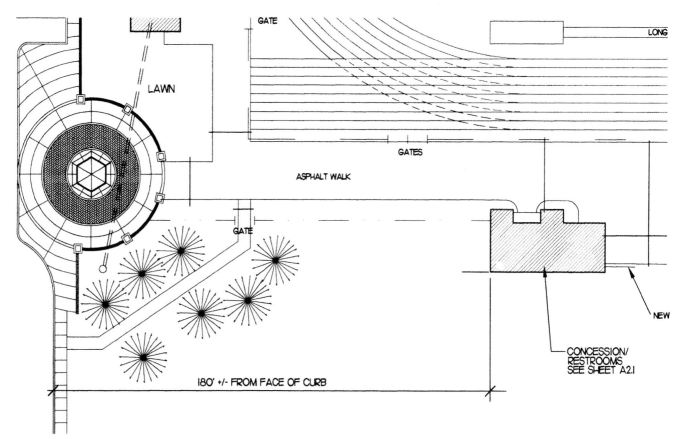

**Figure 3-10.** Site plan for the Concessions building. (The InterDesign Group, Indianapolis, Indiana)

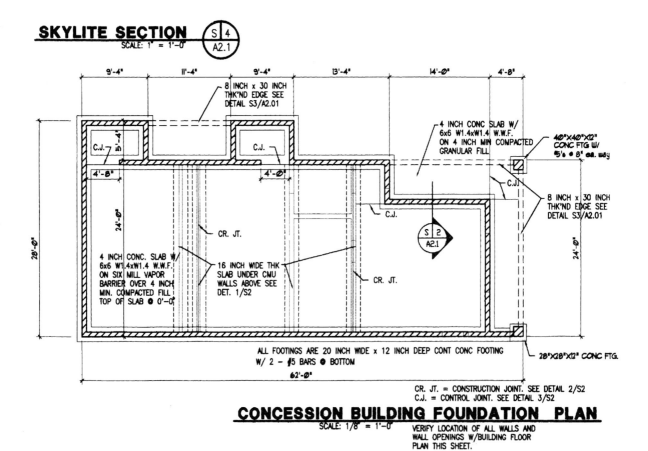

**Figure 3-11.** Foundation plan for the Concessions building. (The InterDesign Group, Indianapolis, Indiana)

The *elevations* shown in Figure 3-13 are drawn for each side of the structure. The drawings show outside features such as placement and height of windows, doors, chimney, and roof lines. Exterior materials are indicated as well as important vertical dimensions.

The *electrical plan* shown in Figure 3-14 is drawn from the floor plan. It locates switches, convenience outlets, ceiling outlet fixtures, TV jacks, service entrance location, and panel box. It also gives general information concerning circuits and special installations.

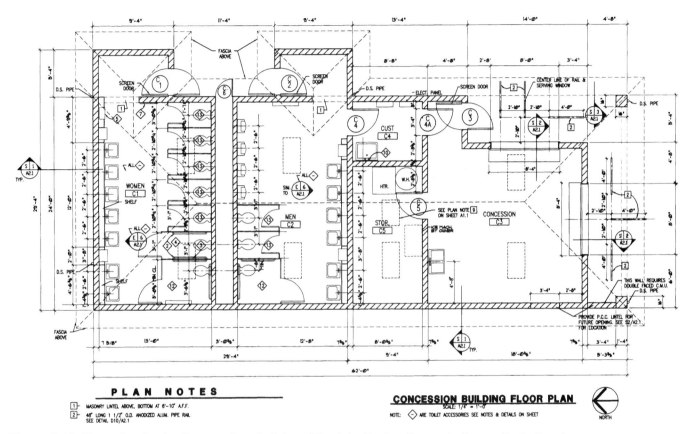

**Figure 3-12.** Floor plan for the Concessions building. (The InterDesign Group, Indianapolis, Indiana)

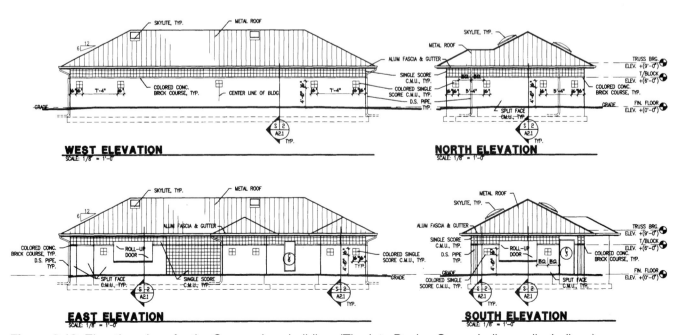

**Figure 3-13.** Elevation plans for the Concessions building. (The InterDesign Group, Indianapolis, Indiana)

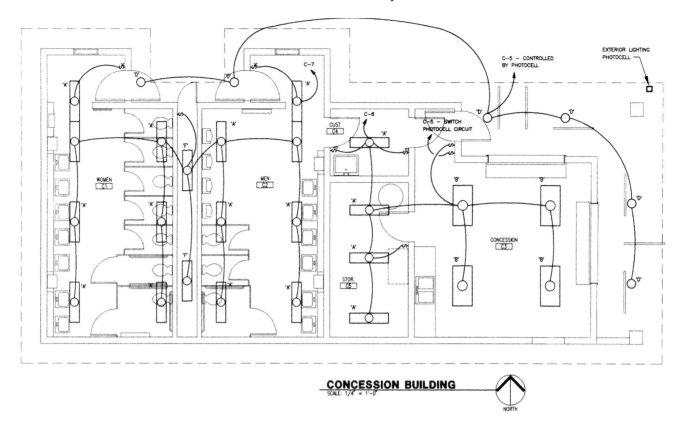

**CONCESSION BUILDING**
SCALE: 1/4" = 1'-0"
NORTH

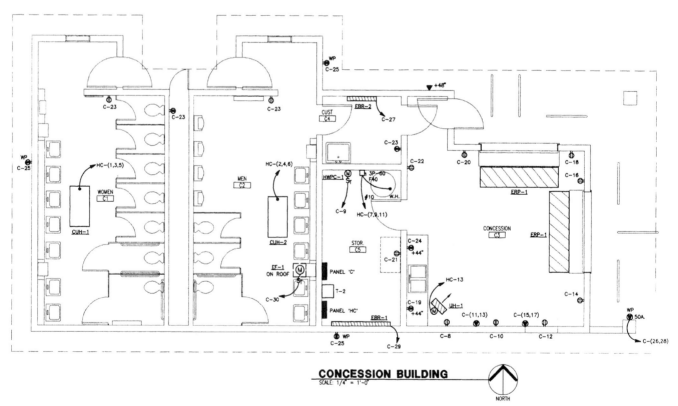

**CONCESSION BUILDING**
SCALE: 1/4" = 1'-0"
NORTH

**Figure 3-14.** Electrical and lighting plans for the Concessions building. (The InterDesign Group, Indianapolis, Indiana)

The **mechanical plan** shown in Figure 3-15 shows the plumbing, heating, and/or cooling systems. Sometimes each of these systems is shown on a separate plan. They would be designated as a **plumbing plan**, **heating plan**, and **cooling plan**. Heating, cooling, humidification, dehumidification, and air cleaning might be drawn on a single **climate control plan**.

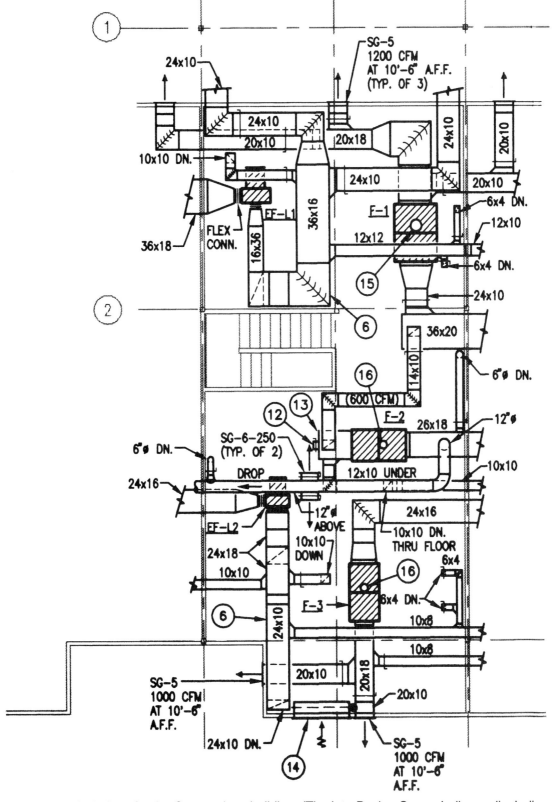

**Figure 3-15.** Mechanical plans for the Concessions building. (The InterDesign Group, Indianapolis, Indiana)

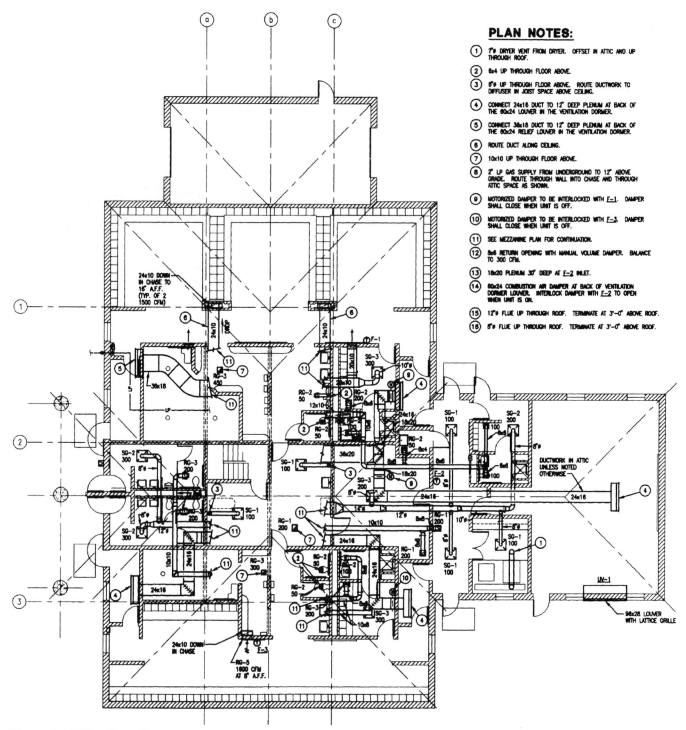

Figure 3-15 (Continued).

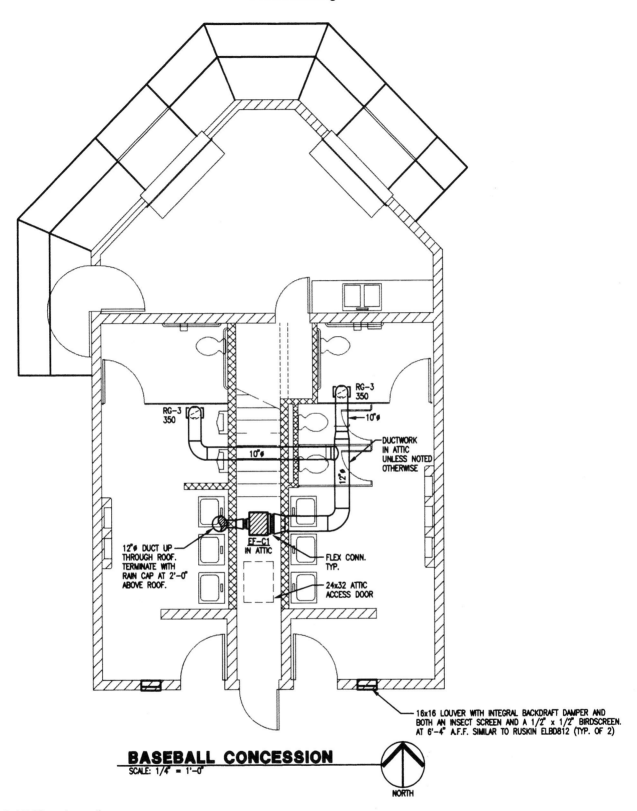

RG-3
350

RG-3
350

←10"∅

RG-3
350

10"∅

DUCTWORK
IN ATTIC
UNLESS NOTED
OTHERWISE

12"∅

12"∅ DUCT UP
THROUGH ROOF.
TERMINATE WITH
RAIN CAP AT 2'-0"
ABOVE ROOF.

EF-C1
IN ATTIC

FLEX CONN.
TYP.

24x32 ATTIC
ACCESS DOOR

16x16 LOUVER WITH INTEGRAL BACKDRAFT DAMPER AND
BOTH AN INSECT SCREEN AND A 1/2" x 1/2" BIRDSCREEN.
AT 6'-4" A.F.F. SIMILAR TO RUSKIN ELBD812 (TYP. OF 2)

## BASEBALL CONCESSION
SCALE: 1/4" = 1'-0"

NORTH

**Figure 3-15 (Continued).**

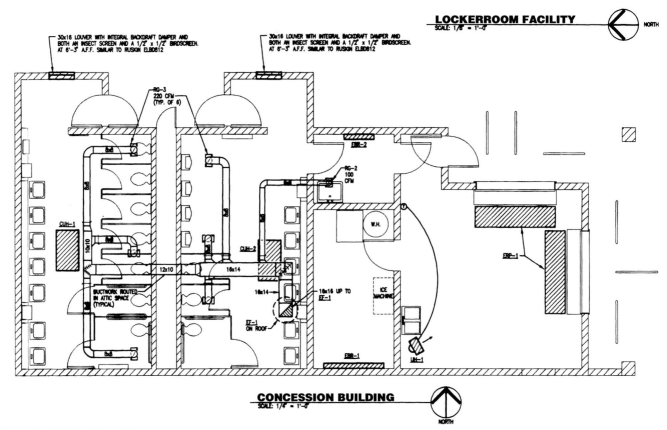

**Figure 3-15 (Continued).**

*Construction details*, as shown in Figure 3-16, are generally drawn of features when more information is needed on how to build them. Typical details include wall sections, sections through the entire structure, stairs, fireplaces, special masonry, or other unique construction.

In addition to these drawings just described, a roof plan, framing plan, landscaping plan, presentation plan, or shop drawings may be needed. A complex structure usually requires more detailed drawings.

A *roof plan,* as shown in Figure 3-17, is included if the roof is complex or not clearly shown on other drawings. It may be included in the site plan.

A *framing plan,* as shown in Figure 3-18, is designed to show, in detail, the framing required for a roof, floor, or other framed area

A *landscaping plan* is sometimes combined with the site plan. It locates and identifies plants and other elements included in landscaping.

A *presentation drawing* shows how the finished structure will appear. The two-point perspective is commonly used. Refer to Figure 3-9.

*Shop drawings* provide additional information about such things as reinforcing steel, complex cabinetwork, and electronic systems. This information may be furnished by suppliers or the various trades that have a part in the construction or installation.

*Drawing notes* are frequently included in a set of construction drawings to clarify details that might otherwise cause confusion. Figure 3-19 shows the drawing notes associated with the School Concessions building.

# Specifications

Each set of drawings is ordinarily accompanied by a set of written specifications. They are called "specs" in the trades. A set of specifications may be only a few pages long or several large volumes if the structure is large and complex.

*Specifications* are written descriptions of the construction project, and are prepared by the architect and/or engineer. The specifications, together with the working drawings, become the *construction documents* upon which an agreement is made between all parties concerned in the project.

Many aspects of a building cannot be adequately shown or explained on the drawings. Such things must be included in the specifications. For example, quality of workmanship, responsibility for various aspects of the construction, and quality of materials and fixtures are explained in the specifications.

Information in the specifications is meant to amplify (enlarge) the drawings. They should agree with the drawings point-by-point. Should a discrepancy occur between the specifications and the drawings, then the specs usually are accepted (take precedence) over the drawings.

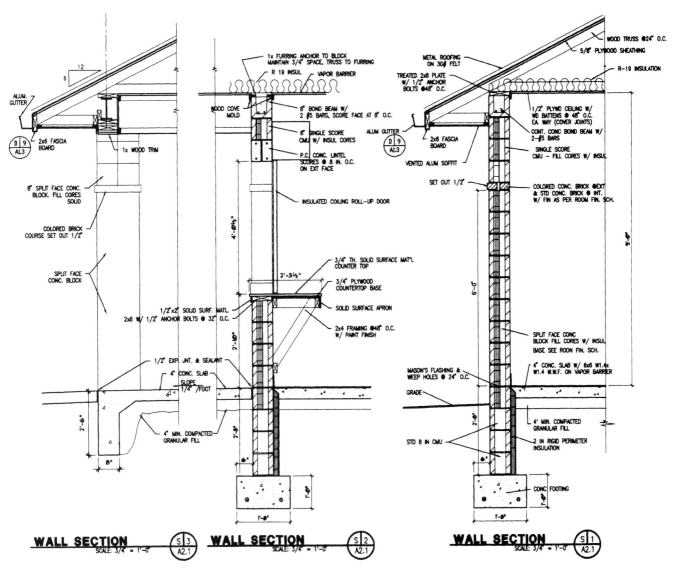

**Figure 3-16.** Construction details of wall sections; skylight sections, railing details, and soffit details. (The InterDesign Group, Indianapolis, Indiana)

Specifications are generally divided into major sections:

1. General conditions
2. Technical sections

The **general conditions** are specifications that deal with such things as terms of the agreement, insurance, responsibility for permits, supervision, payment for the work, and temporary utilities.

The **technical sections** are specifications that deal with the actual building of the structure and the tradework involved in the building. Each section is designed to include information on work to be done by a single subcontractor. Also, the arrangement of the technical sections is roughly similar to the sequence of work on the building. For example, the **excavation section** would be placed before the **finished carpentry** and **millwork section**. The following outline of headings is typical of smaller jobs such as a residence or small commercial building:

1. General Conditions
2. Excavation and Grading
3. Concrete and Masonry Work
4. Waterproofing
5. Miscellaneous Metals
6. Rough Carpentry

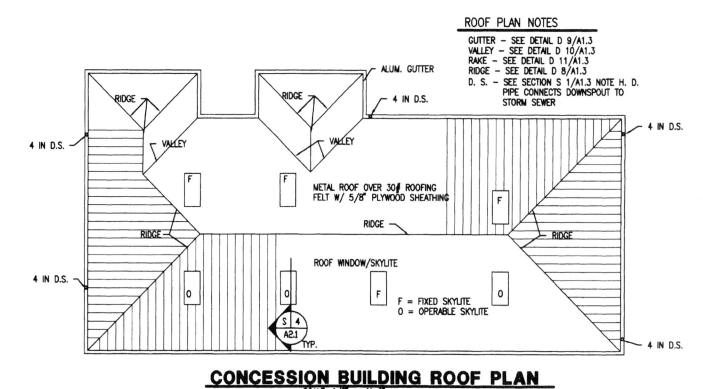

**Figure 3-17.** Roof plan showing ridges, valleys, and skylights. (The InterDesign Group, Indianapolis, Indiana)

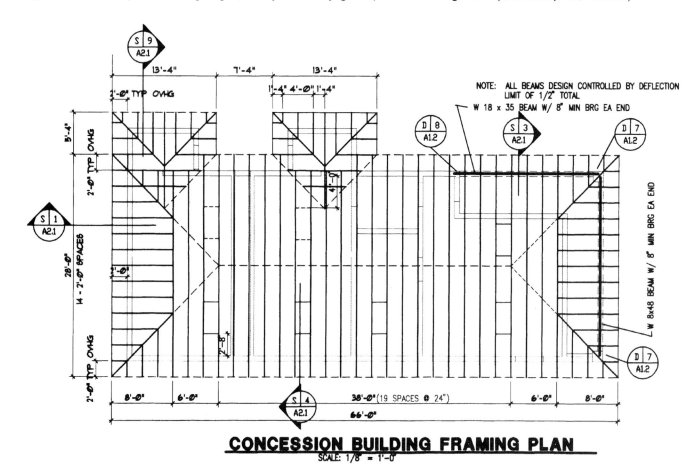

**Figure 3-18.** Framing plan including beam sizes and locations, trusses, section locations, notes and scale. (The InterDesign Group, Indianapolis, Indiana)

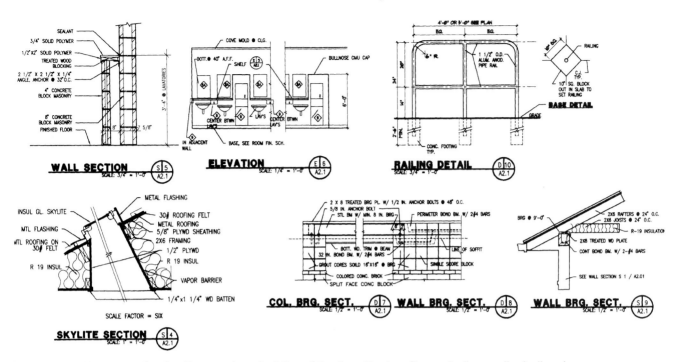

**Figure 3-19.** Plan notes for the Concessions building. (The InterDesign Group, Indianapolis, Indiana)

7. Finish Carpentry and Millwork
8. Roofing and Sheetmetal
9. Weatherstripping, Insulation, Caulking, and Glazing
10. Plastering or Drywall Construction
11. Ceramic Tile
12. Flooring
13. Painting and Decorating
14. Plumbing
15. Mechanical Equipment
16. Electrical
17. Landscaping

A standard preprinted specification form having fill-in blanks to be completed for a specific job is shown in Figure 3-20. This form is used by the **Federal Housing Administration** (**FHA**) and the **Veterans Administration** (**VA**). It includes most of the areas required for small buildings.

○               ○

FHA Form 2005
VA Form 26-1852
Rev. 3/68

For accurate register of carbon copies, form
may be separated along above fold. Staple
completed sheets together in original order.

Form approved.
Budget Bureau No. 63-R055.11.

# DESCRIPTION OF MATERIALS

No. _____
(To be inserted by FHA or VA)

□ Proposed Construction

□ Under Construction

Property address _____ City _____ State _____

Mortgagor or Sponsor _____ _____
                                  (Name)                             (Address)

Contractor or Builder _____ _____
                                 (Name)                             (Address)

## INSTRUCTIONS

1. For additional information on how this form is to be submitted, number of copies, etc., see the instructions applicable to the FHA Application for Mortgage Insurance or VA Request for Determination of Reasonable Value, as the case may be.

2. Describe all materials and equipment to be used, whether or not shown on the drawings, by marking an X in each appropriate check-box and entering the information called for in each space. If space is inadequate, enter "See misc." and describe under item 27 or on an attached sheet.

3. Work not specifically described or shown will not be considered unless required, then the minimum acceptable will be assumed. Work exceeding minimum requirements cannot be considered unless specifically described.

4. Include no alternates, "or equal" phrases, or contradictory items. (Consideration of a request for acceptance of substitute materials or equipment is not thereby precluded.)

5. Include signatures required at the end of this form.

6. The construction shall be completed in compliance with the related drawings and specifications, as amended during processing. The specifications include this Description of Materials and the applicable Minimum Construction Requirements.

**1. EXCAVATION:**

Bearing soil, type _____

**2. FOUNDATIONS:**

Footings: concrete mix _____; strength psi _____ Reinforcing _____

Foundation wall: material _____ Reinforcing _____

Interior foundation wall: material _____ Party foundation wall _____

Columns: material and sizes _____ Piers: material and reinforcing _____

Girders: material and sizes _____ Sills: material _____

Basement entrance areaway _____ Window areaways _____

Waterproofing _____ Footing drains _____

Termite protection _____

Basementless space: ground cover _____; insulation _____; foundation vents_____

Special foundations _____

Additional information: _____

**3. CHIMNEYS:**

Material _____ Prefabricated (make and size) _____

Flue lining: material _____ Heater flue size _____ Fireplace flue size _____

Vents (material and size): gas or oil heater _____; water heater _____

Additional information: _____

**4. FIREPLACES:**

Type: □ solid fuel; □ gas-burning; □ circulator (make and size) _____ Ash dump and clean-out _____

Fireplace: facing _____; lining _____; hearth _____; mantle _____

Additional information: _____

**5. EXTERIOR WALLS:**

Wood frame: wood grade, and species _____ □ Corner bracing. Building paper or felt _____

    Sheathing _____; thickness _____; width _____; □ solid; □ spaced _____" o. c.; □ diagonal; _____

    Siding _____; grade _____; type _____; size _____; exposure _____"; fastening _____

    Shingles _____; grade _____; type _____; size _____; exposure _____"; fastening _____

    Stucco _____; thickness _____"; Lath _____; weight _____ lb.

    Masonry veneer _____ Sills _____ Lintels _____ Base flashing _____

Masonry: □ solid □ faced □ stuccoed; total wall thickness _____"; facing thickness _____"; facing material _____

    Backup material _____; thickness _____"; bonding _____

    Door sills _____ Window sills _____ Lintels _____ Base flashing _____

    Interior surfaces: dampproofing, _____ coats of _____; furring _____

Additional information: _____

Exterior painting: material _____; number of coats _____

Gable wall construction: □ same as main walls; □ other construction _____

**6. FLOOR FRAMING:**

Joists: wood, grade, and species _____; other _____; bridging _____; anchors _____

Concrete slab: □ basement floor; □ first floor; □ ground supported; □ self-supporting; mix _____; thickness _____";

    reinforcing _____; insulation _____; membrane _____

Fill under slab: material _____; thickness _____". Additional information: _____

**7. SUBFLOORING:** (Describe underflooring for special floors under item 21.)

Material: grade and species _____; size _____; type _____

Laid: □ first floor; □ second floor; □ attic _____ sq. ft.; □ diagonal; □ right angles. Additional information: _____

**8. FINISH FLOORING:** (Wood only. Describe other finish flooring under item 21.)

| LOCATION | ROOMS | GRADE | SPECIES | THICKNESS | WIDTH | BLDG. PAPER | FINISH |
|---|---|---|---|---|---|---|---|
| First floor | | | | | | | |
| Second floor | | | | | | | |
| Attic floor | sq. ft. | | | | | | |
| Additional information: | | | | | | | |

FHA Form 2005
VA Form 26-1852

1

DESCRIPTION OF MATERIALS

**Figure 3-20.** Specification forms used by Federal Housing Administration and Veterans Administration.

DESCRIPTION OF MATERIALS

**9. PARTITION FRAMING:**
Studs: wood, grade, and species _____ size and spacing _____ Other _____
Additional information: _____

**10. CEILING FRAMING:**
Joists: wood, grade, and species _____ Other _____ Bridging _____
Additional information: _____

**11. ROOF FRAMING:**
Rafters: wood, grade, and species _____ Roof trusses (see detail): grade and species _____
Additional information: _____

**12. ROOFING:**
Sheathing: wood, grade, and species _____ ; ☐ solid; ☐ spaced _____" o.c.
Roofing _____ ; grade _____ ; size _____ ; type _____
Underlay _____ ; weight or thickness _____ ; size _____ ; fastening _____
Built-up roofing _____ ; number of plies _____ ; surfacing material _____
Flashing: material _____ ; gage or weight _____ ; ☐ gravel stops; ☐ snow guards
Additional information: _____

**13. GUTTERS AND DOWNSPOUTS:**
Gutters: material _____ ; gage or weight _____ ; size _____ ; shape _____
Downspouts: material _____ ; gage or weight _____ ; size _____ ; shape _____ ; number _____
Downspouts connected to: ☐ Storm sewer; ☐ sanitary sewer; ☐ dry-well. ☐ Splash blocks: material and size _____
Additional information: _____

**14. LATH AND PLASTER:**
Lath ☐ walls, ☐ ceilings: material _____ ; weight or thickness _____ Plaster: coats _____ ; finish _____
Dry-wall ☐ walls, ☐ ceilings: material _____ ; thickness _____ ; finish _____ ;
Joint treatment _____

**15. DECORATING:** *(Paint, wallpaper, etc.)*

| ROOMS | WALL FINISH MATERIAL AND APPLICATION | CEILING FINISH MATERIAL AND APPLICATION |
|---|---|---|
| Kitchen _____ | | |
| Bath _____ | | |
| Other _____ | | |

Additional information: _____

**16. INTERIOR DOORS AND TRIM:**
Doors: type _____ ; material _____ ; thickness _____
Door trim: type _____ ; material _____ Base: type _____ ; material _____ ; size _____
Finish: doors _____ ; trim _____
Other trim *(item, type and location)* _____
Additional information: _____

**17. WINDOWS:**
Windows: type _____ ; make _____ ; material _____ ; sash thickness _____
Glass: grade _____ ; ☐ sash weights; ☐ balances, type _____ ; head flashing _____
Trim: type _____ ; material _____ Paint _____ ; number coats _____
Weatherstripping: type _____ ; material _____ Storm sash, number _____
Screens: ☐ full; ☐ half; type _____ ; number _____ ; screen cloth material _____
Basement windows: type _____ ; material _____ ; screens, number _____ ; Storm sash, number _____
Special windows _____
Additional information: _____

**18. ENTRANCES AND EXTERIOR DETAIL:**
Main entrance door: material _____ ; width _____ ; thickness _____". Frame: material _____ ; thickness _____"
Other entrance doors: material _____ ; width _____ ; thickness _____". Frame: material _____ ; thickness _____"
Head flashing _____ Weatherstripping: type _____ ; saddles _____
Screen doors: thickness _____"; number _____ ; screen cloth material _____ Storm doors: thickness _____"; number _____
Combination storm and screen doors: thickness _____"; number _____ ; screen cloth material _____
Shutters: ☐ hinged; ☐ fixed. Railings _____ , Attic louvers _____
Exterior millwork: grade and species _____ Paint _____ ; number coats _____
Additional information: _____

**19. CABINETS AND INTERIOR DETAIL:**
Kitchen cabinets, wall units: material _____ ; lineal feet of shelves _____ ; shelf width _____
Base units: material _____ ; countertop _____ ; edging _____
Back and end splash _____ Finish of cabinets _____ ; number coats _____
Medicine cabinets: make _____ ; model _____
Other cabinets and built-in furniture _____
Additional information: _____

**20. STAIRS:**

| STAIR | TREADS | | RISERS | | STRINGS | | HANDRAIL | | BALUSTERS | |
|---|---|---|---|---|---|---|---|---|---|---|
| | Material | Thickness | Material | Thickness | Material | Size | Material | Size | Material | Size |
| Basement _____ | | | | | | | | | | |
| Main _____ | | | | | | | | | | |
| Attic _____ | | | | | | | | | | |

Disappearing: make and model number _____
Additional information: _____

2

**Figure 3-20 (Continued).** Standard VA and FHA specification form.

**21. SPECIAL FLOORS AND WAINSCOT:**

| | LOCATION | MATERIAL, COLOR, BORDER, SIZES, GAGE, ETC. | THRESHOLD MATERIAL | WALL BASE MATERIAL | UNDERFLOOR MATERIAL |
|---|---|---|---|---|---|
| FLOORS | Kitchen | | | | |
| | Bath | | | | |
| | | | | | |
| | | | | | |

| | LOCATION | MATERIAL, COLOR, BORDER, CAP. SIZES, GAGE, ETC. | HEIGHT | HEIGHT OVER TUB | HEIGHT IN SHOWERS (FROM FLOOR) |
|---|---|---|---|---|---|
| WAINSCOT | Bath | | | | |
| | | | | | |
| | | | | | |

Bathroom accessories: ☐ Recessed; material _____; number _____; ☐ Attached; material _____; number _____
Additional information: _____

**22. PLUMBING:**

| FIXTURE | NUMBER | LOCATION | MAKE | MFR'S FIXTURE IDENTIFICATION No. | SIZE | COLOR |
|---|---|---|---|---|---|---|
| Sink | | | | | | |
| Lavatory | | | | | | |
| Water closet | | | | | | |
| Bathtub | | | | | | |
| Shower over tub△ | | | | | | |
| Stall shower△ | | | | | | |
| Laundry trays | | | | | | |
| | | | | | | |
| | | | | | | |
| | | | | | | |

△☐ Curtain rod    △☐ Door    ☐ Shower pan: material _____
Water supply: ☐ public; ☐ community system; ☐ individual (private) system.★
Sewage disposal: ☐ public; ☐ community system; ☐ individual (private) system.★
★Show and describe individual system in complete detail in separate drawings and specifications according to requirements.
House drain (inside): ☐ cast iron; ☐ tile; ☐ other _____   House sewer (outside): ☐ cast iron; ☐ tile; ☐ other _____
Water piping: ☐ galvanized steel; ☐ copper tubing; ☐ other _____   Sill cocks, number _____
Domestic water heater: type _____; make and model _____; heating capacity _____
_____ gph. 100° rise. Storage tank: material _____; capacity _____ gallons.
Gas service: ☐ utility company; ☐ liq. pet. gas; ☐ other _____   Gas piping: ☐ cooking; ☐ house heating.
Footing drains connected to: ☐ storm sewer; ☐ sanitary sewer; ☐ dry well.  Sump pump; make and model _____
_____; capacity _____; discharges into _____

**23. HEATING:**
☐ Hot water. ☐ Steam. ☐ Vapor. ☐ One-pipe system. ☐ Two-pipe system.
   ☐ Radiators. ☐ Convectors. ☐ Baseboard radiation. Make and model _____
   Radiant panel: ☐ floor; ☐ wall; ☐ ceiling. Panel coil: material _____
   ☐ Circulator. ☐ Return pump. Make and model _____; capacity _____ gpm.
   Boiler: make and model _____   Output _____ Btuh.; net rating _____ Btuh.
   Additional information: _____
Warm air: ☐ Gravity. ☐ Forced.  Type of system _____
   Duct material: supply _____; return _____  Insulation _____, thickness _____ ☐ Outside air intake.
   Furnace: make and model _____   Input _____ Btuh.; output _____ Btuh.
   Additional information: _____
☐ Space heater; ☐ floor furnace; ☐ wall heater.  Input _____ Btuh.; output _____ Btuh.; number units _____
   Make, model _____   Additional information: _____
Controls: make and types _____
Additional information: _____
Fuel: ☐ Coal; ☐ oil; ☐ gas; ☐ liq. pet. gas; ☐ electric; ☐ other _____; storage capacity _____
   Additional information: _____
Firing equipment furnished separately: ☐ Gas burner, conversion type. ☐ Stoker: hopper feed ☐; bin feed ☐
   Oil burner: ☐ pressure atomizing; ☐ vaporizing _____
   Make and model _____   Control _____
   Additional information: _____
Electric heating system: type _____   Input _____ watts; @ _____ volts; output _____ Btuh.
   Additional information: _____
Ventilating equipment: attic fan, make and model _____; capacity _____ cfm.
   kitchen exhaust fan, make and model _____
Other heating, ventilating, or cooling equipment _____

**24. ELECTRIC WIRING:**
Service: ☐ overhead; ☐ underground.  Panel: ☐ fuse box; ☐ circuit-breaker; make _____ AMP's _____ No. circuits _____
Wiring: ☐ conduit; ☐ armored cable; ☐ nonmetallic cable; ☐ knob and tube; ☐ other _____
Special outlets: ☐ range; ☐ water heater; ☐ other _____
☐ Doorbell. ☐ Chimes. Push-button locations _____   Additional information: _____

**25. LIGHTING FIXTURES:**
Total number of fixtures _____   Total allowance for fixtures, typical installation, $ _____
Nontypical installation _____
Additional information: _____

3    DESCRIPTION OF MATERIALS

**Figure 3-20 (Continued).** Standard VA and FHA specification form.

DESCRIPTION OF MATERIALS

**26. INSULATION:**

| Location | Thickness | Material, Type, and Method of Installation | Vapor Barrier |
|---|---|---|---|
| Roof | | | |
| Ceiling | | | |
| Wall | | | |
| Floor | | | |

**HARDWARE:** (make, material, and finish.) _____

**SPECIAL EQUIPMENT:** (State material or make, model and quantity.   Include only equipment and appliances which are acceptable by local law, custom and applicable FHA standards.   Do not include items which, by established custom, are supplied by occupant and removed when he vacates premises or chattles prohibited by law from becoming realty.)_____

**27. MISCELLANEOUS:** (Describe any main dwelling materials, equipment, or construction items not shown elsewhere; or use to provide additional information where the space provided was inadequate.   Always reference by item number to correspond to numbering used on this form.) _____

**PORCHES:**

**TERRACES:**

**GARAGES:**

**WALKS AND DRIVEWAYS:**

Driveway: width _____ ; base material _____ ; thickness _____"; surfacing material _____ ; thickness _____"
Front walk: width _____ ; material _____ ; thickness _____". Service walk: width _____ ; material _____ ; thickness _____"
Steps: material _____ ; treads _____"; risers _____". Cheek walls _____

**OTHER ONSITE IMPROVEMENTS:**

(Specify all exterior onsite improvements not described elsewhere, including items such as unusual grading, drainage structures, retaining walls, fence, railings, and accessory structures.)

**LANDSCAPING, PLANTING, AND FINISH GRADING:**

Topsoil _____" thick: ☐ front yard; ☐ side yards; ☐ rear yard to _____ feet behind main building.
Lawns (seeded, sodded, or sprigged): ☐ front yard _____ ; ☐ side yards _____ ; ☐ rear yard _____
Planting: ☐ as specified and shown on drawings; ☐ as follows:

| | | |
|---|---|---|
| _____ Shade trees, deciduous, _____" caliper. | _____ Evergreen trees. _____' to _____', B & B. |
| _____ Low flowering trees, deciduous, _____' to _____' | _____ Evergreen shrubs, _____' to _____', B & B. |
| _____ High-growing shrubs, deciduous, _____' to _____' | _____ Vines, 2-year _____ |
| _____ Medium-growing shrubs, deciduous, _____' to _____' | |
| _____ Low-growing shrubs, deciduous, _____' to _____' | |

IDENTIFICATION.—This exhibit shall be identified by the signature of the builder, or sponsor, and/or the proposed mortgagor if the latter is known at the time of application.

Date_____ Signature _____

Signature _____

FHA Form 2005
VA Form 26–1852

4

GPO 1968 o48—16—80081-1 296–152

**Figure 3-20 (Continued).** Standard VA and FHA specification form.

# REVIEW QUESTIONS
# CHAPTER 3

Write all answers on a separate sheet of paper. Do not write in this book.

1. What are four other names for construction drawings?
2. Name five line symbols which appear in the Alphabet of Lines.
3. What kind of line is used to represent the visible features of a building?
4. If a part is hidden, it is represented with a(n) _____ line symbol.
5. A symmetrical object usually has a(n) _____ line drawn through the center.
6. What kind of line is used to show how long a wall is?
7. When is a cutting plane line used?
8. What is the purpose of symbols and abbreviations?
9. Symbols may be shown in elevation or section. Which type is usually easiest to recognize?
10. If a plan is drawn at 1/4" = 1'-0" scale, how long on the drawing would a 40'-0" wall be?
11. How long would the same 40'-0" wall be if it were drawn at 1/4 size?
12. Dimensions locate _____ and _____ of building elements.
13. What is the most common method of dimensioning interior frame walls?
14. An exterior brick veneer wall should be dimensioned _____.
    A. To the center of the wall.
    B. To the outside of the brick.
    C. To the outside of the stud wall.
    D. To the inside of the stud wall.
    E. None of the above.
15. In the metric system, what is the basic unit of measure in the construction field?

16. Notes on a drawing are really not very important because they amplify information which is already given on the drawing. True or False?
17. Who usually prepares the working drawings for a building?
18. Identify the seven drawings which usually compose a set of working drawings for a building.
19. Which working drawing, or plan, shows the location of the structure on the site?
20. The plan that shows exterior walls, interior walls, doors and windows, and patios is the _____ plan.
21. A(n) _____ plan shows the footings, foundation size, and materials.
22. The front view of a structure would be called a(n) _____.
23. Which drawing shows the location of switches, convenience outlets, and TV jacks?
24. Why are construction details necessary?
25. Many aspects of a building cannot be adequately shown or explained on the drawings and must be included in the _____.
26. The Construction Documents upon which an agreement is made between all parties concerned in a project is composed of the _____ _____ and the _____.
27. Specifications are generally divided into _____ main sections.
28. Place the following specification outline headings in proper order.
    A. Miscellaneous Metals
    B. Rough Carpentry
    C. Concrete and Masonry Work
    D. Excavating and Grading
    E. Finish Carpentry and Millwork
    F. General Conditions
    G. Waterproofing
    H. Roofing and Sheetmetal

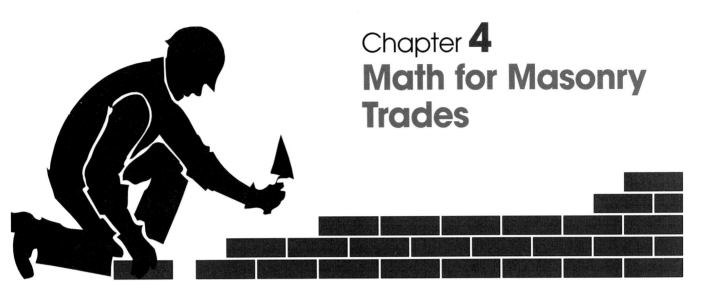

# Chapter 4
# Math for Masonry Trades

A skilled mason or apprentice must be able to take measurements and perform various calculations related to the trade. These calculations require an understanding of certain basic mathematical concepts. The purpose of this chapter is to review some of those basic concepts and give examples of their application using trade terms and figures.

## Whole Numbers ▬▬▬

A *whole number* is any one of the natural numbers such as 1, 2, 5, etc. Numbers represent quantities of anything. They can be added, subtracted, multiplied, or divided.

### Addition

*Addition* is the process of combining two or more quantities (numbers) to find a total. The total in addition is the *sum*. Addition is indicated by the plus (+) sign and may be written as 2 + 2. The sum may be indicated by using the equal (=) sign. For example: 2 + 2 = 4. Another way of writing this problem showing the sum of 4 is:

$$\begin{array}{r} 2 \\ +\,2 \\ \hline 4 \ \textit{sum} \end{array}$$

### Problems—addition

The following problems are included to refresh your memory of basic addition in trade terms.

1. Three bricklayers working together on a job each laid the following number of bricks in one day. The first bricklayer laid 887, the second bricklayer laid 1123, and the third bricklayer laid 1053 bricks. How many bricks did all three lay that day?

**Answer: 887 + 1123 + 1053 = 3063 bricks**

2. A mason was paid the following sum of money for each of five days of work: $96, $105, $87, $97, and $85. How much money was he paid for the five day's work?

**Answer: $96 + $105 + $87 + $97 + $85 = $470**

3. Weight slips on four loads of gravel delivered to one job were 6272 lb., 6098 lb., 6138 lb., and 6193 lb. What was the total weight of the gravel delivered to the job?

**Answer: 6272 + 6098 + 6138 + 6193 = 24,701 lb.**

### Subtraction

*Subtraction* is the process of taking one number away from another number. The number to be subtracted from is the *minuend*, and the number subtracted from it is the *subtrahend*. The portion that is left after the subtraction is the *difference*. The sign that indicates that one quantity (number) is to be subtracted from another is the minus (–) sign. Example: 6 – 4. In this example, 4 is being subtracted from 6. The difference is 2 or 6 – 4 = 2. Another way of writing the same thing is:

$$\begin{array}{r} 6 \quad \textit{minuend} \\ -\,4 \quad \textit{subtrahend} \\ \hline 2 \quad \textit{difference} \end{array}$$

### Problems—subtraction

The following problems are included to refresh your memory of basic subtraction in trade terms.

1. A mason ordered 75 bags of cement and used 68 bags on the job. How many bags of cement are left?

**Answer: 75 – 68 = 7 bags**

2. If a 36" wide chimney had two flues which were 8" and 16" wide and the thickness of the chimney walls were 4", how much space would be between the flues?

**Answer: 36 – (8 + 16 + 4 + 4) = 36 – 32 = 4"**

3. A mason contracts a job for $700. Labor cost is $316 and material cost $203. Other expenses total $78. How much profit is made on the job?

**Answer: 700 – (316 + 203 + 78) = 700 – 597 = $103**

## Multiplication

*Multiplication* is the process of repeated addition using the same numbers. The number to be multiplied is the *multiplicand*, and the number that it is multiplied by is the *multiplier*. The answer to multiplication is the *product*. For example, if 2 + 2 + 2 + 2 + 2 were to be added, the shortest method would be to multiply 5 times 2 to get the product of 10. The sign used to indicate multiplication is the times (×) sign. In the previous example, 5 times 2 equals 10, would be written 5 × 2 = 10. This may also be written as:

|       |               |
|-------|---------------|
| 2     | *multiplicand* |
| × 5   | *multiplier*   |
| 10    | *product*      |

### Problems—multiplication

The following problems are included to refresh your memory of basic multiplication in trade terms.

1. If a bricklayer can lay 170 bricks an hour, how many bricks would he lay in four hours?

**Answer: 170 × 4 = 680 brick**

2. One type of brick costs $9 per hundred. If 14,000 bricks were ordered, how much would they cost?

**Answer: $9 × 140 = $1260. Note: The bricks were $.09 each, $9 per hundred or $90 per thousand. Therefore, the answer could have been determined by multiplying $.09 x 14,000 or $90 × 14, or $9 × 140.**

3. If five bricks covered 1 sq. ft. of wall space, how many bricks would be required to cover a wall 8 ft. high by 10 ft. long?

**Answer: First find the wall area by multiplying 8 × 10. The wall area is 80 sq. ft. The number of bricks required is 5 × 80 = 400 brick.**

## Division

*Division* is the process of finding how many times one number (*divisor*) is contained within another number (*dividend*). The division symbol is (÷). For example, when we wish to find how many times 3 is contained in 9, we say 9 divided by 3 equals 3 or 9 ÷ 3 = 3. The answer of a division problem is the *quotient*. If a number is not contained in another number an equal number of times, the amount left over is the *remainder*. The problem is generally written as follows:

$$\textbf{divisor}\ 4\overline{)\ \begin{array}{c} 2 \\ 9 \\ \underline{8} \\ 1 \end{array}}\quad \textbf{dividend}$$

$$\textbf{remainder}$$

or 2 1/4 **quotient**

Notice that the remainder is placed over the divisor to form a fraction. This fraction is part of the quotient as shown in the problem in Figure 4-1.

## Problems—Division

The following problems are included to refresh your memory of basic division in trade terms.

1. If a set of steps had 8 risers and the total height of all the steps (total rise) was 56", what would the height of each step be?

$$8\overline{)\ \begin{array}{c} 7 \\ 56 \\ \underline{56} \\ 0 \end{array}}$$

**Answer: 7"**

2. Four bricklayers laid 4160 bricks on a job one day. What was the average number of bricks laid by each bricklayer that day?

$$4\overline{)\ \begin{array}{c} 1040 \\ 4160 \\ \underline{4} \\ 01 \\ \underline{0} \\ 16 \\ \underline{16} \\ 00 \\ \underline{0} \\ 0 \end{array}}$$

**Answer: 1040 bricks per bricklayer**

3. If a brick veneer wall requires five bricks to lay up 1 sq. ft., how many square feet would 587 cover?

$$5\overline{)\ \begin{array}{c} 117 \\ 587 \\ \underline{5} \\ 08 \\ \underline{5} \\ 37 \\ \underline{35} \\ 2 \end{array}}$$

**Answer: 117 2/5 sq. ft. of wall**

# Fractions

A *fraction* is one or more parts of a whole. See Figure 4-1. Fractions are written with one number over the other, such as 1/4, 1/2, or 3/4. The top number of a fraction is the *numerator*. The numerator indicates the number of parts of the whole which is of concern. The bottom number of a fraction is called the *denominator*. The denominator identifies the number of parts into which the whole is divided.

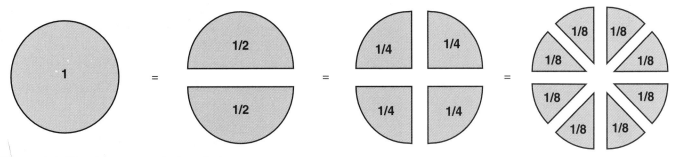

**Figure 4-1.** Fractions are part of a whole.

In reading a fraction, the top number is always read first. For example, 1/2 would be read "one-half;" 3/4 would be read "three-fourths;" and 3/8 would be read "three-eighths."

A fraction should always be reduced to its lowest denominator. For instance, 3/2 is not in the lowest, or reduced, form. It should be 1 1/2 because 2/2 = 1 and 1 + 1/2 = 1 1/2. The 1 1/2 is a **mixed number**. When the numerator and denominator are the same number, such as 1/1, 2/2, 3/3, etc., they are equal to 1.

## Adding fractions

The easiest fractions to add are those that have the same denominators, such as 1/8 + 3/8. Simply add the numerators together and keep the same denominator. For example, 1/8 + 3/8 = 4/8, which reduces to 1/2. Reducing the fraction to its lowest denominator is preferred. Another example of reducing to the lowest denominator is 8/24 = 1/3, because 24 may be divided by 8 three times.

When fractions to be added have different denominators, multiply both numerator and denominator of each fraction by a number that will make the denominators equal. For example, in 1/3 + 3/5 multiply 1/3 by 5, and multiply 3/5 by 3. The result is 5/15 + 9/15. Observation indicated that 15 was the smallest number that could be divided evenly by both denominators. To complete the example, 5/15 + 9/15 = 14/15. Therefore, the sum of 1/3 and 3/5 is 14/15.

## Problems—addition of fractions

1. What is the height of one stretcher course of brick if the bricks are 2 1/4" high and the mortar joint is 3/8 in?

**Answer:** $2\frac{1}{4} + \frac{3}{8}$
$= 2\frac{2}{8} + \frac{3}{8}$
$= 2\frac{5}{8}$" **height for one course**

2. A mason estimated the following amounts of mortar required for a job: 5 1/2 cu. yd., 11 1/3 cu. yd., and 6 1/4 cu. yd. What is the total amount of mortar required for the job?

**Answer:** $5\frac{1}{2} + 11\frac{1}{3} + 6\frac{1}{4}$
$= 5\frac{6}{12} + 11\frac{4}{12} + 6\frac{3}{12}$
$= 22\frac{13}{12}$
$= 23\frac{1}{12}$ **cu. yd. of mortar**

3. A brick wall on a residence was 8 1/16" thick, the furring was 3/4", plaster board was 3/8", and the plaster was 1/2". What was the total thickness of the wall.?

**Answer:** $8\frac{1}{16} + \frac{3}{4} + \frac{3}{8} + \frac{1}{2}$
$= 8\frac{1}{16} + \frac{12}{16} + \frac{6}{16} + \frac{8}{16}$
$= 8\frac{27}{16}$
$= 9\frac{11}{16}$"

## Subtracting fractions

When subtracting fractions, change all fractions to the same common denominator as was done for adding fractions. When the denominators are the same, subtract the numerators.

## Problems—subtraction of fractions

1. What is the dimension of "A" in Figure 4-2?

**Answer:** $2' - 6\frac{1}{2}" - 8\frac{1}{4}"$
$2' - 6\frac{1}{2}" = 30\frac{1}{2}"$
**Therefore,** $30\frac{1}{2}" - 8\frac{1}{4}"$
$= 30\frac{2}{4}" - 8\frac{1}{4}"$
$= 22\frac{1}{4}"$ **(or** $1' - 10\frac{1}{4}"$**)**

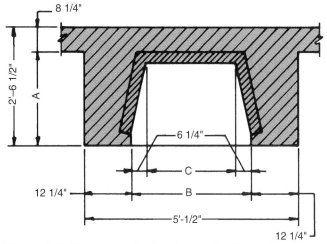

**Figure 4-2.** Plan view of a fireplace with dimensions.

2. What is the dimension of "B" in Figure 4-2?

**Answer:** $5' - \frac{1}{2}" - (12\frac{1}{4}" + 12\frac{1}{4}")$

$5' - \frac{1}{2}" = 60\frac{1}{2}"$

Therefore, $60\frac{1}{2}" - (12\frac{1}{4}" + 12\frac{1}{4}")$

$= 60\frac{1}{2}" - 24\frac{1}{2}"$

$= 36"$ (or 3' - 0")

3. What is the dimension of "C" in Figure 4-2?

**Answer:** $36"$ (from No. 2 above ) $- (6\frac{1}{4}" + 6\frac{1}{4}")$

$= 36" - 12\frac{1}{2}"$

$= 23\frac{1}{2}"$ or 1' - $11\frac{1}{2}"$

## Multiplying fractions

The procedure for multiplying fractions is to multiply the numerators together to find the numerator for the answer. Then, multiply the denominators together to find the denominator for the answer. If the product is a fraction, it should be reduced to its lowest form. For example, in 1/2 × 3/4, multiply the 1 of 1/2 by the 3 in 3/4 to get 3. Then, multiply the 2 of 1/2 by the 4 of 3/4 to get 8. The product of this problem is 3/8.

Whole numbers must be changed to fractions by placing the whole number over a 1. For example, in 4 × 5/8, the 4 is changed to 4/1. Then, 4/1 × 5/8 equals 20/8. The fraction 20/8 reduces to 2 4/8, which reduces to 2 1/2.

## Problems—multiplication of fractions

1. If standard bricks that are 2 1/4" thick are used to lay a wall with 3/8" mortar joints, what will the height of the wall be after nine courses?

**Answer: First, add the thickness of one mortar joint to the thickness of one brick. Then multiply that sum times 9 to find the height.**

$2\frac{1}{4}" + \frac{3}{8}"$

$= 2\frac{2}{8} + \frac{3}{8}$

$= 2\frac{5}{8}$

$2\frac{5}{8} \times 9$

$= \frac{1}{8} \times \frac{9}{1}$

$= \frac{189}{8}$

$= 23\frac{5}{8}"$

2. If a set of steps are five risers high and each riser is 7 1/4", what is the total rise of the steps?

**Answer:** $7\frac{1}{4}" \times 5$

$= \frac{29}{4} \times \frac{5}{1}$

$= \frac{145}{4}$

$= 36\frac{1}{4}"$

3. What is the length of a 28" stretcher wall if each stretcher is 7 1/2" and the mortar joint is 1/2"?

**Answer: 7 1/2" + 1/2" = 8"**

8" × 28" = 224"

224" ÷ 12 = 18 2/3';

convert 2/3' to inches,

2/3 × 12/1 = 24/3 = 8"

= 18'-8"

## Dividing fractions

The process of dividing fractions is accomplished by inverting (turning upside down) the divisor and then multiplying. For example, in 3/8 ÷ 3/4 change the 3/4 to 4/3, and then multiply 3/8 by 4/3. Therefore, 3/8 ÷ 3/4 = 3/8 × 4/3 = 12/24 = 1/2.

## Problems—division of fractions

1. How many risers 7 1/2" high would be required to construct a flight of concrete steps 3'-1 1/2" high?

**Answer: Change 3' - $1\frac{1}{2}"$ to inches. Divide that number by $7\frac{1}{2}"$.**

$3' - 1\frac{1}{2}" = 37\frac{1}{2}"$

$37\frac{1}{2}" \div 7\frac{1}{2}"$

$= \frac{75}{2} \div \frac{15}{2}$

$= \frac{75}{2} \times \frac{2}{15}$

$= \frac{150}{30}$

**= 5 risers**

2. If a brick mantel is corbeled out 4 1/2" in six courses, how much does each course project past the previous course?

**Answer:** $4\frac{1}{2} \div 6$

$= \frac{9}{2} \div \frac{6}{1}$

$= \frac{9}{2} \times \frac{1}{6}$

$= \frac{9}{12}$

$= \frac{3}{4}"$

3. If a story pole was 8' -11 1/4" long and divided into 39 equal spaces, what is the length of each space?

**Answer:** $8' - 11\frac{1}{4}" \div 39$

$= 107\frac{1}{4}" \div 39$

$= \frac{429}{4} \div \frac{39}{1}$

$= \frac{429}{4} \times \frac{1}{39}$

$= \frac{429}{156}$

$= 2\frac{117}{156}$

$= 2\frac{3}{4}"$

# Decimals

*Decimals* are types of fraction that only have numerators. The denominators are implied by the place of the last digit to the right of the decimal point. Therefore, 0.1 is equal to 1/10, 0.01 is equal to 1/100, 0.001 is equal to 1/1000, etc.

The zeros that may follow the significant digits in a decimal number do not increase the value of the fraction. Thus, 0.25 = 0.250 = 0.2500 and so on. The added zeros indicate an increased accuracy, but not an increase in size. However, zeros to the left of the number and before the decimal are very significant. Note: 0.25 is equal to 25/100, 0.025 is equal to 25/1000, 0.0025 is equal to 25/10,000, 0.00025 is equal to 25/100,000, etc.

When there is one number to the right of the decimal, it is read as so many tenths (0.5 is five-tenths). Two numbers to the right of the decimal are read as hundredths (0.05 is five hundredths). Three numbers to the right of the decimal indicate thousandths, four indicate ten thousandths, etc.

When numbers appear on both sides of the decimal, this is an example of a **decimal mixed number**. The word *and* is generally used to represent the decimal when reading a decimal mixed number. For example, 6.125 is read as "six and one hundred twenty-five thousandths." The word *point* is also used to denote the decimal. In this case, 6.125 would be read as "six point one two five."

## Problems—decimals

1. The cost of materials for a masonry job was as follows: $35.50 for sand, $162.38 for cement, $9.63 for one steel lintel, and $338.19 for brick. What was the total cost of the materials for the job?

**Answer: $35.50 + $162.38 + $9.63 + $338.19**
**= $545.70**
```
        35.50
       162.38
         9.63
     +  338.19
      $545.70
```

2. A mason needs to determine if the total cross-sectional area of three pieces of reinforcing steel met the specifications for a reinforced masonry lintel. They were 0.11 sq. in., 0.3 sq. in., and 1.000 sq.in. What was the total cross-sectional area of the three pieces of reinforcing steel?

**Answer:**
```
      0.11
      0.3
    +1.000
     1.410 sq. in.
```

## Problems—subtraction of decimals

1. A mason placed a bid on a job for $1247.38. He allowed $413.75 for materials and $500.00 for labor. How much did he allow for profit and miscellaneous cost?

**Answer: $1247.38 − ($413.75 + $500.00)**
**= 1247.38 − 913.75**
**= $333.63**

2. A cement contractor was to stake out a foundation on a site which was 125.00' wide and 200.00' deep. The plot plan showed that the front of the house (placed along the width of the lot) faced south and was 50.00' from the lot line. The east side of the house was 28.35' from the lot line and the west side of the house was 36.65' from the lot line. How wide is the front of the house?

**Answer: 125.00' − (28.35' + 36.65')**
**= 125.00 − 65.00**
**= 60.00'**
**or 60'-0" as it is written on a construction drawing**

## Problems—multiplication of decimals

1. What is the cost of 20 bags of cement if each cost $3.00?

**Answer: Add up the total number of spaces to the right of the decimal point in both parts of the problem and count over the same number in the answer and place the decimal there.**
```
    $ 3.00
    ×  20
   $60.00
```

2. How much should a mason charge for laying 400 bricks if each brick costs $.09 each and the mason got $.18 for laying each brick? Include the cost of the bricks in the problem.

**Answer: (400 bricks × $0.09)**
**+ (400 bricks × $0.18)**
```
   400          400            36.00
 × .09        × .18          +72.00
 $36.00        3200         $108.00
               400
             $72.00
```

## Problems—division of decimals

1. If a mason charged $2100.00 for laying 14,000 bricks, how much was received per brick?

**Answer: $2100.00 ÷ 14,000**
**= $0.15 ($.15 each)**
```
              .15
   14000 ) 2100.00
           14000
           70000
           70000
               0
```

2. Clay flue lining is sold in 2' lengths. What is the cost of one 2' length if 56' of lining costs $98.00?

**Answer: 56 ÷ 2 = 28 pieces of lining**
**$98.00 ÷ 28 = $3.50 each**
```
           3.50
    28 ) 98.00
         84
         140
         140
          00
           0
           0
```

## Converting fractions to decimals

Common fractions may be changed to decimals by adding one or more zeros to the numerator and dividing by the denominator. See Figure 4-3. Thus, the fraction 3/4 may be changed to a decimal by dividing 3.00 by 4. The answer is .75. Therefore, .75 is equal to 3/4.

| Fraction–Decimal Equivalents | |
|---|---|
| **Fraction** | **Decimal** |
| 1/16 | .0625 |
| 1/8 | .1250 |
| 1/6 | .1600 |
| 3/16 | .1875 |
| 1/4 | .2500 |
| 5/16 | .3125 |
| 1/3 | .3333 |
| 3/8 | .3750 |
| 7/16 | .4375 |
| 1/2 | .5000 |
| 9/16 | .5625 |
| 5/8 | .6250 |
| 2/3 | .6667 |
| 11/16 | .6875 |
| 3/4 | .7500 |
| 13/16 | .8125 |
| 5/6 | .8300 |
| 7/8 | .8750 |
| 15/16 | .9375 |

**Figure 4-3.** Decimal equivalents of fractions.

If the division does not come out even (with no remainder), it is carried out as far as desired. For example, the fraction 1/3 is changed to a decimal by dividing 1.000 by 3. The answer is .333...

When a mixed number such as 3 1/2 must be changed to a decimal form, the mixed number is first changed to a fraction (3 1/2 = 7/2) and then the division is performed shown below:

$$
\begin{array}{r}
3.50 \\
2\,\overline{)\,7.00} \\
\underline{6\phantom{.00}} \\
10\phantom{.} \\
\underline{10\phantom{.}} \\
00 \\
\underline{0} \\
0
\end{array}
$$

## Problems—converting fractions to decimals

1. What is the labor cost of laying 12 3/4 cu. yd. of rubble stone if the cost to lay rubble stone is $9.50 per cu. yd.?
   **Answer: 12 3/4 = 12.75**
   **12.75 × $9.50 = $121.13**
2. What will the cost be for 24 1/4 cu. yd. of fill if it sells for $2.25 per cu. yd.?

**Answer: 24 1/4 = 24.25**
**24.25 × $2.25 = $54.56**

## Converting decimals to fractions

A decimal may be changed to a fraction by dropping the decimal point and using the number as the numerator of the fraction. The denominator is written with a "1" and as many zeros after it as there were decimal places in the original decimal number. Thus, .125 may be changed to a fraction by writing 125 over 1000, or 125/1000. When reduced, 125/1000 equals 1/8.

## Problems—converting decimals to fractions

1. A mason read on a drawing that the reinforcing steel required for a certain situation was .375" in diameter. What is the fractional size of this steel?
   **Answer: 375/1000 = 3/8" in diameter**
2. What is the width of a piece of steel angle which is specified as 3.875"? Answer should be a fraction rather than a decimal.
   **Answer: 375/1000 = 3/8" in diameter**

# Percentage and Interest ▬

*Percentage* is a number used to denote parts of one hundred. Interest is a type of percentage. It represents the cost of borrowing money.

## Percentage

Percentage is indicated by the % sign. For example, 3% of a number is three hundredths of the number. It may be indicated as 3/100, .03, or 3%.

Finding percentage is done by multiplying. The number should be written in the decimal form (.03) when computing the answer. If 4% of 1452 were desired, then 1452 would be multiplied by .04. The answer or product is 58.08.

Study the following conversions before trying the problems:

| | | | | | |
|---|---|---|---|---|---|
| .25 | = 25% | 1/4 | = .25 | = 25% |
| .75 | = 75% | 3/8 | = .375 | = 37.5% |
| .2 | = 20% | | | |
| .7 | = 70% | | | |
| .364 | = 36.4% | | | |
| .5748 | = 57.48% | | | |

## Problems—percentage

1. A mason calculated that 5 cu. yd. of sand was needed for a bricklaying job. Because of waste the mason added 5%. How much sand should be ordered?
   **Answer: .05 × 5 cu. yd. = .25 cu. yd.**
   **5 + .25 = 5.25 cu. yd. should be ordered**
2. It was estimated that 12,000 bricks would be required for a job but an additional 2% were

added because of breakage, bats, and salmon brick. How many bricks should be ordered?

**Answer: .02 × 12,000 = 240**
**12,000 + 240 = 12,240 brick**

## Interest

*Interest* is usually considered to be the cost of borrowing money. The amount borrowed is the *principal.* The percentage charged per year is the *rate of interest.* Interest may be figured by the year, the month, or the day.

To illustrate, if $500 was borrowed for one year (12 months) at 8% interest, the interest to be paid would be $500.00 × .08 = $40.00. The same amount of money and the same rate of interest would cost $6.66 for two months. ($500.00 × .08) ÷ 6. The total is divided by 6 because two months is 1/6 of a year.

## Problems—interest

1. If a mason deposited $1000 in a savings account at 5% interest per year, how much interest would he earn at the end of 12 months?

**Answer: $1000 × .05 = $50.00**

2. A contractor purchased a new piece of equipment for $2500 and agreed to pay 8% interest per year. How much interest would be owed at the end of the year?

**Answer: $2500 × .08 = $200.00 in interest for**
**12 months**

How much interest for 6 months?

**Answer: $200.00 ÷ 2 = $100.00 in interest**
**for 6 months**

# Applied Geometry

The properties of points, lines, and planes are a definite concern of the mason. See Figure 4-4. This section is designed to clarify terms commonly used in geometry that have a direct application to the masonry trades.

## Points

*Points* are basic elements of lines that have no width, length, or height, but merely indicate a location. Figure 4-4 shows recommended symbols for a point. Points may be identified with a name, number, letter, or other symbol.

## Lines

A *line* has only length and is the distance between two or more points. A line may be either straight or curved. A horizontal line is a level line. A vertical line is a plumb line and is perpendicular to a horizontal line. *Perpendicular* means that the line forms a 90 degree (°) angle, or right angle, with respect to another line. The symbol for a perpendicular line is shown in Figure 4-4.

A diagonal line is one joining two opposite angles. Parallel lines remain a constant distance apart and never cross. Curved lines may be either arcs with a single center or irregular curves with many centers. A circle is an exam-

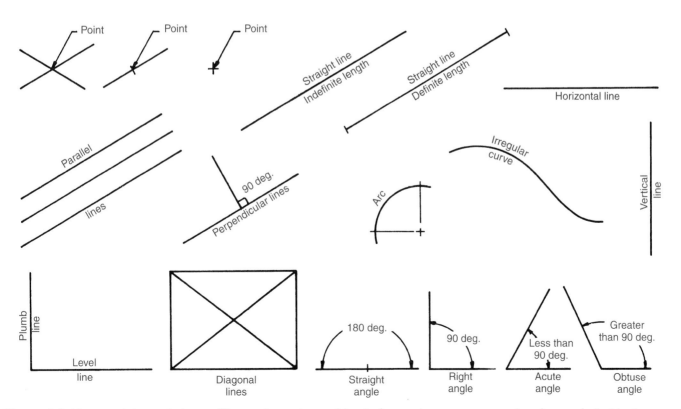

**Figure 4-4.** Lines, points, and planes. These elements combine to form edges, corners, and surfaces of all objects.

ple of a curve with a single center. It is a 360° arc. An irregular or free curve is represented in Figure 4-4.

## Angles

An *angle* is formed by two intersecting lines. An *obtuse angle* is an angle greater than 90°. A *straight angle* is an angle of 180°. A *right angle* is an angle with exactly 90°. An *acute angle* is an angle less than 90°.

Angles are measured in degrees(°), minutes('), and seconds("). Each degree is divided into 60 minutes and each minute is divided into 60 seconds. For example, an angle measurement of 125 degrees, 35 minutes, 40 seconds is written as 125° 35'40".

## Circles

A *circle* is a continuous curve maintaining an equal distance from its center at all times. Figure 4-5 identifies the parts of a circle. The *circumference* of a circle is the distance around it. A *semi-circle* is equal to half the circumference. An *arc* is any portion of the circumference. The *diameter* of a circle is the distance across a circle through its center. The *radius* is the distance from the center to any point on the circumference or one-half the diameter. A *sector* is the portion of a circle between two radii (more than one radius). A *segment* is the portion of a circle contained by a straight line and the circumference which it cuts off. A *tangent line* is a line which touches the circle, but does not cut it and is at right angles to a straight line from the center.

The circumference of a circle has a direct relationship to the circle's diameter. The circumference is equal to a constant of 3.1416 multiplied by the diameter. This constant is called "pi" and is represented by the symbol ($\pi$). Thus, the formula is C = $\pi$D, where C is circumference, $\pi$

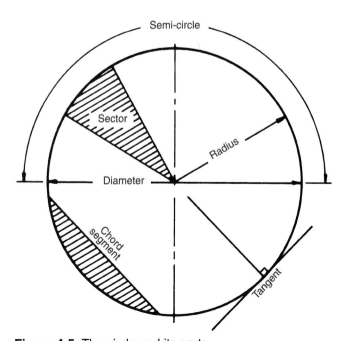

**Figure 4-5.** The circle and its parts.

Thus, the formula is C = $\pi$D, where C is circumference, $\pi$ is pi or 3.1416, and D is diameter.

## Triangular relationships

A *triangle* is a geometric shape that has three angles that total 180°. When two angles are known, the third may be found by subtracting the sum of the two angles from 180°.

A *right triangle* has one 90° angle. Therefore, the sum of the other two is 90°. The longest side of a right triangle is called the *hypotenuse*. The hypotenuse is always opposite of the 90° angle of the right triangle. See Figure 4-6.

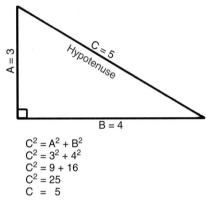

$$C^2 = A^2 + B^2$$
$$C^2 = 3^2 + 4^2$$
$$C^2 = 9 + 16$$
$$C^2 = 25$$
$$C = 5$$

**Figure 4-6.** The length of the sides of a right triangle always have a 3:4:5 proportion.

The square of the hypotenuse of a right triangle is equal to the sum of the squares of the other two sides. For example, if a right triangle had two sides equal to 3' and 4', then the other side (hypotenuse) would equal 5' (Hypotenuse$^2$ = Side A$^2$ + Side B$^2$. Therefore, Hyp$^2$ = 3$^2$ + 4$^2$ = 9 + 16 = 25.) Five multiplied by itself is equal to 25. Thus, the hypotenuse is 5'. This is also called a 3:4:5 triangle and is very useful for laying out square corners for foundations.

# Areas and Volumes ▬▬▬

Much of the math presented thus far will be used to find areas and/or volumes related to a job. Areas are two-dimensional (length × width) and are shown in square measure such as 1296 sq. in., 9 sq. ft., or 1 sq. yd. Volumes are three-dimensional quantities (length × width × height) and are shown in cubic measure (46,656 cu. in., 27 cu. ft., or 1 cu. yd.). Figure 4-7 shows several plane and solid figures with the formulas for calculating their areas or volumes.

## Problems—calculating areas

1. A large window opening is 72 3/8" × 60 3/8". What is the area of the opening?

Areas of Plane Figures

Volumes of Solid Figures

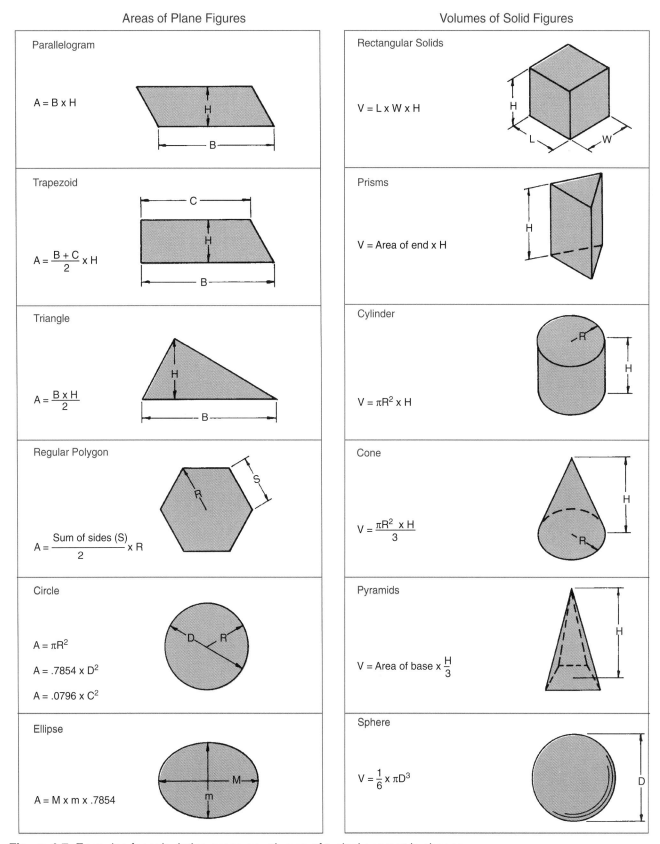

**Parallelogram**

$A = B \times H$

**Trapezoid**

$A = \dfrac{B + C}{2} \times H$

**Triangle**

$A = \dfrac{B \times H}{2}$

**Regular Polygon**

$A = \dfrac{\text{Sum of sides (S)}}{2} \times R$

**Circle**

$A = \pi R^2$

$A = .7854 \times D^2$

$A = .0796 \times C^2$

**Ellipse**

$A = M \times m \times .7854$

**Rectangular Solids**

$V = L \times W \times H$

**Prisms**

$V = \text{Area of end} \times H$

**Cylinder**

$V = \pi R^2 \times H$

**Cone**

$V = \dfrac{\pi R^2 \times H}{3}$

**Pyramids**

$V = \text{Area of base} \times \dfrac{H}{3}$

**Sphere**

$V = \dfrac{1}{6} \times \pi D^3$

**Figure 4-7.** Formulas for calculating areas or volumes of typical geometric shapes.

**Answer: Area for a rectangle is length × height**

**A = L × H**

**A = 72 3/8" × 60 3/8"**

**A = 72.375" × 60.375"**

**A = 4369.64 sq. in. Divide by 144 to find sq. ft.**

**4369.64 ÷ 144 = 30.34 sq. ft.**

2. What is the area of a circular slab whose radius is 6'?

**Answer: A = πR² ( π is a constant 3.1416 or 22/7)**

**A = 3.1416 × 6² (6² means 6 × 6)**

**A = 3.1416 × (6 × 6)**

**A = 3.1416 × 36**

**A = 113.10 sq. ft. of slab**

## Problems—calculating volumes

1. A common brick is 8" × 2 1/4" × 3 3/4". What is its volume in cubic inches?

**Answer: Volume = L × W × H**

**(length × width × height)**

**V = 8" × 2 1/4" × 3 3/4"**

**V = 8" × 2.25" × 3.75"**

**V = 67.5 cu. in.**

2. How many cubic yards of concrete would be required to fill a cylindrical form 12' long and 18" in diameter?

**Answer: Volume = π R² × H**

**(radius is 1/2 the diameter)**

**V = 3.1416 × 9² × 144" (must change feet to inches)**

**V = 3.1416 × 81 sq. in. × 144"**

**V = 36644.788 cu. in. or 21.21 cu. ft. or 0.79 cu. yd.**

**Divide cubic inches by 1728 to determine cubic feet and divide by 27 cu. ft. to determine cubic yards.**

**1 cu. yd. = 27 cu. ft. = 46,656 cu. in.**

# Metric Measurement and U.S. Conversions ▬▬▬▬▬

Through the years more and more countries have begun using the metric system. The United States is changing from the FPS (Foot-Pound-Second) system to SI metrics. It is, therefore, important that we become familiar with the metric units and their relationship to the familiar U.S. Customary units.

## Metrics in construction

The *Metric Conversion Act* was passed in 1975. It called for voluntary compliance, and volunteers were few. The *Omnibus Trade and Competitiveness Act* of 1988 declared that the metric system was to be the preferred system for U.S. trade and commerce. It was also mostly ignored by the public.

U.S. officials, unwilling to risk the political fallout of demanding a complete switch to metrics, launched a covert operation within. The *Omnibus Act* ordered that all U.S. government agencies go metric and, one-by-one, they have complied.

The CIA, NASA, the military, and the State Department have long been metric. But, the covert conversion was hardly noticed in the private sector until the General Services Administration, which oversees all federal building construction, began rejecting bids that did not have metric specifications. All federally funded projects costing more than $10 million must be done according to metric measure. In addition, metric units must be used in all federally assisted highway construction begun after September 30, 1996. On those projects that fall under the law, the architects, engineers, ironworkers, masons, carpenters, and other tradespeople will have to convert to a system of measures in meters, newtons, pascals, kelvins, kilograms, etc.

## SI units of measure

The official name of the new metric system is *Système international d' unités*. Its abbreviation is SI. SI metric is a very convenient system that uses a base of 10 to form the various units of measure. SI metric units of measure include the following:

1. Base units
2. Supplemental units
3. Derived units which have been identified by name
4. Non-SI units

## Base units

There are seven base units to which all other units are related. These units, with the exception of the kilogram, are based on natural phenomena that may be duplicated in the laboratory and require no international prototypes that must be stored by National Standards Organizations. The seven base units are:

| Quantity | SI Unit | SI Symbol |
|---|---|---|
| Length | meter | m |
| Mass (weight) | kilogram | kg |
| Time | second | s |
| Temperature | degree Kelvin | K |
| Electric current | ampere | A |
| Luminous intensity | candela | cd |
| Amount of substance | mole | mol |

## Derived units

Derived units are formed by multiplication or division of two or more SI base units. Some of the SI derived units are as shown in the next column:

| Quantity | SI Unit | SI Symbol |
|---|---|---|
| Electrical charge | coulomb | C |
| Electrical potential (voltage) | volt | V |
| Electrical resistance | ohm | Ω |
| Electrical conductance | siemens | S |
| Electrical capacitance | farad | F |
| Electrical inductance | henry | H |
| Energy (heat or work) | joule | J |
| Force | newton | N |
| Frequency | hertz | Hz |
| Illumination | lux | lx |
| Luminous flux | lumen | lm |
| Magnetic flux | weber | Wb |
| Magnetic flux density | tesla | T |
| Power | watt | W |
| Pressure | pascal | Pa |

## SI prefixes

SI prefixes are formed by multiplying or dividing base units by powers of 10. This is an advantage because it eliminates insignificant digits and decimals. The accepted SI prefixes are:

| Multiplication Factors | Prefix | Symbol |
|---|---|---|
| 1 000 000 000 000 = $10^{12}$ | tera | T |
| 1 000 000 000 = $10^9$ | giga | G |
| 1 000 000 = $10^6$ | mega | M |
| 1 000 = $10^3$ | kilo | k |
| 100 = $10^2$ | hecto | h |
| 10 = $10^1$ | deka | da |
| BASE UNITS 1 = $10^0$ | | |
| 0.1 = $10^{-1}$ | deci | d |
| 0.01 = $10^{-2}$ | centi | c |
| 0.001 = $10^{-3}$ | milli | m |
| 0.000 001 = $10^{-6}$ | micro | μ |
| 0.000 000 001 = $10^{-9}$ | nano | n |
| 0.000 000 000 001 = $10^{-12}$ | pico | p |
| 0.000 000 000 000 001 = $10^{-15}$ | femto | f |
| 0.000 000 000 000 000 001 = $10^{-18}$ | atto | a |

Note: Multiples of 1000 of the base unit are to be used in preference to others.

## Non-SI units

Certain units of measure that are not part of the SI system are so widely used and accepted, they have been recognized. They include the following:

| Quantity | SI Name | SI Unit | Base Unit |
|---|---|---|---|
| Liquid volume | liter | L | meter |
| Area | square meter | m² | meter |
| Solid volume | cubic meter | m³ | meter |
| Temperature | celsius | C | kelvin |
| Angle | degree | ° | radian |
| | minute | ' | radian |
| | second | " | radian |
| Time | minute | min | second |
| | hour | h | second |
| | day | d | second |
| Mass | tonne | t | kilogram |

## Converting U.S. and metric units

Eventually the SI system of measurement will be adopted worldwide. However, until then, the need to convert U.S. to metric and metric to U.S. will exist. Figure 4-8 shows factors for converting from U.S. to metric units.

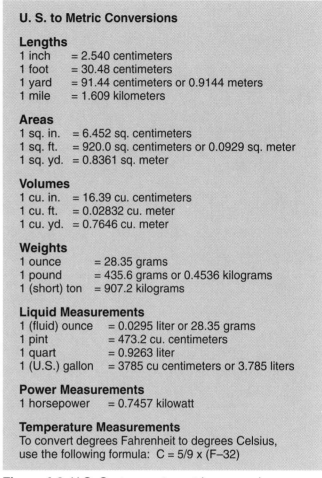

**U. S. to Metric Conversions**

**Lengths**
1 inch   = 2.540 centimeters
1 foot   = 30.48 centimeters
1 yard   = 91.44 centimeters or 0.9144 meters
1 mile   = 1.609 kilometers

**Areas**
1 sq. in.  = 6.452 sq. centimeters
1 sq. ft.  = 920.0 sq. centimeters or 0.0929 sq. meter
1 sq. yd.  = 0.8361 sq. meter

**Volumes**
1 cu. in.  = 16.39 cu. centimeters
1 cu. ft.  = 0.02832 cu. meter
1 cu. yd.  = 0.7646 cu. meter

**Weights**
1 ounce     = 28.35 grams
1 pound     = 435.6 grams or 0.4536 kilograms
1 (short) ton  = 907.2 kilograms

**Liquid Measurements**
1 (fluid) ounce   = 0.0295 liter or 28.35 grams
1 pint        = 473.2 cu. centimeters
1 quart       = 0.9263 liter
1 (U.S.) gallon   = 3785 cu centimeters or 3.785 liters

**Power Measurements**
1 horsepower   = 0.7457 kilowatt

**Temperature Measurements**
To convert degrees Fahrenheit to degrees Celsius, use the following formula:  $C = 5/9 \times (F-32)$

**Figure 4-8.** U.S. Customary to metric conversions.

To convert a quantity from U.S. Customary to metric units follow the procedure listed below.
1. U.S. customary measurements in fractional form should be changed to decimal form.
2. Multiply the quantity by the factor shown in Figure 4-8.
3. Round off the result to the degree of accuracy required.

Remember: The meter is the standard of length for the metric system and all divisions or multiples relate to it. For example:

1 meter = 1000 *milli* meters (thousandths)
1 meter = 100 *centi* meters (hundredths)
1 meter = 10 *deci* meters (tenths)
10 meters = 1 *deka* meter (tens)
100 meters = 1 *hecto* meter (hundreds)
1000 meters = 1 *kilo* meter (thousands)

These prefixes are used for every weight and measure in the metric system.

Figure 4-9 shows factors for converting from metric to U.S. Customary units. Follow the procedure below to convert a quantity from metric to U.S. Customary units.

1. Multiply the quantity by the factor shown in Figure 4-9.
2. Round off the result to the degree of accuracy required.
3. Convert the decimal part of the number to the nearest common fraction for most trade applications.

### Metric to U. S. Conversions

**Lengths**
1 millimeter (mm)  = 0.03937 in.
1 centimeter (cm)  = 0.3937 in.
1 meter (m)        = 3.281 ft. or 1.0937 yd.
1 kilometer (km)   = 0.6214 miles

**Areas**
1 sq. millimeter   = 0.00155 sq. in.
1 sq. centimeter   = 0.155 sq. in.
1 sq. meter        = 10.76 sq. ft. or 1.196 sq. yd.

**Volumes**
1 cu. centimeter   = 0.06102 cu. in.
1 cu. meter        = 35.31 cu. Ft. or 1.308 cu. yd.

**Weights**
1 gram (G)         = 0.03527 oz.
1 kilogram (kg)    = 2.205 lb.
1 metric ton       = 2205 lb.

**Liquid Measurements**
1 cu. centimeter (cm$^3$)   = 0.06102 cu.iln.
1 liter (1000 cm$^3$) = 1.057 quarts, 2.113 pints, or
                        61.02 cu. in.

**Power Measurements**
1 kilowatt (kW)    = 1.341 horsepower (hp)

**Temperature Measurements**
To convert degrees Celsius to degrees Fahrenheit, use the following formula:  F = (9/5 x C) + 32)

**Figure 4-9.** Metric to U.S. Customary conversions.

## Problems—U.S. and metric conversions

1. A residential structure is 40' × 60'. What are its dimensions in meters?

**Answer: 40' × 30.48 cm = 1219.2 cm**
**1 m = 100 cm; therefore, divide by 100**
**1219.2 cm ÷ 100 = 12.192 m**
**40' = 12.192 m**
**60' × 30.48 cm = 1828.8 cm ÷ 100**
**= 18.288 m**
**60' = 18.288 m**
**40' × 60'  = 12.192 m × 18.288 m**

2. A metric drawing showed a recess to be 10.16 cm deep by 20.32 cm wide. What are its dimensions in inches?

**Answer: 2.54 cm = 1 in.; therefore, divide each**
**dimension by 2.54 cm**
**10.16 cm ÷ 2.54 = 4"**
**20.32 cm ÷ 2.54 = 8"**
**The recess is 4" deep by 8" wide**

# Weights and Measures ▬▬▬

Various weights and measures are used in estimating proportioning and ordering materials. The mason must be familiar with common weights and measures so that estimates will be accurate. Some of the more frequently used lengths, quantities, and their equivalents are presented below as a ready source for the mason. At some point in the future, these measurements will be made in metric.

144 sq. in. = 1 sq. ft.
9 sq. ft. = 1 sq. yd.
1296 sq. in. = 1 sq. yd.
1728 cu. in. = 1 cu. ft.
27 cu. ft. = 1 cu. yd.
1 cu. ft. of water weighs 62.5 lb.
1 cu. ft. of damp loose sand equals 80 lb. of dry sand
1 cu. ft. of cement weighs approximately 100 lb.
1 cu. ft. of cement equals 1 bag; 4 bags equal 1 barrel
1 cu. ft. of hydrated lime weighs approximately 100 lb.
1 1/4 cu. ft. of hydrated lime equals one 50 lb. sack
1 cu. ft. of brickwork equals 120 lb.
1 cu. ft. of concrete weighs approximately 150 lb.
1 cu. ft. of packed earth weighs approximately 100 lb.
1 cu. ft. of gravel weighs approximately 110 lb.

## Problems—weights and measures

1. What part of a square foot does the side of a clay tile represent? The side is 8" × 12".

**Answer: 8" × 12" = 96 sq. in.**
**(1 sq. ft. = 144 sq. in.)**
**96 ÷ 144 = .66 sq. ft. = 2/3 sq. ft.**

2. If brickwork weighs 112 lb. per cu. ft., how many tons would a 16" × 16" × 12' pier weigh?

**Answer: 16" × 16" = 256 sq. in.**
**256 sq. in. × 144" = 36,864 cu. in.**
**(1 cu. ft. = 1728 cu. in.)**
**36,864 cu. in. ÷ 1728 cu. in.**
**= 21.33 cu. ft.   (1 ton = 2000 lb.)**
**112 lb. × 21.33 cu. ft. = 2389.33 lb.**
**2389.33 lb. ÷ 2000 = 1.19 tons**

# Estimating Brick Masonry ▬▬▬

According to the *Brick Institute of America*, except for nonmodular "standard bricks" (3 3/4" × 2 1/4" × 8") and some oversize bricks (3 3/4" × 2 3/4" × 8"), almost all of the bricks produced in the United States fit the 4" modular grid system. Modular size masonry units greatly simplify the job of estimating the number of units required for a given wall area.

The most widely used procedure for estimating brick masonry is the "wall-area" method. It consists of multiplying the quantity of materials needed per square foot by the net wall area. **Net wall area** is the gross area less (minus) all wall openings.

There are three standard modular joint thicknesses: 1/4", 3/8", and 1/2". For a given nominal size masonry unit, the number of modular masonry units per square foot of wall will be the same regardless of the mortar joint thickness. This is true if the units are laid with the joint thickness for which they were designed. However, the number of nonmodular standard bricks required per square foot of wall will vary with the thickness of the mortar joint.

When estimating, determine the net quantities of all materials before adding any allowances for waste and breakage. As a general rule, 5% is added for bricks and 10% to 25% for mortar. These factors may vary considerably with job conditions.

Figure 4-10 shows the net quantities of brick and mortar necessary to construct walls one wythe in thickness with various modular and nonmodular brick sizes using either 1/2" or 3/8" mortar joints. A **wythe** is each continuous vertical section of masonry one unit in thickness.

Quantities shown in Figure 4-10 are for running or stack bond which contain no headers. Bonds which require full headers must use the correction factor shown in Figure 4-11. Multiwythe walls also require a correction factor for the collar joint between the wythes. See Figure 4-12.

## Mortar yield

Accurate mortar yield calculations are based on absolute volume. They are complex and require extensive data about the materials being used. Most masons do not have the data nor do most jobs require critical mortar yield calculations. Therefore, the following *rule-of-thumb* may generally be used: For each 1 cu. ft. of damp loose sand, the mortar yield will be 1 cu. ft.

## Problems—estimating brick masonry

1. How many bricks (nominal size of 4" × 2 2/3" × 8") will be required to veneer a wall 24'-0" long and 4'-0" high? A 3/8" mortar joint is to be used. The wall has no openings.

**Answer: First, find the wall area: 24' × 4'**
**= 96 sq. ft.**
**Second, find the number of bricks per sq. ft. in Figure 4-10**
**The table shows 675 bricks/100 sq. ft. or 6.75 bricks/sq. ft.**
**Next, multiply the number of bricks/sq. ft. by the wall area**
**6.75 × 96 = 648 brick**
**Note that 5.28 cu. ft. of mortar will be required to lay these bricks (96% of 5.5)**

| Nominal Size of Brick (In.) T H L | Number of Bricks per 100 Sq. Ft. | Cubic Feet of Mortar | | | |
|---|---|---|---|---|---|
| | | Per 100 Sq. Ft. | | Per 1000 Bricks | |
| | | 3/8" Joints | 1/2" Joints | 3/8" Joints | 1/2" Joints |
| **Modular Brick** | | | | | |
| 4 x 2 2/3 x 8 | 675 | 5.5 | 7.0 | 8.1 | 10.3 |
| 4 x 3 1/5 x 8 | 563 | 4.8 | 6.1 | 8.6 | 10.9 |
| 4 x 4 x 8 | 450 | 4.2 | 5.3 | 9.2 | 11.7 |
| 4 x 5 1/3 x 8 | 338 | 3.5 | 4.4 | 10.2 | 12.9 |
| 4 x 2 x 12 | 600 | 6.5 | 8.2 | 10.8 | 13.7 |
| 4 x 2 2/3 x 12 | 450 | 5.1 | 6.5 | 11.3 | 14.4 |
| 4 x 3 1/5 x 12 | 375 | 4.4 | 5.6 | 11.7 | 14.9 |
| 4 x 4 x 12 | 300 | 3.7 | 4.8 | 12.3 | 15.7 |
| 4 x 5 1/3 x 12 | 225 | 3.0 | 3.9 | 13.4 | 17.1 |
| 6 x 2 2/3 x 12 | 450 | 7.9 | 10.2 | 17.5 | 22.6 |
| 6 x 3 1/5 x 12 | 375 | 6.8 | 8.8 | 18.1 | 23.4 |
| 6 x 4 x 12 | 300 | 5.6 | 7.4 | 19.1 | 24.7 |
| **Nonmodular Brick** | | | | | |
| 3 3/4 x 2 1/4 x 8 | 655 | 5.8 | 8.8 | | |
| | 616 | | 7.2 | | 11.7 |
| 3 3/4 x 2 3/4 x 8 | 551 | 5.0 | | 9.1 | |
| | 522 | | 6.4 | | 12.2 |

**Figure 4-10.** Modular and nonmodular bricks and mortar required for single-wythe walls in running or stack bond. These figures do not allow for waste and breakage.

| Bond | Correction Factor |
| --- | --- |
| Full headers every fifth course only | 1/5 |
| Full headers every sixth course only | 1/6 |
| Full headers every seventh course only | 1/7 |
| English bond | |
|    (Full headers every second course) | 1/2 |
| Flemish bond (alternate full headers | |
|    and stretchers every course) | 1/3 |
| Flemish headers every sixth course | 1/18 |
| Flemish cross bond (Flemish headers | |
|    every second course) | 1/6 |
| Double-stretcher, garden wall bond | 1/5 |
| Triple-stretcher, garden wall bond | 1/7 |

**Figure 4-11.** Correction factors for walls in Figure 4-10 with full headers. Add to facing and subtract from backing. Note: Correction factors are applicable only to those bricks which are twice as long as they are wide

| Cubic Feet of Mortar per 100 Sq. Ft. of Wall | | |
| --- | --- | --- |
| 1/4" Joint | 3/8" Joint | 1/2" Joint |
| 2.08 | 3.13 | 4.17 |

**Figure 4-12.** Correction factors for collar joints. Quantities are given in cu. ft. of mortar for collar joints. Note: cu. ft/1000 units = 10 × 11  cu. ft/100 sq. ft. of wall ÷ the number of units/sq. ft. of wall.

2. How many bricks and how much mortar will be needed to lay a wall 40'-0" long and 8'-0" high using 6" × 4" × 12" SCR bricks with 1/2" mortar joints? The wall has three window openings 3'-0" wide by 4'-0" high.

**Answer: Gross wall area = 40' × 8' = 320 sq. ft.**
      **Net wall area = 320 sq. ft. − (3' × 4' × 3)**
      **= 320 − 36 = 284 sq. ft.**
      **Brick/sq. ft. (Figure 4-10) = 300  per 100**
      **sq. ft. or 3.0/sq. ft.**
      **Therefore, 284 × 3 = 852 brick**
      **Mortar required = 7.4 per 100 sq. ft. or**
      **.074/sq. ft.**
      **.074 × 1284 = 21.01 cu. ft. of mortar**

# REVIEW QUESTIONS
# CHAPTER 4 ▬▬▬▬▬

Write all answers on a separate sheet of paper. Do not write in this book.

1. Four pieces of stone trim weighed 10.5 lb., 17.4 lb., 23.8 lb., and 5.6 lb. What is the total weight of the four pieces of trim?
2. If a mason ordered 5 1/2 cu. yd. of sand and used 3 1/4 cu. yd., how much was left?
3. If four masons each averaged laying 358 bricks a day, how many bricks would they average laying in five

3. If four masons each averaged laying 358 bricks a day, how many bricks would they average laying in five days?
4. A wall area of 23 sq. ft. required 115 bricks to face it. How many bricks were needed for each square foot of area?
5. What is the height of one stretcher course of concrete block if the block are 7 5/8" high and the mortar joint is 3/8"?
6. If the total thickness of a frame wall is 5", the studs 3 1/2", the wall board 3/8", and the weatherboard 1/2", how thick is the siding?
7. If one isolated footing required 1/8 yd. of concrete, how much concrete would be needed for a footing 3/4 as large?
8. If 1 1/2 yd. of mortar was divided among three masons, how much mortar would each get?
9. Which of the following quantities is the largest?
   A.   0.25
   B.   0.025
   C.   0.0025
   D.   2.2500
10. What is the sum of the following quantities; 0.12, 2.03, 12.004, 9.0?
11. A four day masonry job required 13.5 cu. yd. of mortar. Records show that 6.8 cu. yd. were used the first two days. How many yards were used the last two days of the job?
12. A mason purchased 5.5 cu. yd. of washed masonry sand for $7.50 per yd. How much was paid for the sand?
13. If a mason charged $0.25 per brick and the total bill was $3500.00, how many bricks were laid?
14. What is the decimal equivalent of 3 11/16"?
15. What is the fractional size of a piece of steel which is 2.375"?
16. If 3% were allowed for breakage on 10,000 bricks, how many bricks could be used?
17. How much money would be earned on a $1000 savings certificate in one year at 6 3/4% interest per year?
18. The length of the hypotenuse of a right triangle is desired when the other two legs are 6' and 8'. What is the length of the hypotenuse?
19. The circumference of a circle that has a diameter of 6.54" is:
   A.   134.4"
   B.   20.55"
   C.   18.7"
   D.   2.05"
20. The area of a 20' × 20' garage floor is _____ sq. ft.
21. The volume of a cylindrical pipe 6' long with an inside diameter of 4" is:
   A.   904.8 cu. ft.
   B.   75.4 cu. ft.
   C.   75.4 cu. in.
   D.   904.8 cu. in.

22. One meter is equal to:
    A.   1000 millimeters
    B.   100 centimeters
    C.   10 decimeters
    D.   All of the above.
    23. One inch is equal to:
    A.   2.54 cm
    B.   36 m
    C.   1/38 m
    D.   100 mm
24. How many cubic inches does 1 cu. ft. contain?
25. One cubic yard is equal to _____ cubic feet.
26. How many 4" × 2 2/3" × 8" bricks will be required to
    lay a 200 sq. ft. wall area if 675 bricks are needed for
    each 100 sq. ft.?

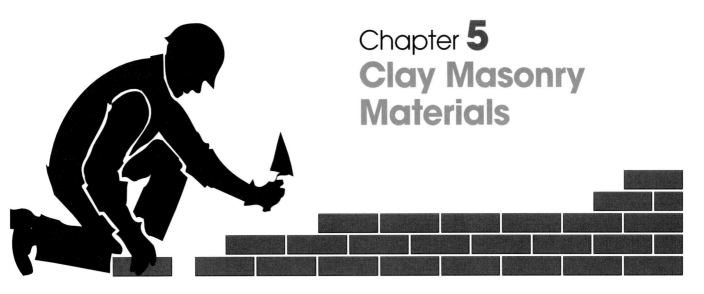

# Chapter 5
# Clay Masonry Materials

The variety of masonry materials available for use in construction has greatly increased. There are many new shapes, textures, colors, and applications. No doubt, this is due, in part, to the variety, versatility, durability, and beauty of masonry materials. See Figure 5-1. For instance, it is estimated that brick alone is made in at least 10,000 different combinations of colors, textures, and shapes.

Masonry construction provides permanence. Few, if any, materials are as maintenance free and long-lasting as brick, block, and stone.

Masonry units and other materials may be grouped under several basic categories or products as follows:

1. Structural clay products (brick, clay tile, and terra cotta)

**Figure 5-1.** These clay masonry applications show only a few of the many textures, colors, and shapes possible in masonry construction. (General Shale, Glen-Gery Corporation, Robinson Brick Company)

2. Concrete masonry units
3. Sand-lime brick
4. Glass block
5. Stone
6. Mortar
7. Masonry anchors, ties, and joint reinforcement

Each of these basic materials will be discussed in terms of classification, characteristics, size, and applications.

# Building Codes for Masonry Structures

The **American Concrete Institute (ACI)** and the **American Society of Civil Engineers (ASCE)** have adopted a consensus standard for the design of masonry structures that also include a standard construction specification. These documents, known as ACI 530—88/ASCE 5-88 *Building Code Requirements for Masonry Structures,* and ACI 530.1—88/ASCE 6-88 *Specification for Masonry Structures,* were developed to replace the existing standards for the design and construction of masonry structures.

## History and development

The development of this sole masonry standard for the design and construction industry began in 1977. At that time there were several design standards for masonry in use. These standards did not have consistent requirements. Concerned individuals representing masonry materials and the design profession saw the need for a single, consensus document for the design and construction of all types of masonry.

ACI and ASCE agreed to jointly develop a consensus standard for masonry design with the support of the masonry industry. Public review began in 1988 with the final approval of the code and specifications in August, 1989. The **Standard Building Code Congress International (SBCCI)** adopted the *Standard Building Code* in October, 1988. The **Building Official and Code Administrators (BOCA),** who publish the *National Building Code*, adopted the *Standard Building Code* in June, 1989.

The new standards are intended to replace the existing masonry design standards. For clay brick masonry, the previous standards are: ANSI A41.1 *Empirical Design of Reinforced Masonry*; ANSI A41.2 *Design for Reinforced Masonry*; and BIA-69 *Building Code Requirements for Engineered Brick Masonry*. Other masonry standards affected by the adoption of the new standards are NCMA TR-75B, ACI 531 and ACI 531.1, which deal only with concrete masonry design and construction. The new standards contribute to consistent design of masonry and a higher degree of quality in construction. These documents incorporate the latest research on material properties and performance. They reduce design procedures that are arbitrary and limiting to masonry construction.

# Structural Clay Products

Structural clay products are classified as:
1. Solid masonry units (brick)
2. Hollow masonry units (tile)
3. Architectural terra cotta

These products are made from clay or shale. See Figure 5-2. These materials are found throughout the country. Clays and shales suitable for the manufacture of brick fall into three groups:
1. Surface clays
2. Shales
3. Fire clays

Clays are produced naturally as rocks weather. Pure clay is mainly silica and alumina. It is made up of very small crystals of certain rocks, chiefly feldspar. **Feldspar** is a crystalline mineral made up of aluminum silicates with either potassium, sodium, calcium, or barium.

Shales are a compressed clay. They are more dense than clay. Fire clays have higher percentages of aluminum silicates, flint, and feldspar. The different characteristics of these materials make possible the great variety of clay products that we have today.

# Manufacture of Brick

The manufacturing process includes six general phases:
1. Mining and storage of raw materials
2. Preparing raw materials
3. Forming units
4. Drying
5. Firing and cooling
6. Drawing and storing finished products

Figure 5-3 shows a diagrammatic representation of this manufacturing process.

Bricks are commonly manufactured by three methods.
1. The soft mud process
2. The stiff mud process
3. The dry-press mud process

## Soft mud process

There are two variations of the soft mud process. They are the water-struck and sand-struck methods. Both methods prevent the clay from sticking to the mold. The clay is mixed wet and forced into a mold. The thickness of the brick is formed by scraping off the top of the mold.

In the water-struck process, the brick face is usually rather smooth, Figure 5-4. A sand-struck brick has a sandpaper-like surface. See Figure 5-5. Particles of sand stick to the brick material during removal from the mold and create this roughness.

**Classifications of Structural Clay Products**

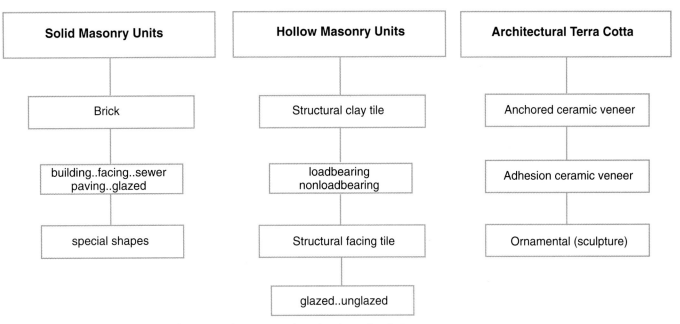

| Solid Masonry Units | Hollow Masonry Units | Architectural Terra Cotta |
|---|---|---|
| Brick | Structural clay tile | Anchored ceramic veneer |
| building..facing..sewer paving..glazed | loadbearing nonloadbearing | Adhesion ceramic veneer |
| special shapes | Structural facing tile | Ornamental (sculpture) |
|  | glazed..unglazed |  |

**Figure 5-2.** Structural clay products can be grouped under three basic types.

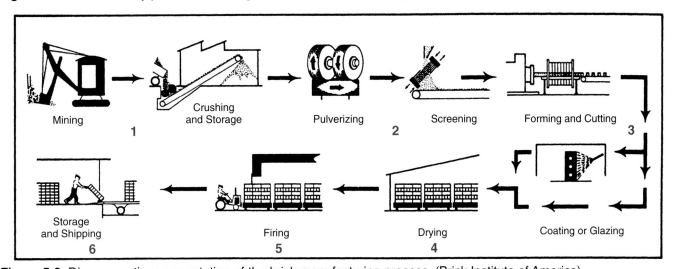

Mining  1  Crushing and Storage  Pulverizing  2  Screening  Forming and Cutting  3

Storage and Shipping  6  Firing  5  Drying  4  Coating or Glazing

**Figure 5-3.** Diagrammatic representation of the brick manufacturing process. (Brick Institute of America)

## Stiff mud process

In the stiff mud process, the clay is just wet enough to stick together. It is forced through a die and extruded (shaped) in the form of a ribbon or column. The thickness and width of the ribbon is the same as the dimensions of the brick being produced. Wires are used to cut the ribbon to lengths desired. Figure 5-6 shows the brick produced by this method.

## Dry-press process

In the dry-press process the clay is dry and loose. It is forced under high pressure into molds and forms the brick. These bricks are very dense.

**Figure 5-4.** The surface of this water-struck brick is smooth.

**Figure 5-5.** This sand-struck brick has a sandpaper-like surface.

**Figure 5-6.** Wire-cut brick has a rough surface.

# Drying and Burning

*Green bricks* are bricks that have not been fired. They will contain from 5% to 30% moisture as they come from the molding or extruding machines. Most of this moisture must be evaporated before the bricks are burned. Therefore, the bricks are placed in a drying kiln for 24 hours to 48 hours. On the other hand, they may be allowed to air dry.

When the moisture conditions are right, the bricks are put into a kiln designed to maintain a certain temperature. Even distribution of heat is very important for uniform results.

The most popular types of kilns used are the periodic and tunnel kilns. The primary difference between the two is that the brick remain stationary in the periodic kiln and the temperature is changed for each stage of the burning. In the tunnel kiln, the brick are placed on moving cars that pass through different temperature zones.

Burning time may be as little as 50 hours in a tunnel kiln and as long as 150 hours in a periodic kiln. Cooling may require over 50 hours. This prevents checking (small cracks) and maintains the right color.

# Classification of Brick

Bricks are usually classified as:
1. Building brick (very often referred to as common brick). ASTM C 62
2. Facing brick. ASTM C 216, A82.1
3. Hollow brick. C 652, A82
4. Paving brick. C 902
5. Ceramic glazed brick. C 126
6. Thin brick veneer units. C 1088
7. Sewer and manhole brick. C 32

Fire brick is a special type of brick made from a refractory ceramic material that is highly heat resistant. It is used mainly in the construction of fireplace linings and hearths. It is not considered a main classification type.

## Building brick

Building or common bricks are used chiefly as structural material where durability and strength are of more importance than appearance. Three grades of building bricks are available.

*Grade SW brick* is brick that is highly resistant to frost action. It is used where bricks are exposed to freezing weather. An example is brick used for foundation courses or retaining walls.

*Grade MW brick* is brick that is used where there may be exposure to temperatures below freezing, but where they are not likely to be permeated with water. An example is brick used in the face of a wall aboveground.

*Grade NW brick* is brick designed to be used as a backup on interior masonry.

## Facing brick

*Facing bricks* are used on exposed surfaces where appearance is an important consideration. Such brick must also be strong and durable. Appearance is covered in three grades:
1. *Type FBX.* These bricks are for general use in exposed exterior and interior walls and partitions where a high degree of perfection is required. They must be uniform in color and size.
2. *Type FBS.* This type is for general use in exposed exterior and interior masonry walls and partitions where wider color ranges and greater variation in sizes are permitted.
3. *Type FBA.* These bricks are manufactured and selected to produce architectural effects resulting from nonuniformity in size, color, and texture.

## Hollow brick

*Hollow bricks* are identical to facing bricks but have a larger core area. Most hollow bricks are used in the same application as facing bricks. Hollow bricks with very large cores are used in walls that are reinforced with steel and grouted solid. Larger cores or cells in hollow bricks allow reinforcing steel and grout to be placed in these units, whereas, it would be difficult to do so with building bricks, facing bricks, or some hollow bricks.

## Paving brick

*Paving bricks* are intended for use as a paving material to support pedestrian and light vehicular traffic. They must meet additional requirements for abrasion resistance or alternate performance requirements. Paving bricks are assigned a type by the traffic or abrasion expected. Type I pavers are exposed to extensive abrasion, such as driveways or public entries. Type II pavers are exposed to high levels of pedestrian traffic, such as in stores, restaurant floors, or exterior walkways. Type III pavers are exposed to pedestrian traffic, such as floors, or patios in homes.

## Ceramic glazed brick

*Ceramic glazed bricks* are units with a ceramic glaze fused to the body and used as facing brick. The body may be either facing brick or other solid masonry units. Requirements for properties of the finish include imperviousness, opacity, resistance to fading, resistance to crazing, flame spread, fuel contribution and smoke density, toxic fumes, hardness, and abrasion resistance.

## Thin brick

*Thin brick veneer* units are fired clay units with normal face dimensions, but a reduced thickness. They are used in adhered veneer applications.

## Sewer and manhole brick

*Sewer* and *manhole bricks* are intended for use in drainage structures for the conveyance of sewage, industrial wastes, and storm water; and related structures, such as manholes and catch basins.

# Brick Sizes and Nomenclature

For many years, bricks were manufactured in only a few sizes—Standard, Roman, and Norman. But today, there are many more sizes available. Until now, a given brick size may have been known by several names due to regional variations. A joint committee of the **Brick Institute of America** and the **National Association of Brick Distributors** recently developed standard nomenclature for bricks which represent roughly 90% of all sizes currently manufactured. They can be separated into two groups: modular and nonmodular.

A *modular unit* is based on a measurement of 4". It could include a measurement of 2" (one-half of the module) or 8" which is twice the module.

However, in bricks, this size is only nominal. This means that it is not exactly 4". Some allowance has to be made for the thickness of the mortar joints. Thus, a nominal 4" may be actually 3 7/8" or some other dimension.

Bricks are available in sizes ranging in thickness from a nominal 3" to 12". Heights vary from a nominal 2" to 8" and lengths up to 16" are common.

Figure 5-7 shows the most typical sizes of nonmodular brick. Sizes of modular brick are shown in Figure 5-8.

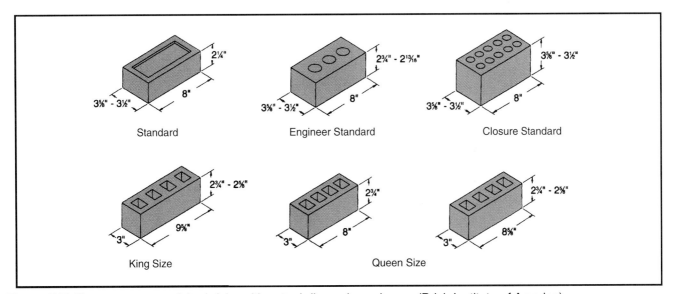

**Figure 5-7.** These are nonmodular bricks with actual dimensions shown. (Brick Institute of America)

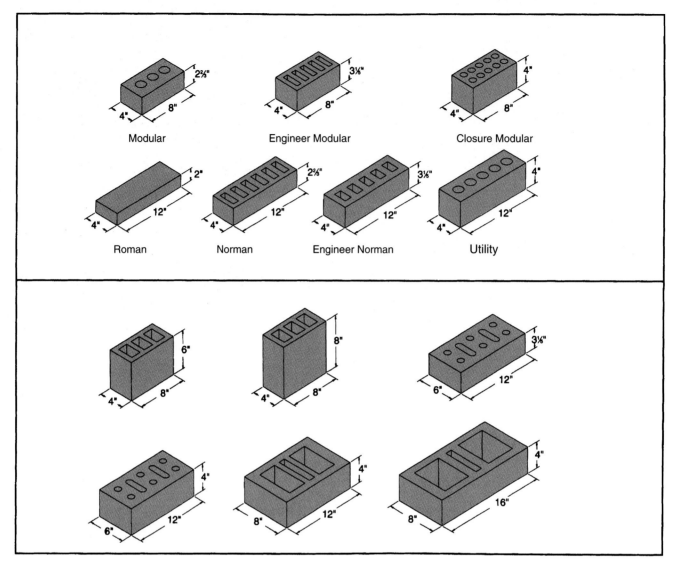

**Figure 5-8.** Modular brick sizes are shown in nominal dimensions. (Brick Institute of America)

Few manufacturers produce all the sizes shown and other sizes are produced by some manufacturers. As design requirements change, new sizes may be added and less popular sizes dropped. For these reasons, it is recommended that the sizes available in a given region be checked before purchasing.

With the exception of the *nonmodular standard, oversize,* and *3"* units, most bricks are produced in modular sizes. The nominal dimensions of modular bricks are equal to the manufactured dimensions plus the thickness of the appropriate mortar joint. See Figure 5-9. Usually the mortar joint is 3/8" or 1/2" thick. The standard joint thickness (3/8" or 1/2") is subtracted from the length, height, and thickness of the masonry unit to get the standard or specified dimensions of the unit. However, the actual dimensions may vary from the standard, usually by the amounts at the top of the next column:

1. Plus or minus 1/16" in thickness
2. 1/8" in width
3. 1/4" in length

### Why modular bricks are popular

Modular brick sizes became popular because they provided a quick and simple way of estimating the number of bricks needed. Since the nominal dimensions used in modular brick is the actual dimension of the brick plus the thickness of one mortar joint, the estimator can measure the area to be bricked and divide it by the nominal dimensions of the specific bricks to be used.

For example, suppose that the "standard modular" brick is 4" × 2 2/3" × 8". If a mortar joint of 1/2" is used, the actual manufactured dimensions would be 3 1/2" × 2 1/4" × 7 1/2". These dimensions should be used when ordering the bricks. Figure 5-10 shows the most common

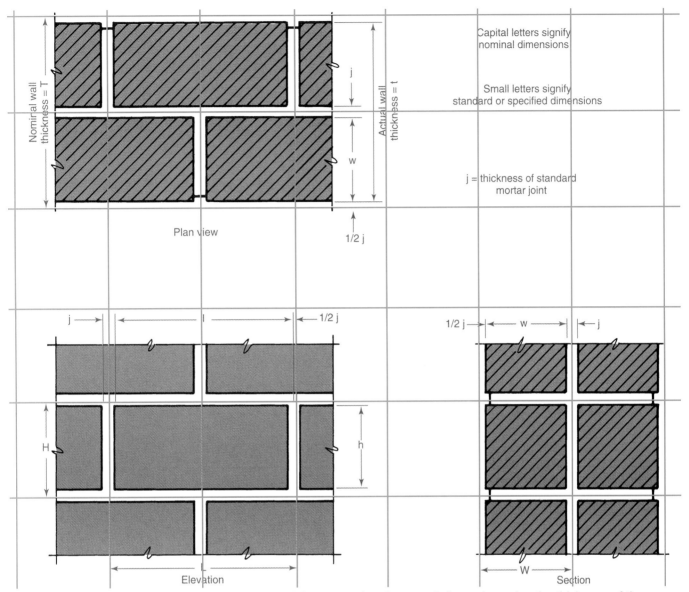

**Figure 5-9.** The nominal dimensions of modular brick are equal to the actual dimensions plus the thickness of the mortar joint.

brick sizes. The standard Brick Institute of America nomenclature for brick sizes is shown in Figure 5-11. Metric sizes are also shown in this illustration.

## Color and finishes

Facing bricks are produced in many colors, textures, and finishes. In fact, no two bricks are exactly alike. For this reason, **American Society for Testing and Materials (ASTM)** specifications for bricks and structural clay facing tile state that texture and color shall conform to an approved sample showing the full range of color and texture. In general, from three to five samples will be required depending on the range of color and texture in the bricks.

About 99% of all bricks produced is in the range of the reds, buffs, and creams. The actual color varies with the chemical makeup of the clay. Usually, this depends on the source. Color is also affected by the method of manufacture and the length and temperature of the burn.

Addition of chemicals and the method of molding and cutting the bricks determine texture. Much of the brick manufactured has smooth or sand finished texture. However, many other textures are produced on bricks. Figure 5-12 shows several brick textures created during the production process.

A large sample of popular bricks are included at the end of this chapter to illustrate the variety of color and texture available. See pages 96 through 101.

## Properties of Brick

The term **structural clay products** is generally used to indicate burned clay units that are used primarily in building construction. They may help support the structure or they may serve only as a decorative finish. Technically, when assembled in a structure, they should support their own weight. They may be loadbearing or nonloadbearing.

**Standard Nomenclature for Brick Sizes[1]**

| Unit Designation | Nominal Dimensions, in. | | | Joint Thickness[2], in. | Specified Dimensions[3], in. | | | Vertical Coursing |
|---|---|---|---|---|---|---|---|---|
| **Modular Brick Sizes** | | | | | | | | |
| | w | h | l | | w | h | l | |
| Modular | 4 | $2^2/_3$ | 8 | $^3/_8$ $^1/_2$ | $3^5/_8$ $3^1/_2$ | $2^1/_4$ $2^1/_4$ | $7^5/_8$ $7^1/_2$ | 3C = 8 in. |
| Engineer Modular | 4 | $3^1/_5$ | 8 | $^3/_8$ $^1/_2$ | $3^5/_8$ $3^1/_2$ | $2^3/_4$ $2^{13}/_{16}$ | $7^5/_8$ $7^1/_2$ | 5C = 16 in. |
| Closure Modular | 4 | 4 | 8 | $^3/_8$ $^1/_2$ | $3^5/_8$ $3^1/_2$ | $3^5/_8$ $3^1/_2$ | $7^5/_8$ $7^1/_2$ | 1C = 4 in. |
| Roman | 4 | 2 | 12 | $^3/_8$ $^1/_2$ | $3^5/_8$ $3^1/_2$ | $1^5/_8$ $1^1/_2$ | $11^5/_8$ $11^1/_2$ | 2C = 4 in. |
| Norman | 4 | $2^2/_3$ | 12 | $^3/_8$ $^1/_2$ | $3^5/_8$ $3^1/_2$ | $2^1/_4$ $2^1/_4$ | $11^5/_8$ $11^1/_2$ | 3C = 8 in. |
| Engineer Norman | 4 | $3^1/_5$ | 12 | $^3/_8$ $^1/_2$ | $3^5/_8$ $3^1/_2$ | $2^3/_4$ $2^{13}/_{16}$ | $11^5/_8$ $11^1/_2$ | 5C = 16 in. |
| Utility | 4 | 4 | 12 | $^3/_8$ $^1/_2$ | $3^5/_8$ $3^1/_2$ | $3^5/_8$ $3^1/_2$ | $11^5/_8$ $11^1/_2$ | 1C = 4 in. |
| **Nonmodular Brick Sizes** | | | | | | | | |
| Standard | | | | $^3/_8$ $^1/_2$ | $3^5/_8$ $3^1/_2$ | $2^1/_4$ $2^1/_4$ | 8 8 | 3C = 8 in. |
| Engineer Standard | | | | $^3/_8$ $^1/_2$ | $3^5/_8$ $3^1/_2$ | $2^3/_4$ $2^{13}/_{16}$ | 8 8 | 5C = 16 in. |
| Closure Standard | | | | $^3/_8$ $^1/_2$ | $3^5/_8$ $3^1/_2$ | $3^5/_8$ $3^1/_2$ | 8 8 | 1C = 4 in. |
| King | | | | $^3/_8$ | 3 3 | $2^3/_4$ $2^5/_8$ | $9^5/_8$ $9^5/_8$ | 5C = 16 in. |
| Queen | | | | $^3/_8$ | 3 | $2^3/_4$ | 8 | 5C = 16 in. |

[1]1 in. = 25.4 mm; 1 ft. = 0.3 m
[2]Common joint sizes used with length and width dimensions. Joint thicknesses of bed joints vary based on vertical coursing and specified unit height.
[3]Specified dimensions may vary within this range from manufacturer to manufacturer.

**Figure 5-10.** The nominal and manufactured dimensions are shown for most common modular and nonmodular bricks. Metric common bricks will have a nominal size of 215 mm × 102.5 mm × 65 mm. Mortar joints will be 10 mm. (Brick Institute of America)

Because brick masonry is bonded into an integral mass by mortar and grout, it is considered to be a homogeneous construction. It is the behavior of the combination of materials that determines the performance of the masonry as a structural element. However, the performance of a structural masonry element is dependent upon the properties of the constituent materials and the interaction of the materials as an assemblage. Therefore, it is important to first consider the properties of the constituent materials: clay and shale units, mortar, grout, and steel reinforcement.

Structural clay products must meet the following standards established by the Federal Trade Commission:
1. The composition is primarily of clay or shale or mixtures.
2. The ingredients have been fused together as a result of the application of heat.

When products do not meet these requirements, manufacturers must clearly show what the product is made of. For example, concrete brick, coral brick, plaster brick, sand-lime brick, and concrete structural tile are not ceramic products. Structural clay products may be separated into

| BIA Standard Nomenclature for Brick Sizes | | | |
|---|---|---|---|
| **Unit Name** | **Unit Dimensions** | | |
| | **Width in. (mm)** | **Height in. (mm)** | **Length in. (mm)** |
| Modular | 3 1/2 - 3 5/8 (89 - 92) | 2 1/4 (57) | 7 1/2 - 7 5/8 (190 - 194) |
| Standard | 3 1/2 - 3 5/8 (89 - 92) | 2 1/4 (57) | 8 (203) |
| Engineer Modular | 3 1/2 - 3 5/8 (89 - 92) | 2 3/4 - 2 13/16 (70 - 71) | 7 1/2 - 7 5/8 (190 - 194) |
| Engineer Standard | 3 1/2 - 3 5/8 (89 - 92) | 2 3/4 - 2 13/16 (70 - 71) | 8 (203) |
| Closure Modular | 3 1/2 - 3 5/8 (89 - 92) | 3 1/2 - 3 5/8 (89 - 92) | 7 1/2 - 7 5/8 (190 - 194) |
| Closure Standard | 3 1/2 - 3 5/8 (89 - 92) | 3 1/2 - 3 5/8 (89 - 92) | 8 (203) |
| Roman | 3 1/2 - 3 5/8 (89 - 92) | 1 5/8 (41) | 11 1/2 - 11 5/8 (292 - 295) |
| Norman | 3 1/2 - 3 5/8 (89 - 92) | 2 1/4 (57) | 11 1/2 - 11 5/8 (292 - 295) |
| Engineer Norman | 3 1/2 - 3 5/8 (89 - 92) | 2 3/4 - 2 13/16 (70 - 71) | 11 1/2 - 11 5/8 (292 - 295) |
| Utility | 3 1/2 - 3 5/8 (89 - 92) | 3 1/2 - 3 5/8 (89 - 92) | 11 1/2 -11 5/8 (292 - 295) |
| King Size | 3 (76) | 2 5/8 - 2 3/4 (67 - 70) | 9 1/2 - 9 5/8 (241 - 244) |
| Queen Size | 3 (76) | 2 3/4 (70) | 7 5/8 - 8 (194 - 203) |

**Figure 5-11.** *Brick Institute of America* nomenclature for brick sizes in U.S. Customary and metric. (Brick Institute of America)

solid masonry units and hollow masonry units. These two groups indicate how much solid material is in them.

## Solid masonry units

The *American Society for Testing and Materials* has set the standard. A solid masonry unit, they say, is "one whose cross-sectional area in every plane parallel to the bearing surface is 75% or more of its gross cross-sectional area measured in the same plane."

To know what this means we must understand some terms. The *bearing surfaces* are the tops and bottoms of the bricks or blocks.

Now we need to know what a "parallel plane of the bearing surface" is. Just imagine that you have placed a brick on its side and have sliced it into a number of thin pieces. The sides of each slice would be parallel planes. If you measured the area of the slices, then subtracted the area of the holes in the brick you could tell what percent is solid material. See Figure 5-13.

Many bricks are solid, but many facing bricks have holes running through them. These cored bricks are usually classified as solid masonry units because 75% of their area is solid material. The cored holes reduce weight, permit easier handling, and make the mortar joint stronger.

## Hollow masonry units

A *hollow masonry unit* is a brick with less than 75% solid material in "its net cross-sectional area in any plane." Structural clay tile are included in this classification.

## Weight of brick

The weight of a brick unit will depend on materials used, method of manufacture, burning, and the size of the brick. Since there are so many variables involved, you will have to check information secured from the manufacturer. Building bricks weigh about 4 1/2 lbs. to 5 lbs. each.

## Durability

Several conditions affect the weathering (durability) of a brick. These include heat, cold, wetting, drying, and the action of soluble salts.

Figure 5-14 shows the areas of the United States in which brick construction is subjected to severe, moderate, and negligible weathering.

The weathering index for any locality is found or estimated from material published by the *Weather Bureau, U.S. Department of Commerce*. This material is in the form of tables of *Local Climatological Data*. A severe weathering region has an index greater than 500. A moderate weathering region has an index of 100 to 500. A

Ruggs Face

Smooth Face

Stippled Face

Sand Mold Face

Sand Struck Face

Matt Face

**Figure 5-12.** These brick textures are made during the production process.

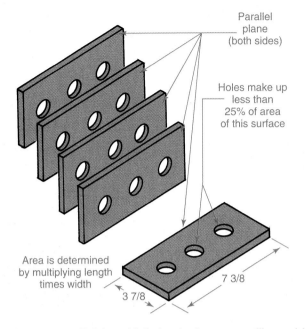

Parallel plane (both sides)

Holes make up less than 25% of area of this surface

Area is determined by multiplying length times width

7 3/8

3 7/8

**Figure 5-13.** Bricks with holes in them are still considered *solid* brick if 75% of the area in any parallel plane is solid material. Imagine that the brick is sliced like the drawing above. Each cut surface is a parallel plane. To find area, multiply length by width. Area of the holes is found by multiplying 3.14 by the square of the radius.

negligible weathering region has an index of less than 100. The only weathering action that has any real effect on burned clay products is freezing and thawing when moisture is present.

## Efflorescence

*Efflorescence* is a white powder or salt-like deposit on masonry walls, which is caused by water-soluble salts. See Figure 5-15. These salts collect on the wall's surface as water evaporates. The salts may have been mixed into the masonry units, mortar, or plaster.

Several things must happen for efflorescence to develop:

1. There must be soluble salts in the masonry wall.
2. Moisture must pass through the wall and carry the salts to the surface.

If either of these conditions are taken away, efflorescence will not occur.

## Compressive strength

*Compressive strength* is the ability of a masonry unit to stand up under heavy weight without crumbling. The material properties of brick and structural clay tile that have the most significant effect upon structural performance of the masonry are compressive strength and those properties affecting bond between the unit and

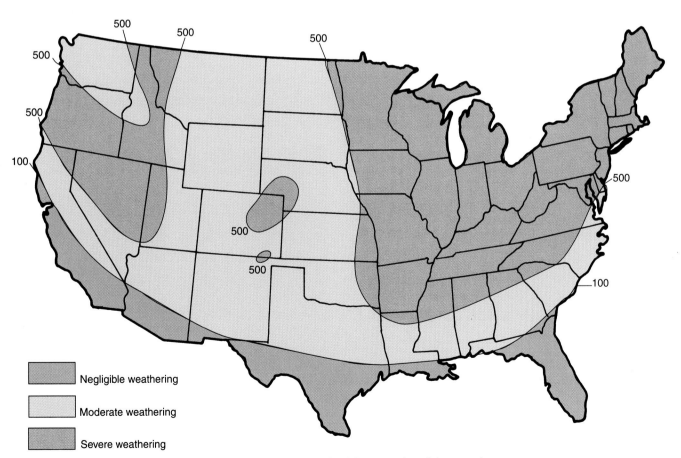

Negligible weathering

Moderate weathering

Severe weathering

**Figure 5-14.** Regions of the United States have been marked for severity of the weather.

Figure 5-15. Efflorescence on masonry walls is caused by soluble salts in wet brick being carried to the surface as water evaporates.

mortar. These properties include such things as rate of water absorption and surface texture. Four things will affect compressive strength:

1. Strength of the units (bricks) themselves
2. Strength of the mortar
3. Quality of workmanship
4. The proportion of gross area that is given a bearing surface at bed joints

High-compressive strength is developed by using high-strength brick and mortars and making well-filled joints. In most construction today, compressive strengths do not exceed 100 psi (pounds per square inch). Most brick have compressive strengths of over 4500 psi. Typical mortars have strengths from 750 psi to 2500 psi. Therefore, we can say that most masonry walls have ample compressive strength for current construction practices.

# Bonds and Patterns

The strength, durability, and appearance of a brick wall depends to a great extent upon the **bond** used to lay up the wall. The word bond, when it refers to masonry, has three meanings:

1. **Structural bond** is the method by which individual units are interlocked or tied together to form a wall. See Figure 5-16.
2. **Pattern bond** is the pattern formed by the masonry units and the mortar joints on the face of a wall. The pattern may be the result of the

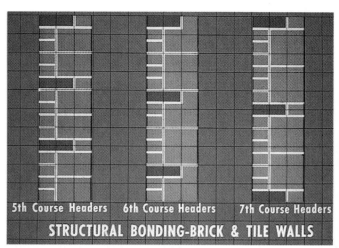

Figure 5-16. Three typical brick and tile walls are shown. They illustrate various structural bonding methods using headers. (Brick Institute of America)

structural bond used or may be purely for decorative purposes.
3. **Mortar bond (joint)** is the adhesion of mortar to the masonry units or reinforcing.

## Structural bonds

Structural bonding in brick masonry may be done in three ways:

1. Arranging the brick in an overlapping fashion.
2. The use of ties embedded (buried) in the mortar joints. See Figure 5-17.
3. The application of grout to adjacent wythes of masonry. A **wythe** is a single vertical tier or stack of masonry.

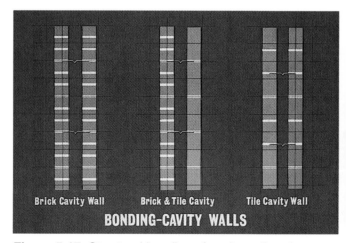

Figure 5-17. Structural bonding of cavity walls using metal ties. (Brick Institute of America)

The overlapping method is best illustrated by two traditional bonds known as **English bond** and **Flemish bond.** See Figure 5-18. The **English bond** consists of alternating courses of headers and stretchers. Alternating is defined as switching back and forth between two things.

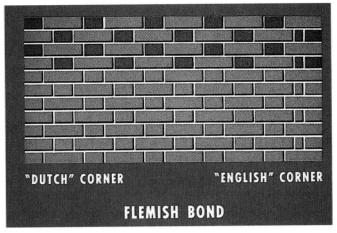

**Figure 5-18.** These two traditional bonds are used in brick masonry. (Brick Institute of America)

A *course* of bricks is a continuous level row of bricks. The *Flemish bond* consists of alternating headers and stretchers in every course, so that the headers and stretchers in every other course appear in vertical lines.

Stretchers are the bricks laid with the length of the wall. They develop longitudinal (lengthwise) bonding strength. Headers, laid across the width of the wall, bond the wall across its thickness. Figure 5-19 identifies the name of six brick positions. Figure 5-20 identifies the various brick positions in a wall.

Names have been given to various sizes of cut or broken bricks. See Figure 5-21. For example, a half brick is called a *bat* or *half.* Three-fourths of a brick is called a *three-quarter closure,* and one-fourth of a brick is called a *quarter closure.* When one corner is cut off, the remaining brick is called a *king closure.* A brick cut lengthwise across the end is a *queen closure.* Finally, a brick cut lengthwise parallel to the bed is called a *split.*

Modern building codes specify that masonry-bonded brick walls must have no less than 4% headers in the wall surface. The distance between nearest headers should not exceed 24" vertically or horizontally.

Structural bonding of masonry walls with metal ties is acceptable for solid wall and cavity wall construction. See Figure 5-22. No less than one 3/16" diameter metal tie should be used for each 4 1/2 sq. ft. of wall surface.

Ties in alternate courses should be staggered with the distance between adjacent (next to) ties not more than 24" vertically or 36" horizontally. Additional bonding ties should be placed around the perimeter of openings. These should be spaced not more than 3' apart and within 12' of the opening. If smaller diameter ties are used, the spacing between them should be reduced.

Solid and reinforced brick masonry walls may be structurally bonded by pouring grout into the cavity between wythes of masonry. *Grout* is a liquid cement mixture used to fill voids.

The method of bonding used will depend on the wall type, use requirements, and other factors. The metal tie is usually recommended for exterior walls. It makes construction easier and provides greater resistance to rain penetration. The metal tie also allows for slight movements of facing and backing, which may prevent cracking.

Unit texture and absorption are properties that affect the bond strength of the masonry assemblage. In general, mortar bonds better to roughened surfaces, such as wire cut surfaces, than to smooth surfaces, such as die skin surfaces. Cores or frogs provide a means of mechanical interlock. The bond strength of sanded surfaces is dependent upon the amount of sand on the surface, the sand's adherence to the unit, and the absorption rate of the unit at the time of laying.

In practically all cases, mortar bonds best to a unit whose suction at the time of laying is less than 30 g/min/30 sq. in. When the suction is greater than this amount, the brick should be wetted to reduce the rate of water absorption of the unit prior to laying.

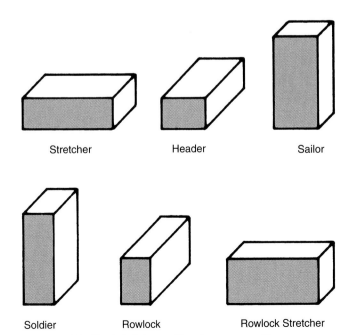

Stretcher    Header    Sailor

Soldier    Rowlock    Rowlock Stretcher

**Figure 5-19.** Each brick position has a specific name.

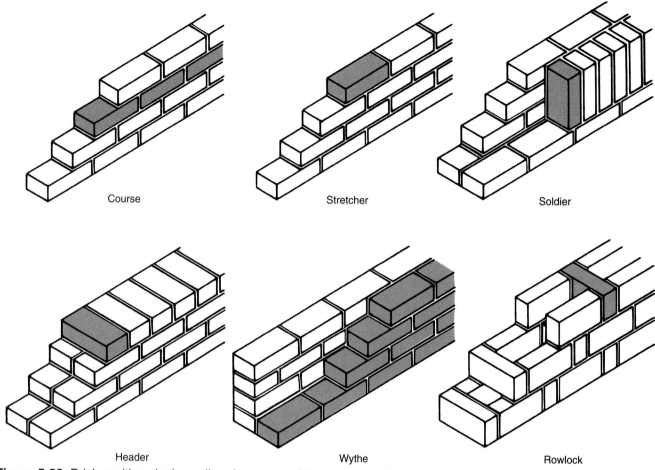

**Figure 5-20.** Brick positions in the wall and terms used in masonry wall construction.

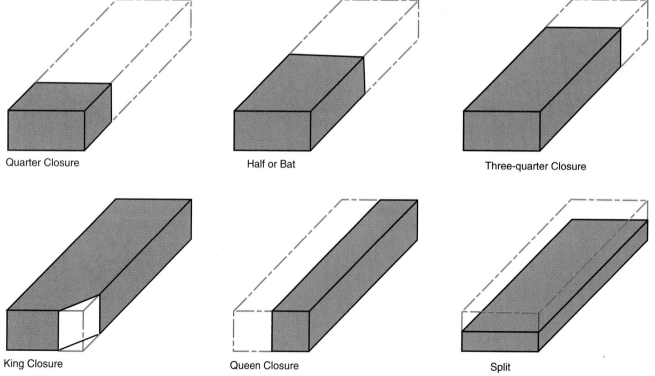

**Figure 5-21.** Names used to describe cut or broken brick pieces.

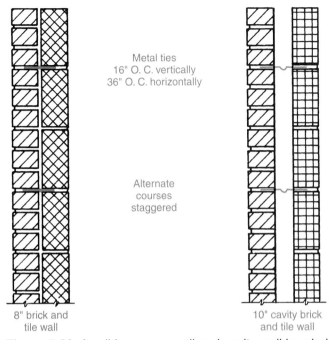

Metal ties
16" O. C. vertically
36" O. C. horizontally

Alternate
courses
staggered

8" brick and
tile wall

10" cavity brick
and tile wall

**Figure 5-22.** A solid masonry wall and cavity wall bonded with metal ties.

## Pattern bonds

Often, structural bonds, such as Flemish or English, are used to create patterns in the face of the wall. Patterns may be created by the way the mortar joint is handled. The arrangement of the brick remains unchanged.

It may also be desirable to produce a pattern bond by allowing certain brick to stick out beyond or sink below the plane (surface) of the wall. Pattern bonds can also be created by varying the texture of brick units.

## Five basic bonds

There are five basic structural bonds. These bonds include the running bond, the common or American bond, the Flemish bond, the English bond, and the block or stack bond.

Color and texture in both brick and joints create endless variety. Figure 5-23 shows how many different patterns can be arranged. Figure 5-24 illustrates the basic structural bonds.

**Running bond** consists of all stretchers and is used largely in cavity wall and veneer wall construction. It is the simplest of the basic pattern bonds. See Figure 5-25.

**Common** or **American bond** is a variation of the running bond that has a course of full-length headers at regular intervals to provide structural bonding as well as pattern. Header courses generally are used every fifth to seventh course. See Figure 5-26.

**Flemish bond** alternate courses of stretchers and headers. The headers in alternate courses are centered over the stretchers in the course between them.

If the headers are not needed for structural bonding, as in veneer wall construction, half bricks may be used. These half bricks are called **clipped** or **snap headers**.

**Figure 5-23.** This wall is decorated with a collection of pattern bonds.

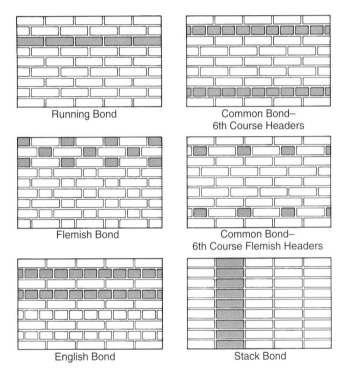

Running Bond

Common Bond–
6th Course Headers

Flemish Bond

Common Bond–
6th Course Flemish Headers

English Bond

Stack Bond

**Figure 5-24.** The basic structural bonds.

**Figure 5-25.** This reclaimed brick is laid in running bond.

**Figure 5-26.** The common or American bond is a variation of the running bond. Two variations are shown here. (Brick Institute of America)

Garden wall bond with units
in dovetail fashion

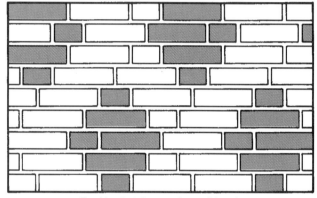

Double stretcher garden wall bond
with units in diagonal lines

**Figure 5-27.** Two variations of the garden wall bond.

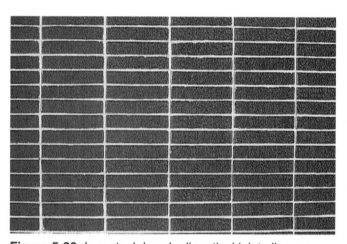

**Figure 5-28.** In a stack bond, all vertical joints line up.

The Flemish bond may be varied by increasing the number of stretchers between headers in each course. Three stretchers alternating with a header is known as a **garden wall bond**. Two stretchers between headers is called a **double stretcher garden wall bond**. Four or five stretchers between headers may be laid to form other garden wall bonds. See Figure 5-27.

**English bond** is made up of alternate courses of headers and stretchers. The headers are centered on the stretchers and joints between stretchers in all courses are aligned vertically. Snap headers may be used when structural bonding is not necessary.

**English cross** or **Dutch bond** is a variation of English bond. The only difference is that vertical joints between the stretchers in alternate courses do not align vertically. These joints center on the stretchers themselves in the courses above and below.

**Block** or **stack bond** is purely a pattern bond. All vertical joints are aligned. There is no overlapping of units. See Figure 5-28. Since the bond lacks the strength that other bonds provide, it is usually bonded to the backing with rigid steel ties. In loadbearing construction or large wall areas, it is advisable to reinforce the wall with steel. This is placed in the horizontal mortar joints.

Figure 5-29 shows a pattern that was made through variations in color of the brick used. This is a common way to develop patterns.

## Wall texture

Many new and interesting patterns and effects are created by projecting and recessing units, Figure 5-30, and omitting units to form perforated walls or screens, Figure 5-31. These techniques greatly extend the traditional patterns and add a new dimension to brick masonry. Figure 5-32 shows a modern brick sculptured wall which incorporates a variety of colors and textures.

## Mortar joints

**Mortar joints** use mortar between brick units to perform one or more of the following four functions:

1. Bonds units together and seals spaces between them.
2. Bonds to reinforcing steel and causes it to be part of the wall.
3. It compensates for variations in the size of brick units.
4. It provides a decorative effect.

Mortar joint finishes are either troweled or tooled joints. In the **troweled joint**, the excess mortar is struck (cut off) with a trowel and finished with the same tool. The **tooled joint** is made with a special tool to compress and shape the mortar in the joint. Figure 5-33 shows typical mortar joints used in brick masonry.

**Concave** and **V–shaped** joints are formed by the use of a steel jointing tool. These joints are usually rather small. They are effective in resisting rain penetration. Their use is recommended where heavy rains and high winds are likely to occur. This is the most popular mortar joint.

The **weathered** joint requires care since it must be worked from below. It is a very functional joint because it is compacted and readily sheds water.

**Figure 5-29.** Glazed brick and tile of various colors were used to create this pattern.

**Figure 5-30.** Pattern bonds may be produced by projecting or recessing certain bricks from the plane (surface) of the wall as shown in these illustrations.

**Figure 5-31.** Brick screens are highly decorative. They also help reduce noise and control light.

**Figure 5-32.** A panoramic view created with sculptured brick. (General Shale Brick)

A ***struck*** joint is common for ordinary brickwork. The joint is struck with the trowel and some compacting occurs, but it does not shed water as well as other joints.

The ***rough cut*** or ***flush*** joint is the simplest joint to make. It is made by holding the trowel flat against the brick and cutting in any direction. The joint is not compacted and leaves a small hairline crack where the mortar is pulled away from the brick by the trowel. This joint may not be watertight.

The ***raked*** joint is made by removing some of the mortar while it is still soft. This joint is difficult to make watertight even though slightly compacted. It is not recommended for areas with heavy rains, high winds, and freezing temperatures. However, this joint is very effective with irregular brick. It tends to enhance its character.

The ***extruded*** joint is made by squeezing out the mortar as the unit is laid. It is not trimmed off but is left to harden in its extruded form. This joint is best suited for dry climates. See Figure 5-34.

**Figure 5-34.** The extruded mortar joint is best suited for dry climates because its ledges do not shed water very well. (Portland Cement Association)

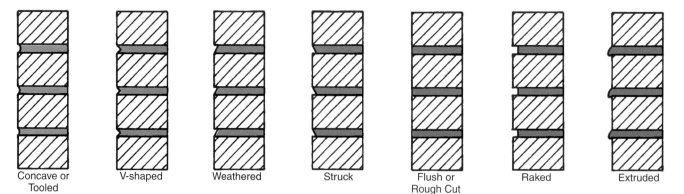

Concave or Tooled   V-shaped   Weathered   Struck   Flush or Rough Cut   Raked   Extruded

**Figure 5-33.** Typical mortar joints used in masonry work.

Colored mortars may be used to accent brick masonry patterns. See Figure 5-35. It may be applied in two ways:

1. The entire mortar joint may be colored.
2. Where a tooled joint is used, tuck-pointing is the preferred method.

When tuck-pointing, the entire wall is finished with a 1" deep raked joint. The colored mortar is filled in later, with special care taken to keep the colored mortar off the brick.

**Figure 5-35.** Colored mortar may be used to accent the brick pattern. (Brick Institute of America)

## Special Shapes

One of the primary reasons for the popularity of brick with architects, designers, and builders is its ability to be adapted to a wide variety of design criteria. See Figure 5-36. It can be shaped and used in traditional and contemporary detailing—from arches to sills, the water table shown in Figure 5-37, copings, columns, and the treads shown in Figure 5-38. Special shapes add a unique and distinctive touch, projecting individual styling and taste.

**Figure 5-36.** Handmade bricks have been used in a variety of shapes to produce this example of classic architecture.

**Figure 5-37.** The water table on this traditional home was formed with specially shaped bricks. (General Shale Brick)

**Figure 5-38.** These traditional steps with rounded nosing were made possible with specially shaped bricks. (General Shale Brick)

Many shapes such as these are manufactured by large brick companies, but other shapes may be desired that do not fall within this range. See Figure 5-39. Custom shapes that conform to the designer's specification can be produced in either extruded, molded, or handmade brick. You should consult with the manufacturer before specifying unique shapes to determine the most practical and cost effective method of achieving the desired shape. Figure 5-40 shows a residential structure that incorporates special shapes in the jack arches, treads, circular arch, and water table.

## Hollow Masonry Units (Tile)

Hollow masonry units are a machine-made product that is extruded (squeezed) through a die and cut to the desired height or length. The raw materials may be surface clay, shale, fire clay, or combinations of these. They are pulverized (crushed), mixed with water, and then extruded.

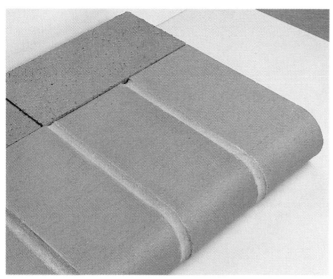

**Figure 5-39.** Specially shaped bricks, such as these pool coping units, are available for specific applications. (Glen-Gery Corporation)

**Figure 5-40.** This traditional home incorporates several examples of specially shaped and custom brick units. (General Shale Brick)

After the moisture content is reduced to a satisfactory level, they are fired in a kiln at temperatures from 1750°F to 2500°F.

Unlike bricks, tile is hollow. Hollow in this instance means that each masonry unit has a net cross-sectional area in any plane parallel to the bearing surface less than 75% of its gross cross-sectional area measured in the same plane (Standards of the **American Society for Testing and Materials**). Hollow clay tile is classified as either structural clay tile or structural clay facing tile.

## Structural clay tile

Structural clay tile is produced as loadbearing and nonloadbearing types. Both loadbearing and nonloadbearing tile is produced with vertical or horizontal cells, which are the open spaces through the tile.

When the tile is to be laid with the cells in a horizontal plane, the unit is called **side construction** or **horizontal cell tile**. When the tile is to be laid with the cells in a vertical plane, the unit is called **end construction** or **vertical cell tile**. The size, number, shape, and thickness of cells will vary from one manufacturer to another.

## Loadbearing tile

Structural clay loadbearing tile is suitable for use in walls as backup or in partitions carrying superimposed loads (weight of other parts). Loadbearing wall tiles are produced in two grades—LB and LBX.

**Grade LBX tile** is suitable for use in areas exposed to frost action. **Grade LB tile** is designed for areas not exposed to frost action.

## Nonloadbearing tile

Structural clay nonloadbearing tile is suitable for use as fireproofing, furring, or the construction of partitions that do not support superimposed loads. Nonloadbearing tile is available in one grade—NB.

## Sizes and shapes

Structural clay tile is most often used in actual thickness of 3" and nominal thicknesses of 4", 6", and 8". However, nominal 2", 10", and 12" thick units are available. Heights for clay tile units are 5 1/3", 8", and 12" nominal.

Tile unit length is generally 12" nominal. However, 8" and 16" units are available in some areas.

Figure 5-41 shows some of the many shapes and sizes of structural clay tile. Other shapes include header units, jamb units, closures, and kerfed units for use in reinforced bond beams or lintels.

The most widely available unit shapes and sizes of structural clay tile are shown in Figure 5-42. The units shown are all horizontal cell tile. The same sizes are also manufactured in vertical cell tile units.

All structural clay tile is designed to be used with 1/2" mortar joints. All are modular sizes. The basic group shown in Figure 5-42 provides compatibility (can be used together) with all facing tile and modular brick sizes.

## Properties of structural clay tile

Properties of structural clay tile include color, texture, strength, and variations in dimensions. Structural clay tile is produced in a wide range of colors. The greatest influence upon the color will be the chemical composition of the raw materials and the temperature of the burn. Due to these factors, color alone cannot be used as a measure of quality.

Structural clay tile is produced with surface textures. The two most generally seen are the scored and wire cut. Wire cut is known as "universal finish" in some regions. Both surfaces will readily receive plaster and the wire cut may be painted or left exposed.

The compressive strength of structural clay tile is affected not only by the raw materials and method of manufacture, but also by the design of the unit. ASTM standard

Various Sizes and Shapes of Structural Clay Tile

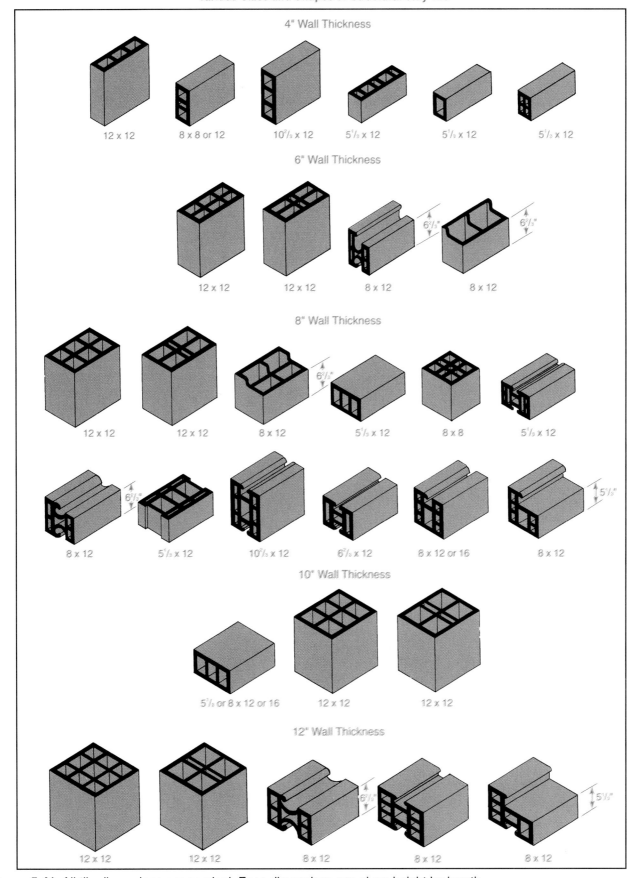

**Figure 5-41.** All tile dimensions are nominal. Face dimensions are given height by length.

Most Widely Available Sizes of Structural Clay Tile

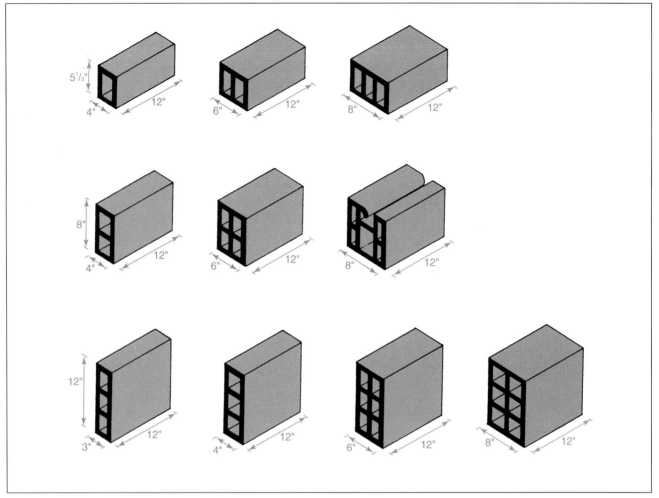

**Figure 5-42.** All units shown are horizontal cell tile, but vertical cell tile is also available. Dimensions are nominal.

C34—62, *Specifications for Structural Clay Loadbearing Wall Tile* requires that the average compressive strength of five units be not less than the following:

Most clay tile units produced today will have compressive strengths of two to four times these requirements.

| Average Minimum Compressive Strength | | |
|---|---|---|
| Grade | End Construction | Side Construction |
| | Tile (in psi) | Tile (in psi) |
| LBX | 1400 | 750 |
| LB | 1000 | 750 |

Variations in sizes are also controlled by ASTM. The biggest allowable variation is 3% more or less than required by the actual dimensions.

### Structural clay facing tile

Structural clay facing tile consists of two broad types. These types are glazed and unglazed. See Figure 5-43.

**Glazed clay facing tile** is produced from high-grade, light-burning fire clay suitable for the application of ceramic glaze. (They may be left unglazed.) There are two grades of glazed facing tile—S and G.

**Grade S (select)** is intended for use with somewhat narrow mortar joints. **Grade G (ground edge)** is used where face dimension must be very exact. An example of this would be a stacked bond, where units are placed in the wall so that both the horizontal and vertical joints are continuous. Both grades may be purchased as single-faced units or two-faced units. A single-faced unit will have one face exposed. In a two-faced unit, two opposite finished faces will be exposed.

### Unglazed facing tile

**Unglazed facing tile** is made from either light-burning fire clay or shale and other darker-burning clays. These tiles are not glazed and have either a smooth or a rough-textured finish. They can be purchased as standard or heavy duty tile. The difference is in the thickness of the face shells.

Another classification for unglazed facing tile is based on factors that control appearance of the finished wall. This classification includes FTX and FTS types.

**FTX unglazed facing tile** is suitable for general use in exposed exterior and interior masonry walls and partitions. It is easily cleaned, resists staining, and is low in absorption. This type is also best where these characteristics are important:

**Figure 5-43.** Structural facing tile is produced with either a glazed or unglazed face.

1. A high degree of mechanical perfection
2. Narrow color range
3. A minimum variation in face dimensions

***FTS unglazed facing tile*** is smooth or rough textured. These tile are suitable for the following:

1. General use in exposed exterior or interior masonry walls and partitions
2. Special conditions where tile of moderate absorption are required
3. Usage where moderate variation in face dimensions and medium color range is acceptable
4. Usage where minor defects in surface finish or small handling chips are acceptable.

***SCR acoustile*** is designed to absorb sound. Typical units are shown in Figure 5-44. Distinguishing characteristics are the perforations through the faces and the fibrous glass pads behind the faces. The perforations (holes) can be circular or slotted, the same or different in size, and of regular or random pattern.

All types of structural clay facing tile, glazed or unglazed, can have an acoustile. These units may be used in wall or ceiling applications.

## Sizes and shapes

Some sizes of structural clay facing tile are listed in Figure 5-45. Some makers also produce these units with a nominal 4" × 12" face size.

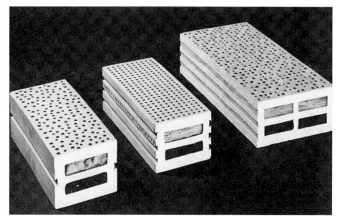

**Figure 5-44.** These are typical SCR acoustile units designed to absorb sound. Note use of insulating blanket behind the face. (Brick Institute of America)

Structural facing tile are designed to be laid with a 1/4" mortar joint. The specified unit dimensions are 1/4" less than the nominal dimensions shown in Figure 5-45 under Available Sizes.

Nominal modular sizes of structural clay facing tile are shown in the bottom part of Figure 5-45. Units shown in nominal modular sizes are usually designed to be laid with either 1/4", 3/8", or 1/2" joints.

The basic shapes of structural clay facing tile are:

1. Stretchers
2. Corners and closures
3. Starters and miters

A wide variety of supplementary shapes are also available for complex wall layouts. These shapes include

Structural Clay Facing Tile
Available Sizes

| Series | Nominal face dimensions in inches | Nominal thickness in inches |
|---|---|---|
| 6T | 5 1/3 by 12 | 2, 4, 6, 8 |
| 4D | 5 1/3 by 8 | 2, 4, 6, 8 |
| 4S | 2 2/3 by 8 | 2, 4 |
| 8W | 8 by 16 | 2, 4 |

| Nominal Modular Sizes* | |
|---|---|
| Face Dimensions | |
| Height by length in inches | Height by length in inches |
| 4 by 8 | 8 by 8 |
| 4 by 12 | 8 by 12 |
| 5 1/3 by 8 | 8 by 16 |
| 5 1/3 by 12 | 12 by 12 |
| Thickness: All of the above are in nominal thicknesses of 4, 6 and 8 inches | |

* Nominal sizes include the thickness of the standard mortar joint for all dimensions.

**Figure 5-45.** The typical sizes of structural clay facing tile. Joint thickness should be deducted from the dimensions above in determining actual size of the units.

the following:

1. Sills
2. Caps and lintels
3. Cove base stretchers and fittings
4. Coved internal corners
5. Octagons
6. Radials

Figure 5-46 shows some of the typical shapes.

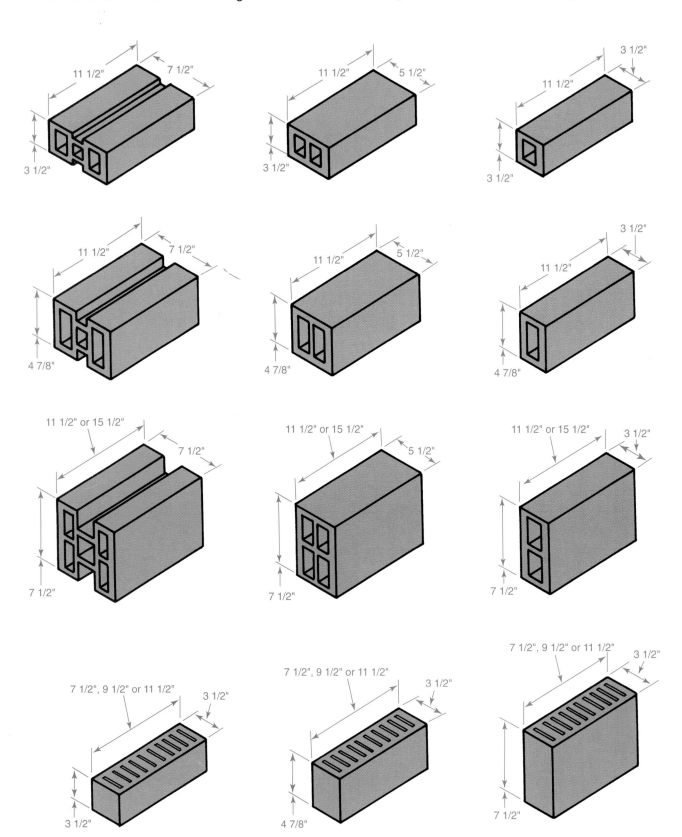

**Figure 5-46.** Typical shapes of structural clay facing tile. Included are stretchers, jamb units, and corner units.

All structural clay facing tile is modular, which means that all units are multiples of 4". They will fit together with each other and with all other modular building materials. This reduces cutting. However, some units have a dimension of 5 1/3", which is 1/3 of 16" rather than a multiple of 4".

Clay tile is also produced in various shapes for use as decorative screens. Figures 5-47 and 5-48 show some of the different patterns possible with tiles.

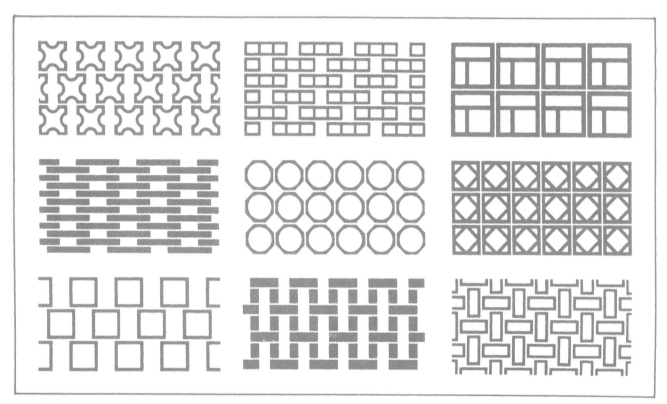

**Figure 5-47.** Clay tile in shapes used for decorative screens.

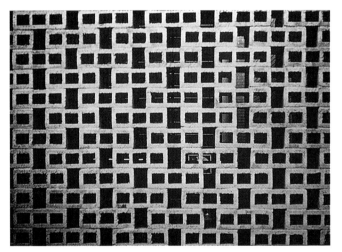

**Figure 5-48.** A wall in one of the patterns shown (top center) in Figure 5-47.

# Architectural Terra Cotta

Architectural terra cotta is a custom (made to order) product. It is produced in sizes and shapes to meet specific requirements. An unlimited color range of high-fired, ceramic vitreous (glass-like) glazes are available. See Figure 5-49. The configurations (arrangement of parts), forms, and radius shapes in which architectural terra cotta can be made are almost unlimited.

## Classification of architectural terra cotta

Architectural terra cotta includes several classes of products. These classes include anchored ceramic veneer, adhesion ceramic veneer, and ornamental or sculptured terra cotta.

**Figure 5-49.** These are examples of the broad color range in which architectural terra cotta is produced. (Interpace Corp.)

## Anchored ceramic veneer

*Anchored ceramic veneer* is usually produced in thicker sections held in place by grout and wire anchors connected to its backing. Typical maximum sizes are shown in Figure 5-50. These are standard die sizes. It is cheaper to use these sizes. Anchored ceramic veneer may be used as facing for exterior and interior walls, column facings, soffits, copings, and sills.

## Adhesion ceramic veneer

*Adhesion ceramic veneer* is usually produced in thin sections. It is held in place without metal anchors by a mortar backing. Thickness is ordinarily 1". Largest lengths and widths are shown in Figure 5-51.

Adhesion ceramic veneer can be installed over concrete or masonry, metal, or wood. It is used primarily as facing for exterior and interior walls and coping units. See Figure 5-52.

Anchored Ceramic Veneer
Maximum Finished Sizes

| A | B | C |
|---|---|---|
| 1' – 5 3/4" | 3' – 0" | 7 3/4" |
| 1' – 6 3/4" | 3' – 0" | 7 3/4" |
| 1' – 7 3/4" | 3' – 6" | 7 3/4" |
| 1' – 8 3/4" | 3' – 6" | 7 3/4" |
| 1' – 9 3/4" | 3' – 6" | 7 3/4" |
| 1' – 11 3/4" | 4' – 3" | 11 3/4" |
| 2' – 1 3/4" | 4' – 3" | 11 3/4" |
| 2' – 3 3/4" | 4' – 3" | 11 3/4" |
| 2' – 5 3/4" | 4' – 3" | 1' – 2 3/4" |
| 2' – 8 3/4" | 4' – 3" | 11 3/4" |

**Figure 5-50.** Anchored ceramic veneer is produced in the maximum finished sizes shown.

Adhesion Ceramic Veneer
Maximum Finished Sizes

| A | B | C |
|---|---|---|
| 5 3/4" | 2' – 0" | 3 3/4" |
| 7 3/4" | 2' – 0" | 3 3/4" |
| 11 3/4" | 2' – 0" | 3 3/4" |
| 1' – 2 3/4" | 2' – 2" | 3 3/4" |
| 1' – 3 3/4" | 2' – 2" | 3 3/4" |
| 1' – 5 3/4" | 2' – 2" | 5 3/4" |
| 1' – 7 3/4" | 2' – 6" | 4 1/2" |

**Figure 5-51.** Adhesion ceramic veneer is produced in these maximum finished sizes.

**Figure 5-52.** Adhesion ceramic veneer is available in a variety of colors for use as facing and coping units. (Interpace Corp.)

## Ornamental or sculptured terra cotta

*Ornamental terra cotta* is produced in sculptured patterns or freestanding sculpture. It is used frequently as cornices and column capitals on large buildings.

Ornamental terra cotta can be either anchored or adhesion veneer. It depends on the application. Molded pieces may also be made in larger dimensions than facing veneer.

# REVIEW QUESTIONS
# CHAPTER 5 ▰▰▰▰▰▰▰▰

Write all answers on a separate sheet of paper. Do not write in this book.

1. What three main types are structural clay products classified as?
2. The three methods of manufacturing brick are the soft mud process, the stiff mud process, and the _____ process.
3. Bricks remain stationary while being fired in a periodic kiln, but are placed on moving cars in the _____ kiln.
4. Identify five of the seven classifications of brick.
5. A special type of brick made from a refractory ceramic material that is highly heat resistant is known as _____ brick.
6. Name the three grades of building brick.
7. When appearance is important as well as durability and strength, _____ bricks are used.
8. What is the difference between hollow brick and facing brick?
9. Paving bricks are assigned a type by the traffic or _____ expected.
10. A brick that has ominal dimensions based on the 4" module is called a _____ brick.
11. Why did the brick type mentioned in Question 10 become popular?
12. The usual mortar joint thickness for brick is _____ " or _____ ".
13. Identify three factors that determine brick color.
14. A masonry unit whose net cross-sectional area in every plane parallel to the bearing surface of 75% or more of its gross cross-sectional area measured in the same plane is known as a(n)_____ masonry unit.
15. What is the weight of a typical standard building brick?
16. Efflorescence is caused by _____ salts.
17. _____ strength is the ability of a masonry unit to stand up under heavy weight without crumbling.
18. The three types of bonds discussed in this chapter were structural bond, pattern bond, and mortar bond. Identify the three methods of producing a structural bond.
19. A(n) _____ of brick is a continuous level row of brick.
20. Three-fourths of a brick is called three-quarter _____.
21. A brick that is laid across the width of a wall is called a(n) _____.
22. When a structural bond is used to create patterns in the face of the wall, you then have a(n) _____ bond.
23. The simplest of the basic pattern bonds is the _____ bond.
24. The two types of mortar joint finishes are troweled and _____ joints.
25. Name four types of mortar joints.
26. Hollow clay tile may be classified into two groups: structural clay tile and structural facing tile. Structural clay tile is produced as _____ and _____ types.
27. Structural clay tile units are generally a nominal _____ " in length.
28. All structural clay tile is designed to be used with _____ " mortar joints.
29. Structural clay facing tile may be divided into the two broad types of _____ and _____.
30. What is the function of SCR acoustile?
31. Identify the three classification types of architectural terra cotta.

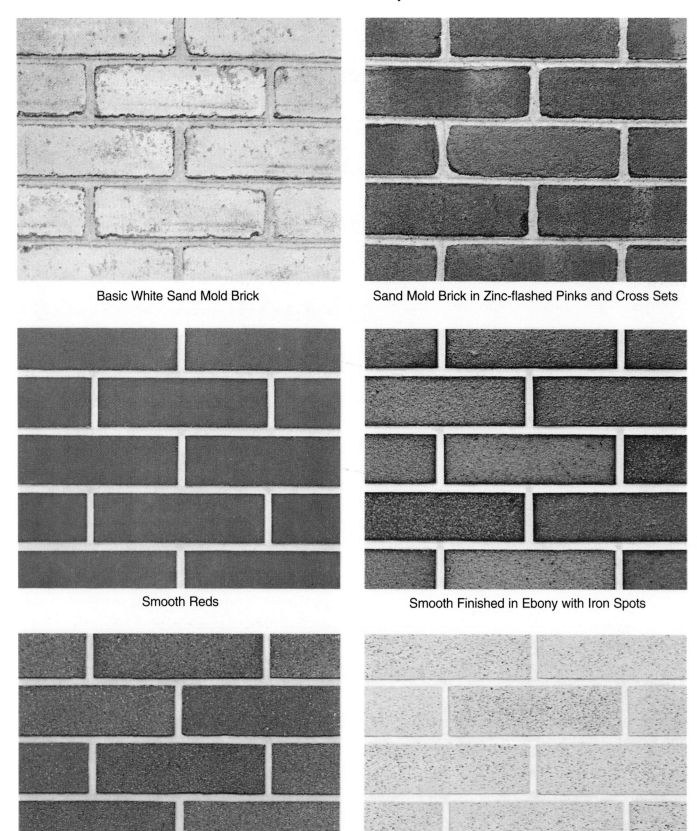

Basic White Sand Mold Brick

Sand Mold Brick in Zinc-flashed Pinks and Cross Sets

Smooth Reds

Smooth Finished in Ebony with Iron Spots

Sand Finished with Soft Pinks and Dusty Rose

Smooth Gray with Manganese Specks

Matt Textured Fireclay Brick

Textured Charcoals

Matt Textured Burgundy

White Velour Texture

Tan Flashed Shades and Cross Sets

Water Struck Brick in Fine Sand-finished Texture

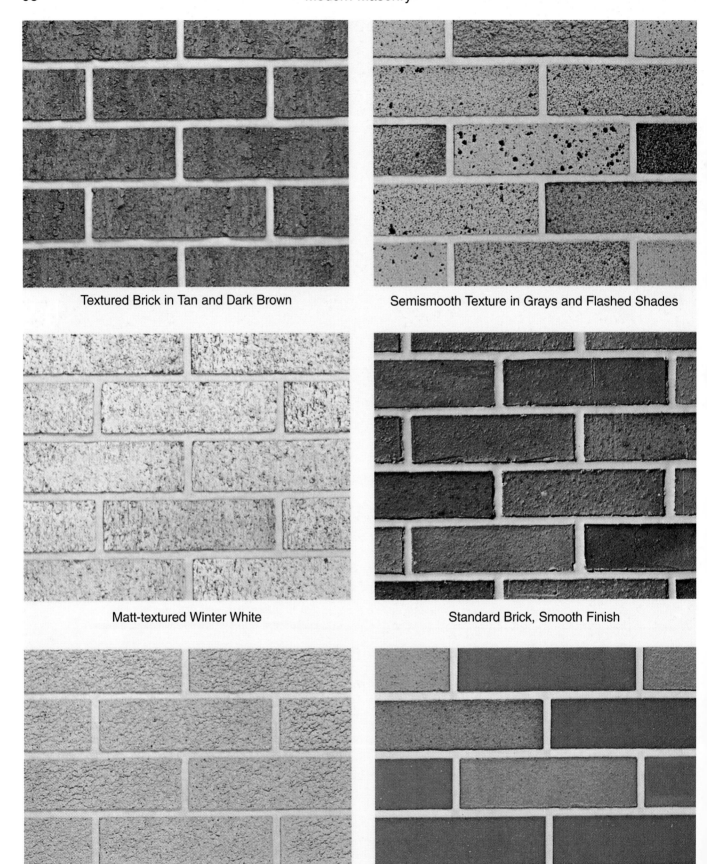

Textured Brick in Tan and Dark Brown

Semismooth Texture in Grays and Flashed Shades

Matt-textured Winter White

Standard Brick, Smooth Finish

Matt Texture in Buff

Semismooth in Reds, Cross Sets, Hearts, Brown, and Tans

Matt Textured Multicolors

Matt Texture in Pinks and Tans

Sand Finished in Buff Pink, Tan, and Flashed Shades

Matt Texture in Gray Shades

Rustic Sand Finish

Antique Finish

Textured Snow White

Sand Finish in Pinks, Reds, and Flashed Shades

Sand Texture in Brown

Double-fired, Glazed, Smooth, and Textured

Sand Mold Brick in Reds, Tans, Cross Sets, and Earth Tones

Velour Texture in Pink

Molded, Textures in Brown

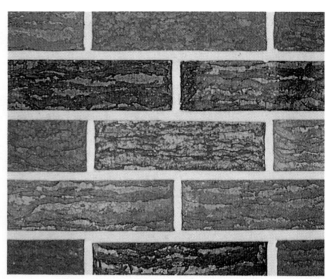

Textured in Grays, Green, and Brown

Sand Molded in Grays and Tans

Sand Molded in Earth Tones

Colonial Russet Brown

Standard Striated in Brown

The Classico paver system, comprising five different shapes, permits the creation of circles, half circles, fans, sweeping curves, geometric, and random patterns. (Uni-Group USA)

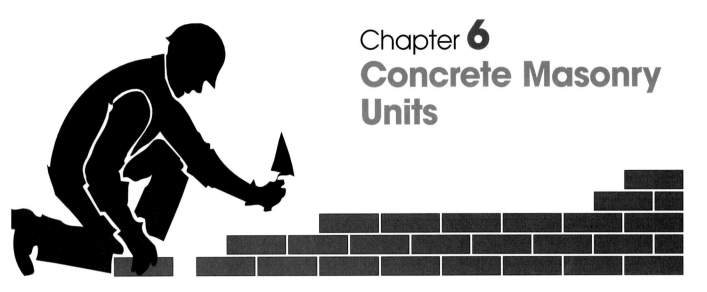

# Chapter 6
# Concrete Masonry Units

The term *concrete masonry* includes all the sizes and kinds of hollow or solid block, brick, and concrete building tile, so long as they are made from concrete and are laid by masons. See Figure 6-1.

There is no doubting the usefulness of concrete masonry. We have only to look at the range of sizes and shapes of units being produced today. A modern block manufacturing plant will offer over 100 different sizes and shapes. Today, over 700 different units are made.

Concrete masonry is popular because it is inexpensive to make and easy to use. It is a good insulator, has good sound reduction properties, and stands up well in all kinds of weather. It is fire resistant and can be bought almost anywhere in many sizes, textures, and shapes.

## How Concrete Masonry Units Are Made

Concrete masonry units are made from a relatively dry mixture of Portland cement, aggregates, water, and

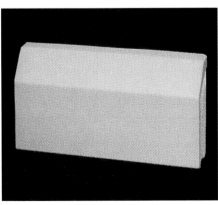

**Figure 6-1.** Concrete block materials—cracked block, screen block, typical concrete blocks, and prefaced blocks. (Trenwyth Industries, Inc.)

often admixtures. Sometimes other cement-like materials are used in place of Portland cement. **Aggregate** means sand and gravel or a suitable substitute. **Admixtures** are such things as coloring agents, air-entraining materials, accelerators, retarders, and water repellents. These materials are mixed, and then the mixture is molded into desired shapes through compaction and vibration. See Figure 6-2. **Compaction** is a squeezing action to make a material more dense. They are then cured under controlled moisture and temperature conditions. After a period of aging they are ready for use.

**Figure 6-2.** In this machine, the concrete mixture is molded into desired shapes through compaction and vibration. (Besser Company)

## Aggregates

Aggregates normally make up about 90% of the block by weight. Desirable properties of aggregates include:
1. Uniform amounts of fine and coarse sizes
2. Toughness, hardness, and strength to resist impact abrasion and loading
3. Ability to resist the forces exerted by freezing, thawing, expansion, and contraction
4. Cleanliness and lack of foreign materials which would reduce strength and/or cause surface roughness

The two classes of aggregates by weight are normal weight (dense) and lightweight. **Normal weight aggregate** includes sand, gravel, crushed limestone, and air-cooled slag. **Lightweight aggregate** is expanded shale or clay, expanded slag, coal cinders, pumice, and scoria. **Scoria** is the scrap left after the melting of metals. The effect of aggregate on weight, strength, and other characteristics is shown in Figure 6-3.

# Classification of Concrete Masonry Units

Concrete masonry units are classified into six main groups based on their intended use, size, and appearance. The groups include:
1. *Concrete Brick*, ASTM C 55-95
2. *Loadbearing Concrete Block*, ASTM 90-95
3. *Nonloadbearing Concrete Block*, ASTM 129-95
4. *Calcium Silicate Face Brick*, ASTM 73-95
5. *Prefaced Concrete Units*, ASTM 139-95
6. *Units for Catch Basins and Manholes*, ASTM 139-95

Currently, these six groups of masonry units are covered by six ASTM standards that apply to units intended primarily for construction of concrete masonry walls, beams, columns, or specialty applications. The first part of an ASTM standard number is the fixed designation for that standard. For example, ASTM 90 is the fixed designation for loadbearing concrete masonry units. The number immediately following the fixed designation indicates the year of the last revision. Units for catch basins and manholes are not covered in the text. The classification of concrete masonry units shown in Figure 6-4 presents another useful categorization.

## Concrete brick

Concrete bricks are solid brick size units, solid concrete veneer, and facing units larger than brick size. See Figure 6-5. They are similar in function to clay brick. They are completely solid or have a shallow depression called a *frog*. The frog is designed to reduce weight and provide for a better bond when the bricks are laid in mortar.

There are two grades of concrete bricks included in ASTM C 55 — Grade N and Grade S. **Grade N** is concrete brick used for architectural veneer and facing units in exterior walls. It is also used where high strength and resistance to moisture penetration and severe frost action are desired. **Grade S** is concrete brick used where moderate strength and resistance to frost action and moisture penetration are required.

Concrete bricks are meant to be laid with a 3/8" mortar joint. The most popular nominal modular dimension is 4" × 8". The thickness of horizontal mortar joints is such that three courses (three brick and three bed joints) equal a height of 8". Some manufacturers produce oversized jumbo and double brick units for special applications.

**Properties of Concrete Block Made of Different Aggregates**

| Aggregate (Graded: 3/8" to 0) | | Weight lb./cu. ft. of concrete | Weight 8" x 8" x 16" unit | Compressive strength (gross area) psi | Water absorption lb./cu. ft. of concrete |
|---|---|---|---|---|---|
| Type | Density (air-dry) lb./cu. ft. | | | | |
| Sand and gravel | 130 – 145 | 135 | 40 | 1200 – 1800 | 7 – 10 |
| Limestone | 120 – 140 | 135 | 40 | 1100 – 1800 | 8 – 12 |
| Air-cooled slag | 100 – 125 | 120 | 35 | 1100 – 1500 | 9 – 13 |
| Expanded shale | 75 – 90 | 85 | 25 | 1000 – 1500 | 12 – 15 |
| Expanded slag | 80 – 105 | 95 | 28 | 700 – 1200 | 12 – 16 |
| Cinders | 80 – 105 | 95 | 28 | 700 – 1000 | 12 – 18 |
| Pumice | 60 – 85 | 75 | 22 | 700 – 900 | 13 – 19 |
| Scoria | 75 – 100 | 95 | 28 | 700 – 1200 | 12 – 16 |

**Figure 6-3.** This chart shows the effect of aggregate type on weight, strength, and other characteristics of concrete block.

**Classification of Concrete Masonry Units**

| Concrete Brick | Concrete Block | Special Units |
|---|---|---|
| Solid brick size units | Solid loadbearing | Ground face units |
| Solid concrete veneer | Hollow loadbearing | Split face block |
| Facing units larger than brick size | Solid nonloadbearing | Prefaced concrete block |
| | Hollow nonloadbearing | Decorative block |

**Figure 6-4.** A diagram of 11 types of concrete masonry products.

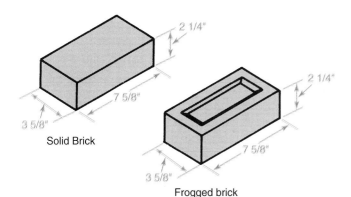

**Figure 6-5.** In size, concrete building bricks are much like clay brick. They may be solid or frogged.

Solid Brick

Frogged brick

## Slump brick

Slump brick or block is produced from a mixture that is wet enough to cause the units to sag or slump when removed from the molds. Resulting faces are irregular.

Height, surface texture, and appearance vary greatly. See Figure 6-6. Slump blocks produce special effects, and they resemble stone. See Figure 6-7.

**Figure 6-6.** This concrete slump brick has the appearance of stone. (Portland Cement Association)

**Figure 6-7.** Slump blocks are used here to produce a very attractive masonry planter and wall. (Portland Cement Association)

## Concrete block

Concrete block units usually are produced in four classes:

1. Solid loadbearing
2. Hollow loadbearing
3. Solid nonloadbearing
4. Hollow nonloadbearing

Each of these classes may be purchased with either heavyweight or lightweight aggregates. Figure 6-8 shows the strength and absorption requirements for concrete masonry units using lightweight concrete and normal weight concrete.

A solid concrete block is one in which the hollow parts in a cross section are not more than 25% of the total cross-sectional area.

A hollow concrete block is one in which the core or hollow area is greater than 25% of its total cross-sectional area. Generally, the core area in hollow units will be from 40% to 50% of the gross area.

In 1990, the basis for compressive strength in ASTM C 90 was changed from gross area to average net area of the unit. This change was mandated by the ever-increasing use of engineered masonry design, which uses net area strength as a basis for allowable stresses. Compressive stress based on gross area of units is used with masonry designed by empirical codes.

## Concrete block grades

In previous editions of ASTM C 90, two grades of block (N and S) were included, distinguished only by differences in compressive strength and water absorption requirements. The lower grade has been discontinued and there is now one grade, which is not given a grade classification.

Two types of concrete masonry units are covered by ASTM C 90—Type I, *Moisture-Controlled Units*, and Type II, *Nonmoisture-Controlled Units*.

*Type I units* are required to comply with the moisture content provisions at the time of delivery to the jobsite. The purpose of these requirements is to provide the specifier with a method of limiting residual drying shrinkage of units in the wall regardless of the shrinkage properties of the units (up to maximum unit shrinkage of 0.065 %). These provisions prescribe maximum allowable moisture content depending on shrinkage properties of the units and typical atmospheric humidity conditions at the point of use.

*Type II units*, which are not required to meet a maximum moisture content, are extensively used. With exposed Type II units, closer control joint spacing or increased horizontal reinforcement may be advisable depending on the potential shrinkage properties of the units. As with Type I units, Type II units are limited to a maximum linear shrinkage of 0.065%.

## Uses of concrete block

Solid loadbearing blocks are most often used where great loads are placed on it. Hollow loadbearing blocks have many uses. They combine high compressive strength with lightweight and flexible design, size, and shape. A large percentage of the block produced is hollow loadbearing block. Hollow nonloadbearing block are thin-shelled and lightweight. They are intended to be used mostly in nonloadbearing partitions. If used on nonloadbearing exterior walls, they should be protected from the weather.

Concrete masonry units are available in such a large number of shapes, sizes, textures, colors, and profiles that they are used in almost all aspects of masonry construc-

| Strength and Absorption Requirements for Concrete Masonry Units | | | | | | | | |
|---|---|---|---|---|---|---|---|---|
| Type of Unit | ASTM Designation | Grade of Unit | Minimum Compressive Strength, psi, on Average Gross Area | | Maximum Water Absorption, pcf, Based on Oven-Dry Unit Weight | | | |
| | | | | | Lightweight Concrete | | Medium-Weight Concrete, 105 to 125 pcf | Normal-Weight Concrete, 125 pcf or more |
| | | | Average of Three Units | Individual Unit | Less than 85 pcf | Less than 105 pcf | | |
| Concrete Brick | C55 | N | 3500* | 3000* | --- | 15 | 13 | 10 |
| | | S | 2500* | 2000* | --- | 18 | 15 | 13 |
| Solid Loadbearing Block | C145 | N | 1800 | 1500 | --- | 18 | 15 | 13 |
| | | S | 1200 | 1000 | 20 | --- | --- | --- |
| Hollow Loadbearing Block | C90 | N | 1000 | 800 | --- | 18 | 15 | 13 |
| | | S | 700 | 600 | 20 | --- | --- | --- |
| Hollow and Solid Nonloadbearing Block | C129 | --- | 600** | 500** | --- | --- | --- | --- |

\* Concrete brick tested flatwise.
\** Minimum compressive strength, psi, on average net area.

**Figure 6-8.** Compressive strength is an important property of concrete masonry and its related applications. (Portland Cement Association)

tion. Further, a great variety of patterns and designs are possible to produce a wide range of wall surfaces and architectural treatments. The following list of applications illustrates the broad use of concrete masonry units.

- Exterior loadbearing walls. See Figure 6-9.

**Figure 6-9.** Concrete blocks are used in this residential structure as exterior loadbearing walls and columns.

- Interior loadbearing walls
- Curtain walls, party walls, and fire walls
- Solar screens, panel walls, and partitions
- Veneer or nonstructural facing for concrete, masonry, wood, or steel
- Backing for brick, stone, stucco, and exterior insulation finish systems. See Figure 6-10.

**Figure 6-10.** This Florida home is made of concrete block with a stucco finish.

- Fire protection of structural steel members
- Pilasters, columns, and piers. See Figure 6-9.
- Bond beams, lintels, and sills
- Firesafe enclosures of stairwells, elevator shafts, or storage vaults
- Chimneys and fireplaces
- Retaining walls and slope protection
- Paving and turf block
- Catch basins

## Finish and appearance

Provisions relating to finish and appearance prohibit defects that would impair the strength or permanence of the construction. However, these provisions permit minor cracks incidental to the usual manufacturing methods.

For units that will be used in exposed wall construction, the presence of objectionable imperfections is based on viewing the face or faces from a distance of 20' under diffused lighting. The specification requires that color and texture be specified by the purchaser. An approved sample consisting of not less than four units, representing the range of color and texture permitted, is used to determine conformance.

## Sizes and shapes

Concrete building units are made in many sizes and shapes to fit different construction needs. It is common practice to specify the width first, height second, and the length last of the stretcher block.

Sizes are usually given in their nominal dimensions. A unit measuring 7 5/8" wide, 7 5/8" high, and 15 5/8" long is known as an 8" $\times$ 8" $\times$ 16" unit. The mortar joint is intended to be 3/8". This fits the modular design based on a 4" module. Concrete blocks are commonly available in widths of 2", 4", 6", 8", 10", and 12". Standard block heights are 4" and 8", but 12" units are also available. Lengths include 8", 12", 16", 20", and 24". Figure 6-11 shows most of the popular shapes and sizes. All are produced in full and half-length units. Some block units have a two-core design rather than three. This design has some advantages:

1. The shell is thicker at the center web. This increases the strength of the unit and reduces the tendency to crack from shrinkage.
2. Heat conduction is reduced by 3% to 4%.
3. Blocks are about 4 lb. lighter.
4. Hollow portions provide more space for placing conduit or other utilities.

The use of some of the units shown in Figure 6-11 is explained in the following descriptions:

- **Stretcher block** is the most common and is used on exterior surfaces.
- **Corner block** has one flush end that may be used for corners, simple windows, and door openings.
- **Double corner** or **pier block** is designed for piers or pilasters or any other place where both ends will be exposed.
- **Bullnose block** is used the same way as the corner block, but has a rounded corner.
- **Jamb** or **sash block** is used to make installation of windows or other openings easier.
- **Solid top block** has a solid top for use as a bearing surface in the finishing course of a foundation wall.

# Modern Masonry

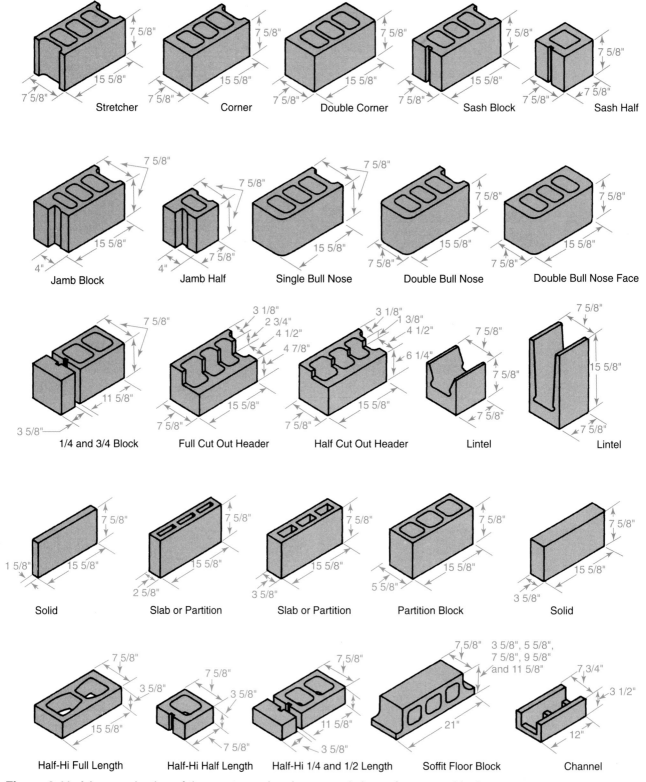

**Figure 6-11.** A large selection of the most popular shapes and sizes of concrete block.

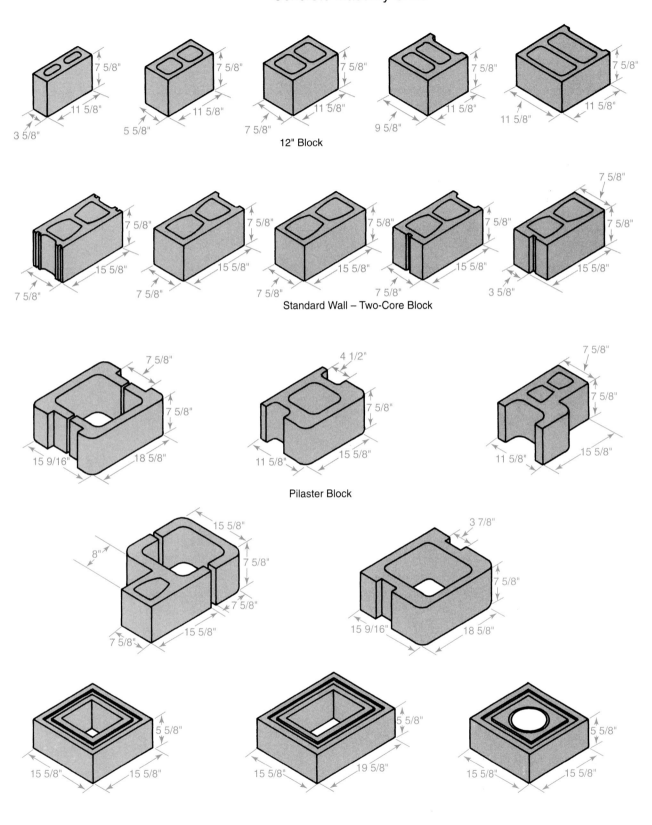

12" Block

Standard Wall – Two-Core Block

Pilaster Block

One-Piece Chimney Block

**Figure 6-11 (Continued).**

- **Header block** has a recess to receive the header unit in a masonry bonded (brick and block) wall.
- **Metal sash block** is used for window openings in which a metal sash is to be used. The slot anchors the jamb.
- **Partition block** (4" or 6") is used in constructing nonloadbearing partition walls.
- **Lintel block** is used to construct horizontal lintels or beams. It is generally U- or W-shaped.
- **Control joint block** is used to construct vertical shear-type control joints.

## Block terminology

Several terms are used to describe parts of standard concrete blocks, Figure 6-12. Blocks may be plain or flat on the end or concave in shape. Plain end blocks are used at corners, piers, or other locations where a plain end is necessary or desirable. Blocks with concave ends have two **ears** or **ends**. Blocks are produced with ears on one or both ends. Blocks with ears on both ends are called **stretchers**. Stretchers are used along the course between the corners.

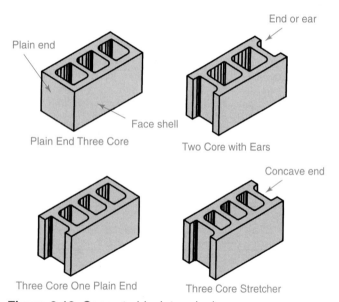

**Figure 6-12.** Concrete block terminology.

The openings in blocks are called **cells, cores** or **voids**. The wall between the cells or cores is called a **cross web**. The outside face of a block is called the **face shell**. Standard blocks are produced with two and three cells. Half blocks have a 4" thickness and are either solid, two-cell, or three-cell blocks.

## Surface texture

Surface texture is important in block. It affects sound absorption, appearance, and painting characteristics. Coarse textures look nicer and absorb sound better than smooth textures. Fine textures may absorb less sound but also take less paint than coarse textures. Some concrete masonry units have face surfaces ground smooth to produce interesting aggregate colors.

## Color

The typical color of concrete masonry units varies from light to dark gray to tints of buff, red, or brown depending on the color of aggregate, cement, and other ingredients used. The method of curing can also affect the color. Variations in color are common.

## Prefaced concrete masonry units

ASTM C 744 is the applicable standard for prefaced concrete masonry units. These masonry units have a specially prepared surface material applied to one or more surfaces of a concrete masonry unit to provide color, pattern, or texture. See Figure 6-13. Shapes such as stretchers, cove bases, bond beam units, caps or sills, and lintel blocks are frequently selected for prefacing. See Figure 6-14.

**Figure 6-13.** A sampling of prefaced concrete units showing a variety of colors, textures, and shapes. (Trenwyth Industries, Inc.)

The most common binders used to face masonry units are resin, resin and fine sand, or Portland cement and inert fillers. Glazed masonry units that use ceramic or mineral glazes are also produced. The facing material can add as much as 1/4" to the block dimension. As a result, mortar joints are reduced to 1/4" to maintain modular dimensions.

Since the facings are resistant to water penetration, cleaning detergents, and abrasion, prefaced units are frequently used in locations where cleanliness, decoration, or low maintenance are desirable. Examples of where prefaced units are used include school corridors, locker rooms, hospitals, and food processing areas.

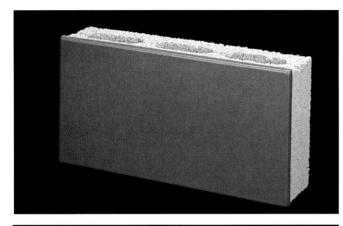

**Figure 6-14.** Top—Concrete block stretcher prefaced unit. Bottom—Concrete block prefaced cove based unit. (Trenwyth Industries, Inc.)

## Ground face units

A recent addition to the types of concrete blocks available are the ground face units. Innovative cutting and molding techniques have brought a new dimension to concrete blocks. These units include oversized and specialty units that may be used to create walls, columns, and arches of monumental proportions. Lengths of 11 5/8", 15 5/8", 17 5/8", and 23 5/8"; widths of 3 9/16" and 7 9/16"; and heights of 7 5/8", 11 5/8", and 15 5/8" are available in hollow and solid shapes. Figure 6-15 shows a close-up view of the surface of a ground face unit.

**Figure 6-15.** This close-up view of a ground face concrete block unit shows the aggregate which gives this product its character.

Special shapes of ground face units are also produced. These shapes include keystone arch units, watertable shapes, cornice units, sculpted sill shapes, column wraps, chamfered sill units, slant units, scored units, chamfer and quoining units, radial units, corbel and medallion units, cap units, and fluted units.

## Decorative blocks

Fluted and scored or ribbed units are shown in Figure 6-16 and Figure 6-17. They provide striations that can be developed into many kinds of patterns. Striations are a series of parallel ridges or grooves. The accuracy that is achieved in machine production makes it possible to produce the effect of long, vertical, straight lines. These blocks can be quite decorative.

**Figure 6-16.** Fluted and ribbed concrete blocks are used to achieve unique textured surfaces.

Blocks with recessed corners and raised patterns add endless numbers of effects to concrete masonry wall construction. See Figure 6-18. A few recessed and raised designs are shown in Figure 6-19.

Units that look like rough quarried stone are produced by splitting solid concrete block lengthwise as seen in Figure 6-20. Several sizes are available. Special hollow block as well as ribbed units may also be split to produce unusual effects.

Metric sizes are also available in glazed units. Typical sizes of stretchers include: lengths of 194 mm and 394 mm; widths of 94 mm, 194 mm, and 198 mm; and heights of 90 mm, 194 mm, and 394 mm. Cove base, cap, jamb, and other special shapes are also produced in metric sizes.

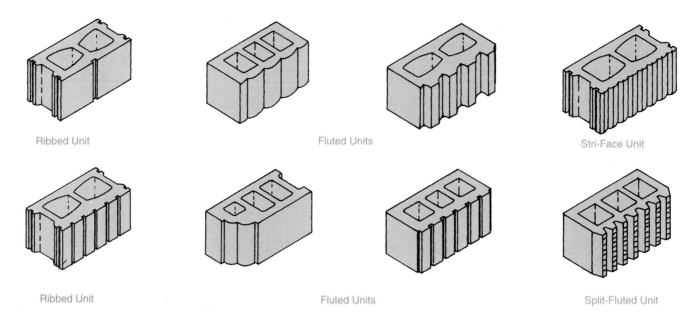

Ribbed Unit          Fluted Units          Stri-Face Unit

Ribbed Unit          Fluted Units          Split-Fluted Unit

**Figure 6-17.** Split-fluted concrete block units are used to form attractive wall patterns.

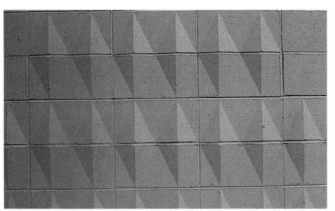

**Figure 6-18.** Even though this sculptured faced concrete block wall is painted a uniform color, it develops the effect of several shades.

## Screen block

Screen blocks are highly decorative. See Figure 6-21. However, they can fill several other useful functions such building facades, ornamental room dividers and partitions, garden fences, and patio screens. They can be used, for example, to provide only limited, directional vision. Or they can produce a balance between privacy and openness from within or without. They also screen out some light and produce a degree of shade.

There are an almost unlimited variety of patterns produced. Some are shown in Figure 6-22. The number of patterns you can find in any given community, however, may be limited.

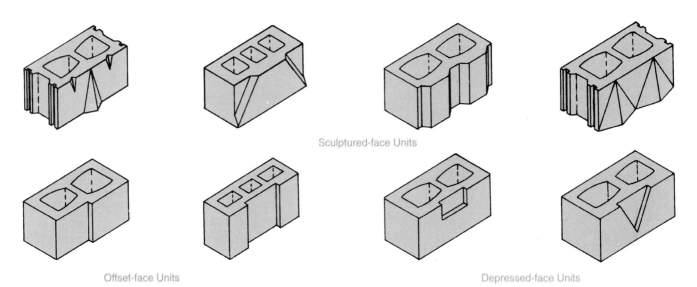

Sculptured-face Units

Offset-face Units          Depressed-face Units

**Figure 6-19.** Concrete block with offset, depressed, or sculptured faces.

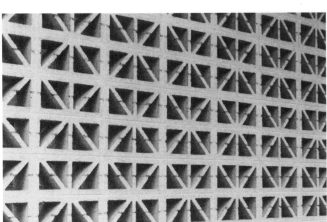

**Figure 6-20.** Split concrete block produce the effect of quarried stone.

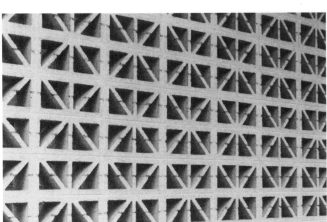

**Figure 6-21.** Concrete block screens are highly decorative and can be used to control light or noise.

Screen blocks, or grille units as they are also called, range in size from 4" to 16" square and 8" × 16" rectangles. Even though a wide variety of patterns of screen block exists, the number available in any given area may be limited, so it is prudent to check with local suppliers before planning to use a particular design.

Some screen blocks can be used in several positions in a wall to produce different patterns. There seems to be no limit as to how these units can be combined to produce interesting designs. They can also be mixed with common block shapes to add variety.

## Sound block

A sound block is made to absorb sound. Molded or sawed openings conduct the sound into the cores and absorb it. See Figure 6-23. This design is patented. Sound block is often used in gymnasiums, plants, subways, bowling alleys, or any place where noise is a problem. See Figure 6-24.

## Insulating block

Concrete blocks are available that have plastic foam inserts that increase their resistance to heat gain and heat loss. Some types have inserts that can be inserted in the cores as the block are laid. Others have the foam inserts cast into the block as it is made. One popular example of the latter is Therma-Lock™ block. See Figure 6-25. The foam insert in these blocks isolates one side of the block from the other, thus creating a better insulator—typical webs are not present to transmit heat or cold from one side to the other.

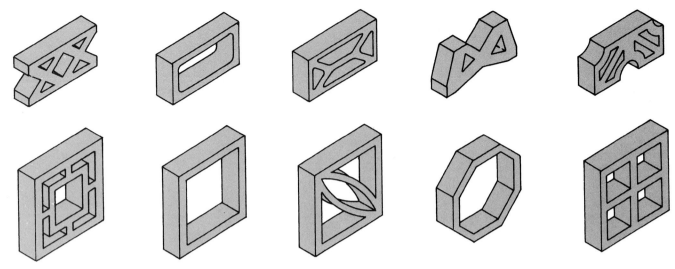

**Figure 6-22.** Typical screen block patterns.

**Figure 6-23.** Sound block absorb sound and are used where noise is a problem. (Trenwyth Industries, Inc.)

**Figure 6-25.** Therma-Lock™ blocks have high thermal performance because no concrete webs extend through the block to conduct heat. (Therma-Lock Products, Inc.)

**Figure 6-24.** Sound block used in a computer lab. (Trenwyth Industries, Inc.)

Therma-Lock™ blocks are laid in the conventional manner with typical mortar. No special tools are required. The blocks are available in 8", 10", and 12" widths. R-values of 15, 19, and 24 respectively are claimed by the manufacturer for Therma-Lock™ blocks. The inserts are slightly taller than the block and thereby eliminate any void at the mortar joint. This feature also increases the insulation value.

Additional information about Therma-Lock™ blocks can be obtained by contacting:
Therma-Lock Products, Inc.
162 Sweeney Street
North Tonawanda, NY 14120

# Pattern bonds

Thus far, we have discussed decorative effects that center around characteristics of the individual block shape, texture, size, or form of the block. Still other decorative effects can be achieved through pattern bonds. See Figure 6-26. Basically, the patterns belong to the following classifications:

1. Running bond
2. Stacked bond
3. Coursed or patterned ashlar
4. Diagonal bond
5. Basket weave
6. Diagonal basket weave

Figure 6-27 shows each of these patterns. Each classification includes several variations and the number may be enlarged further by using various size blocks.

**Figure 6-26.** Projecting concrete blocks produce a dramatic and decorative effect. (Portland Cement Association)

# Calcium silicate face brick (sand-lime brick)

Sand-lime brick, as the name indicates, is manufactured from sand and lime. A small amount of cement is sometimes added.

Among the advantages of sand-lime brick are uniform size, shape, and weight. These bricks are not burned. Shrinkage, warping, and color variation is not as great a problem as in clay and shale brick. To produce various colors the manufacturer adds coloring to the mixture.

Sand-lime bricks are used mainly in back-up work. They are generally produced in three grades—SW, MW, and NW. **Grade SW** includes brick that can be exposed to freezing temperatures and water saturation. This grade is used in foundations, retaining walls, or exposed piers. They can be used almost anywhere that burned-clay bricks are used.

**Grade MW** bricks may be exposed to freezing conditions, but not to water saturation. Uses of this grade include wall faces where the frost conditions are not too severe and where the climate is rather dry.

**Grade NW** includes bricks that may be exposed to light frost action with very little moisture present. This grade can be used in exterior wall construction in warm, dry climates.

# Glass block

Glass blocks are hollow, partially evacuated units of clear pressed glass. See Figure 6-28. Partially evacuated means that the air is partly removed from inside the block. They are produced in two sections and sealed. This provides a dead air space which acts as insulation. Glass blocks are made in square, rectangular, radial, and other special shapes.

Mortar-bearing surfaces of glass block are corrugated (a rippled design) with a gritty, alkali- and moisture-resistant substance. This coating acts as a bond between the block and the mortar. The construction of the block is such that designs may be imprinted on both sides. These designs distribute and control the direction of light rays. This is one of the most important properties of glass block. Some of the other advantages are good insulation value, low cost of maintenance, and the elimination of dust which comes through windows. The three categories of glass block are as follows:

1. Functional
2. General purpose
3. Decorative

The functional type of glass block controls both the distribution and diffusion (scattering) of light. See Figure 6-29. It is used often in schools, hospitals, and basement walls. The general purpose type is installed where direct light transmission is required. Bathrooms, windows in homes such as entrance side panels, and interior partitions would require this type of glass block. When architectural design is most important, decorative block is used. Examples would include the front entrance of an office building, decorative screens, and corridor partitions or stairwell walls.

Glass blocks are manufactured in three popular sizes: 5 3/4 inches square, 7 3/4 inches square, and 11 3/4 inches square. These blocks are all 3 7/8" thick. A 1/4" mortar joint is required. A glass block layout table is shown in Figure 6-30. The maximum panel area, using chase construction, (reinforced masonry frame around the glass block panel), is 144 sq. ft. (maximum height is 20'; maximum width is 25'). The maximum panel area using wall anchor construction is 100 sq. ft. (maximum height is 10', maximum width is 10'). Panel and curtain wall sections may be erected up to 250 sq. ft. if properly braced to limit movement and settlement.

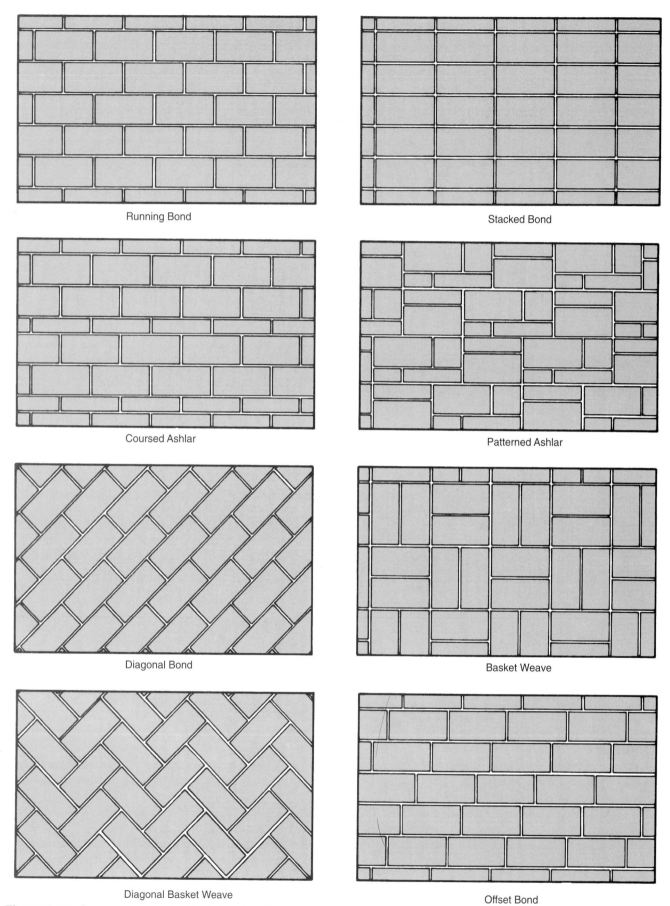

**Figure 6-27.** Common concrete block bond patterns.

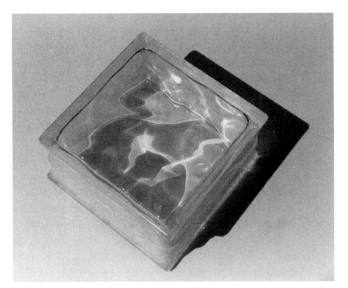

**Figure 6-28.** Glass blocks have rough edges to form a good bond with the mortar.

**Figure 6-29.** Glass blocks are functional as well as decorative.

**Glass Block Layout Table**

| No. of Blocks | 6" 5 3/4 x 5 3/4 × 3 7/8 | 8" 7 3/4 x 7 3/4 × 3 7/8 | 12" 11 3/4 x 11 3/4 × 3 7/8 |
|---|---|---|---|
| 1 | 0' – 6" | 0' – 8" | 1' – 0" |
| 2 | 1' – 0" | 1' – 4" | 2' – 0" |
| 3 | 1' – 6" | 2' – 0" | 3' – 0" |
| 4 | 2' – 0" | 2' – 8" | 4' – 0" |
| 5 | 2' – 6" | 3' – 4" | 5' – 0" |
| 6 | 3' – 0" | 4' – 0" | 6' – 0" |
| 7 | 3' – 6" | 4' – 8" | 7' – 0" |
| 8 | 4' – 0" | 5' – 4" | 8' – 0" |
| 9 | 4' – 6" | 6' – 0" | 9' – 0" |
| 10 | 5' – 0" | 6' – 8" | 10' – 0" |
| 11 | 5' – 6" | 7' – 4" | 11' – 0" |
| 12 | 6' – 0" | 8' – 0" | 12' – 0" |
| 13 | 6' – 6" | 8' – 8" | 13' – 0" |
| 14 | 7' – 0" | 9' – 4" | 14' – 0" |
| 15 | 7' – 6" | 10' – 0" | 15' – 0" |
| 16 | 8' – 0" | 10' – 8" | 16' – 0" |
| 17 | 8' – 6" | 11' – 4" | 17' – 0" |
| 18 | 9' – 0" | 12' – 0" | 18' – 0" |
| 19 | 9' – 6" | 12' – 8" | 19' – 0" |
| 20 | 10' – 0" | 13' – 4" | 20' – 0" |
| 21 | 10' – 6" | 14' – 0" | 21' – 0" |
| 22 | 11' – 0" | 14' – 8" | 22' – 0" |
| 23 | 11' – 6" | 15' – 4" | 23' – 0" |
| 24 | 12' – 0" | 16' – 0" | 24' – 0" |
| 25 | 12' – 6" | 16' – 8" | 25' – 0" |

**Figure 6-30.** This chart shows the course height of glass block in 6", 8", and 12" sizes. For example, 12 courses of 8" glass block with a 1/4" mortar joint will produce an 8'-0" high wall. An 8'-0" long wall will also require 12 blocks.

# Metric Concrete Masonry Units

In 1991 the Executive Order 12770, *Metric Conversion Act*, mandated that metrics be used in the construction of all United States federal buildings. From that date all building designs and construction drawings had to be submitted in metric units, and constructed according to metric specifications.

Metric conversion for some materials, like concrete masonry units, means that the inch-pound dimensions are stated in metric units, but does not require changing the material sizes currently produced. This is called a **soft conversion**. On the other hand, a **hard conversion** requires a physical change in the size of the product.

Because concrete masonry uses soft metric conversion, the metric equivalents of concrete masonry unit dimensions are usually the exact metric conversions of the inch-pound unit dimensions. For example, an 8" × 8" × 16" concrete block is considered to be 200 mm × 200 mm × 400 mm. (These are the nominal dimensions specified by the Brick Institute of America.) However, the exact metric equivalents are 203 mm × 203 mm × 406 mm for this nominal size block. Remember that the nominal size of a block includes the mortar joint.

# REVIEW QUESTIONS CHAPTER 6

Write all answers on a separate sheet of paper. Do not write in this book.

1. The term _____ includes all the sizes and kinds of hollow or solid block, brick, and concrete building tile, so long as they are made from concrete and are laid by masons.
2. Identify four reasons why concrete masonry is popular today.
3. What are the three primary ingredients in concrete?
4. Sand and gravel in mortar or concrete is referred to as _____.
5. Aggregates are classified according to their weight as _____ and _____.
6. Concrete masonry units can be classified in six main groups. Name the groups.
7. What concrete masonry unit is similar in function to clay brick?
8. Coarse-textured block absorb sound poorly. True or False?
9. Concrete block units are grouped into two grades according to their degree of resistance to frost action, and into two types according to the amount of moisture in each block. Identify the four grade types.
10. Concrete blocks are usually specified by their nominal dimensions. True or False?
11. A concrete block whose nominal dimensions are 8" × 8" × 16" has an actual size of _____.
12. Standard concrete blocks are produced in two-core and three-core designs. Identify four advantages of two-core blocks.
13. _____ block is used the same way as the corner block, but has a rounded corner.
14. Blocks with concave ends have two _____ or ends.
15. The openings in blocks are called _____ or voids.
16. Decorative block with a ceramic type glaze on its face is called _____ block.
17. _____ are highly decorative and serve several functions. They provide limited directional vision, screen out some light and produce some shade while giving privacy.
18. A sound block is designed to _____.
19. Concrete blocks that have plastic foam inserts to increase their resistance to heat gain and heat loss are called _____ block.
20. Identify five typical pattern bonds used with concrete block.
21. Name two advantages of sand-lime brick (calcium silicate face brick).
22. Name the three grades of sand-lime brick.
23. A unit that is partially evacuated of air and made of clear pressed glass is known as a(n) _____ block.
24. What are the three categories of the type of block mentioned in Question 23?
25. What are the three popular sizes of the type of block mentioned in Question 23?
26. Which Executive Order (1991) mandated that metrics be used in the construction of all United States federal buildings.
    A. *The National Metrication Act*
    B. *Metric Conversion Act*
    C. *The Federal Construction Act*
    D. *The Metric Construction Act*
27. What is the difference between soft conversion and hard conversion when referring to metrics?
28. What are the nominal dimensions of an 8" × 8" × 16" concrete block in metric units?

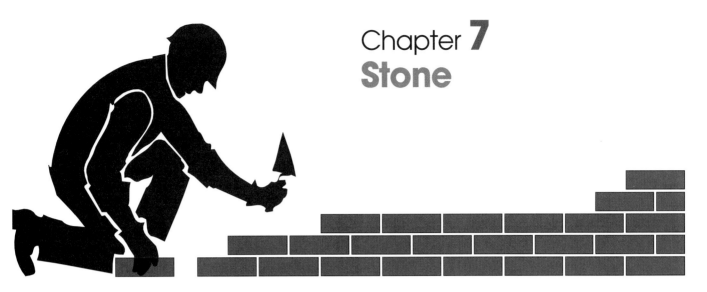

# Chapter 7
# Stone

Stone is one of the oldest building materials. It is used for many things and in many places in construction. Its use as the sole building material for a structure has declined over the years. Today, it is almost always used as a nonstructural material. Generally, it is applied as a facing, a veneer, or for decorative purposes. See Figure 7-1. Other uses such as in sandwich and panel systems are increasing. This trend is expected to continue.

## Classification of Stone

Stone is made up of minerals in various mixtures. Individual samples vary greatly in composition. These variations greatly affect the strength, color, texture, and durability of stone. This is especially true of marble, limestone, and sandstone.

**Figure 7-1.** This collection of photos shows the use of four different types of stone in traditional as well as modern applications. (Eldorado Stone Corporation & Stone Products Corporation)

Stone is divided into three categories based on the process of formation. See Figure 7-2. The three categories are as follows:

1. Igneous (volcanic)
2. Sedimentary
3. Metamorphic

These categories are many times related and many samples are the result of more than one process of formation.

| Classification of Stone | | |
|---|---|---|
| **Igneous** | **Sedimentary** | **Metamorphic** |
| Granite<br>Traprock | Sandstone<br>Limestone | Marble<br>Slate<br>Schist<br>Gneiss<br>Quartzite |

**Figure 7-2.** This classification diagram identifies the three main types of stone. Division is based on the process of formation.

## Igneous stone

Igneous stone is further classified by its texture, mineral content, and origin. It all comes from magmas generally found deep in the earth. *Magma* is a molten mixture of minerals. Igneous stone usually contains ferro-magnesian minerals and feldspar or quartz. When composed primarily of light minerals (quartz and potash feldspar), it is called *acidic*. This type is light in color and weight.

Stone with a greater degree of ferro-magnesian minerals is called *basic*. It is darker and heavier. In texture, igneous stone ranges from those with large crystals to glassy stone with no crystals at all. Two popular types of igneous stone are granite and traprock.

### Granite

*Granite* is an igneous stone that is usually light colored, and formed mainly of potash feldspar, quartz, and mica. Fine granite has a salt-and-pepper pattern, but feldspar may redden it. Granite is hard and takes a good polish. Figure 7-3 shows a fine granite quarried in Canada. Granite is probably the best known of the deeper igneous stone.

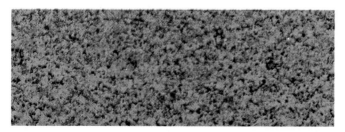

**Figure 7-3.** This granite sample shows the typical grain pattern of most granites. It may be purchased in a broad range of colors, including white, gray, pink, green, and various combinations.

### Traprock

*Traprock* is an igneous stone that is hard, durable, and used in construction, but is limited because of iron minerals it contains. These iron minerals make rusty stains as the stone weathers. It is the quarryman's term for diabase, basalt, or gabbro. Traprock is excellent as a crushed rock.

## Sedimentary stone

Sedimentary stone differs widely in texture, color, and composition. See Figure 7-4. Sedimentary stone is formed in layers.

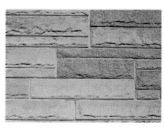

**Figure 7-4.** These photos illustrate the varied color, texture, and composition of sedimentary stone. (Stone Products Corporation)

Deposits of shell, disintegrated stone, or sand have become cemented together under pressure to form sedimentary stone. Two popular types are sandstone and limestone.

### Sandstone

*Sandstone* is sedimentary stone composed mainly of grains of quartz cemented by silica, lime, or iron oxide.

Silica cement can produce a hard, durable sandstone, but other cements are not as resistant. Porous sandstone may not weather well and special treatment may be required in cold regions. Color and texture are available in such varieties as bluestone, brownstone, silica sandstone, lime sandstone, and many mixed varieties. See Figure 7-5

**Figure 7-5.** These thin sandstones form an attractive facing for this fireplace. (Stone Products Corporation)

## Limestone

**Limestone** is sedimentary stone that varies widely in color and texture and consists mainly of the mineral calcite. It is usually marine in origin. Limestone weathers rapidly in humid climates, but very slowly where it is dry. Limestone from Bedford in central Indiana is a well-known building stone. It is widely used in building construction. See Figure 7-6. Bedford limestone is white, even textured, and sometimes packed with small fossils. Many other types of limestone, which include dolomitic, oolitic, crystalline, and travertine marble, are also used for construction.

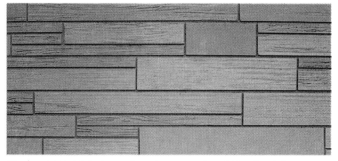

**Figure 7-6.** This is a classic example of Bedford limestone from central Indiana.

*Indiana limestone*, geologically known as Salem limestone, is essentially a monomineralic rock consisting of the calcium carbonate mineral named calcite. The calcite comes from the skeletal material (about 75%) that form the framework grains and from the cementing material (about 20%) that binds the grains together. Porosity (about 5%) and small amounts of non-calcareous material comprise the remainder of the rock.

Indiana limestone was formed in a shallow sea that covered the midwest of the United States that included the Bedford-Bloomington, Indiana quarry area. This formation took place more than 300 million years ago during the Mississippian geological epoch. See Figure 7-7 and Figure 7-8 for relative position of stone deposits.

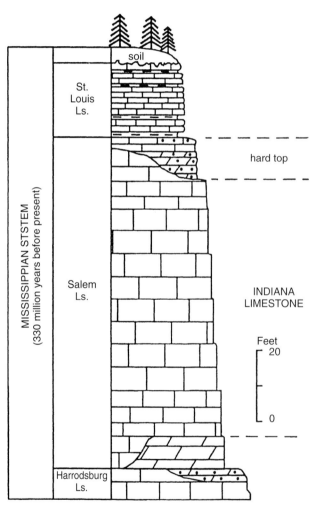

**Figure 7-7.** A stratigraphic section showing Indiana limestone's relative position in the Indiana stone deposit. (Courtesy Indiana Geological Survey)

This limestone is characteristically a freestone, without pronounced cleavage planes, possessing a remarkable uniformity of composition, texture, and structure. It has a high internal elasticity, which helps it adapt to extreme temperature changes without damage.

This stone combines excellent physical properties with a remarkable degree of machinability. This ease of

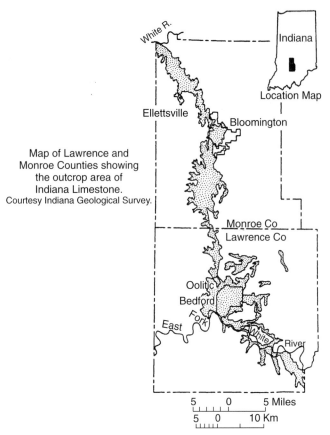

Map of Lawrence and
Monroe Counties showing
the outcrop area of
Indiana Limestone.
Courtesy Indiana Geological Survey.

**Figure 7-8.** A map of Lawrence and Monroe Counties showing the outcrop area of Indiana limestone. (Courtesy Indiana Geological Survey)

machining provides complete flexibility of shape and texture at low cost.

Over the past hundred years, Indiana limestone has proved its ability to resist the forces of weather and pollution. Its qualities of strength and beauty continue to adapt themselves to the needs of contemporary architecture.

Indiana limestone is a sedimentary formation, but the deposition of the minute calcareous seashell is so uniform that no weak cleavage planes occur in the material. It can be machined or cut in any direction without danger of splitting. However, because it is a sedimentary rock it does have a grain running horizontally in the deposit.

Stone set in a building with its grain running horizontally, as it does in the quarry, is said to be set on its *natural bed*. Stone set with the grain running vertically is on *edge*.

## Metamorphic stone

Metamorphic stone is formed through reconstitution due to great heat and pressure. Changes may be barely visible, or may be so great that it is impossible to determine what the original stone once was. Metamorphic stone includes marble, slate, schist, gneiss, and quartzite.

## Marble

*Marble* is recrystallized limestone that can be white, yellow, brown, green, or black. See Figure 7-9. Marble does not often develop the parallel bonding and mineral arrangement seen in slate and schist. Marble is a classic stone used for the finest work. See Figure 7-10. However, it is softer and less resistant to weathering than granite.

## Slate

*Slate* is a metamorphic stone that is frequently a blue-gray color, but may be green, red, or brown. See Figure 7-11. It may be split into sheets used for roofing or flagstones. Most United States slate comes from Vermont, Maine, and Pennsylvania.

## Schist

*Schist* is a rather coarse-grained stone with large amounts of mica in it. Several kinds of schist are classified according to the most characteristic mineral present. Examples of schist include mica schist, hornblende schist, chlorite schist, and quartz schist.

## Gneiss

*Gneiss,* pronounced *nice,* is difficult to define or describe because it is so varied. In general, it is a coarse-textured stone with minerals in parallel streaks or bands. It is relatively rich in feldspar and usually contains mica.

## Quartzite

*Quartzite* is sandstone that has been recrystallized. The grain structure in quartzite is not nearly as clear as in sandstone. It is very hard and durable.

# Grades of Stone

Stone is generally graded into four groups for construction purposes. These groups are statuary, select, standard, and rustic in descending order of quality. Fineness of grain or texture is the basis for this classification of stone.

## Classification of Indiana limestone

The **Indiana Limestone Institute** classifies Indiana limestone into two colors and four grades based on granular texture and other natural characteristics. See Figure 7-12. When specifying Indiana limestone it is necessary to identify both the color and grade required as well as the surface finish to be applied to the stone.

## Color descriptions

There are two color descriptions of Indiana limestone. *Buff* is the color description of Indiana limestone that varies from a light creamy shade to a brownish buff. *Gray* is the color description of Indiana limestone that varies from a light silvery gray to shades of bluish-gray.

## Grade descriptions

The **Indiana Limestone Institute** classifications are based on the degree of fineness of the grain particles and

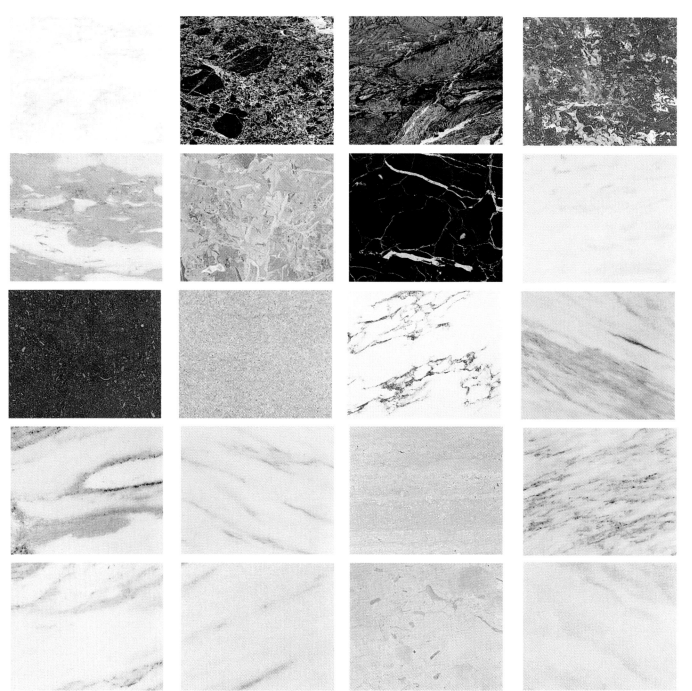

**Figure 7-9.** Marble is noted for its great beauty. These samples show some of the variety, color, and patterns available.

other natural characteristics which make up the stone. The structural soundness of each of the grades is essentially identical.

As a natural product, Indiana limestone contains at least a few distinguishable calcite streaks or spots, fossils or shell formations, pit holes, reed formations, open texture streaks, honeycomb formations, iron spots, travertine-like formations, and grain-formation changes. The four grades recognized by the *Indiana Limestone Institute* are:

1. ***Select.*** Fine to average grained stone having a controlled minimum of the above characteristics.

2. ***Standard.*** Fine to moderately large grained stone permitting an average amount of the above characteristics.

3. ***Rustic.*** Fine to very coarse grained stone permitting an above-average amount of the above characteristics.

4. ***Variegated.*** An unselected mixture of grades 1 through 3 permitting both the buff and gray colors.

Figure 7-10. The pattern displayed in the marble face of this fireplace is unsurpassed by any other stone. (Marco Manufacturing, Inc.)

Figure 7-11. Slate is a favorite flooring material because it is easy to clean, resists soil and stains, and lasts almost forever.

| Grade and Color Classifications | |
|---|---|
| **Buff** | **Gray** |
| Select | Select |
| Standard | Standard |
| Rustic | Rustic |
| Variegated | |

Figure 7-12. The grade and color classifications of Indiana limestone. (Indiana Limestone Institute)

 **Trade Tip.** It is advisable that all stone for each project be furnished from a single quarry. This should result in the best possible color control.

## Stone Surface Finishes

Many surface finishes are possible with stone. Figure 7-13 shows the graphic symbols for typical stone finishes. Some of the more popular ones are:

- **Gang sawed.** This finish has a moderately smooth surface with visible saw marks. It is inexpensive to produce because it requires no further finishing after leaving the saw. It may be used on all types of stone.
- **Shot sawed.** A variable or slightly rough finish according to the amount of shot used and the texture of the stone. It should be used with softer stone.
- **Machined (planer).** Finish is smooth with some texture from tool marks. May be used with most types of stone.
- **Machine tooled.** Has two to ten grooves per inch. Grooves are parallel and concave in shape. It may be used on all types of stone, but is expensive.
- **Plucked.** This surface is obtained by rough planing. Then it is textured by "plucking" out small particles of the stone. Used mainly for limestone.
- **Hand tooled.** Finish may be applied in a regular or random pattern. It is very expensive and should be used only in instances that require special accent.
- **Carborundum.** A very smooth finish attained by using a carborundum machine rather than a planer. The finish is used primarily for limestone.
- **Rubbed and honed.** A very smooth finish—smoother than the carborundum. It is most often used for interior marble and granite
- **Honed and polished.** Finish is the smoothest of all finishes. It has a "high glass" sheen. This surface finish is used on marble and granite.

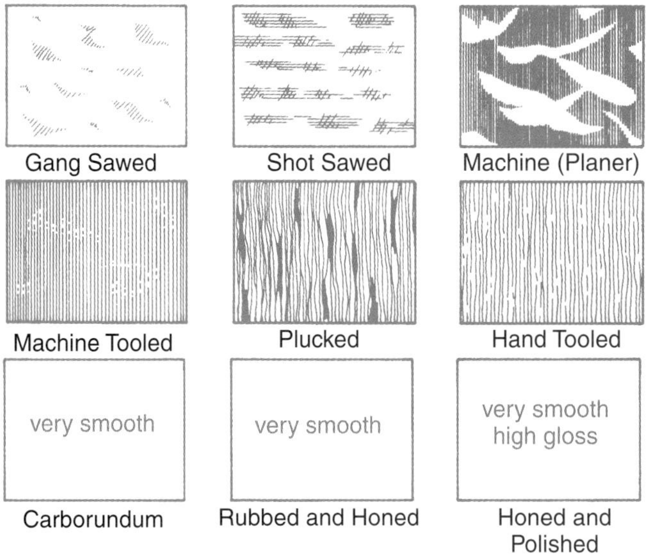

Gang Sawed    Shot Sawed    Machine (Planer)

Machine Tooled    Plucked    Hand Tooled

very smooth    very smooth    very smooth
high gloss

Carborundum    Rubbed and Honed    Honed and
Polished

**Figure 7-13.** Graphic symbols for typical stone finishes.

# Stone Wall Patterns

Stone wall patterns fall roughly into three groups. These groups are rubble, roughly squared stone, and dimensioned or ashlar. They are described in the following sections.

## Rubble

Stone pattern No. 1 in Figure 7-14, illustrates **uncoursed field stone** or random (common) rubble. Stones of many sizes and shapes are used as they are found in fields and streams. Even though no coursing is possible, stones are usually laid in a horizontal position.

Pattern No. 2 in Figure 7-14 shows **uncoursed cobweb** or polygonal rubble. These stones are dressed with relatively straight line edges and are selected to fit a place in the wall that has a particular shape. Mortar joints are approximately the same width.

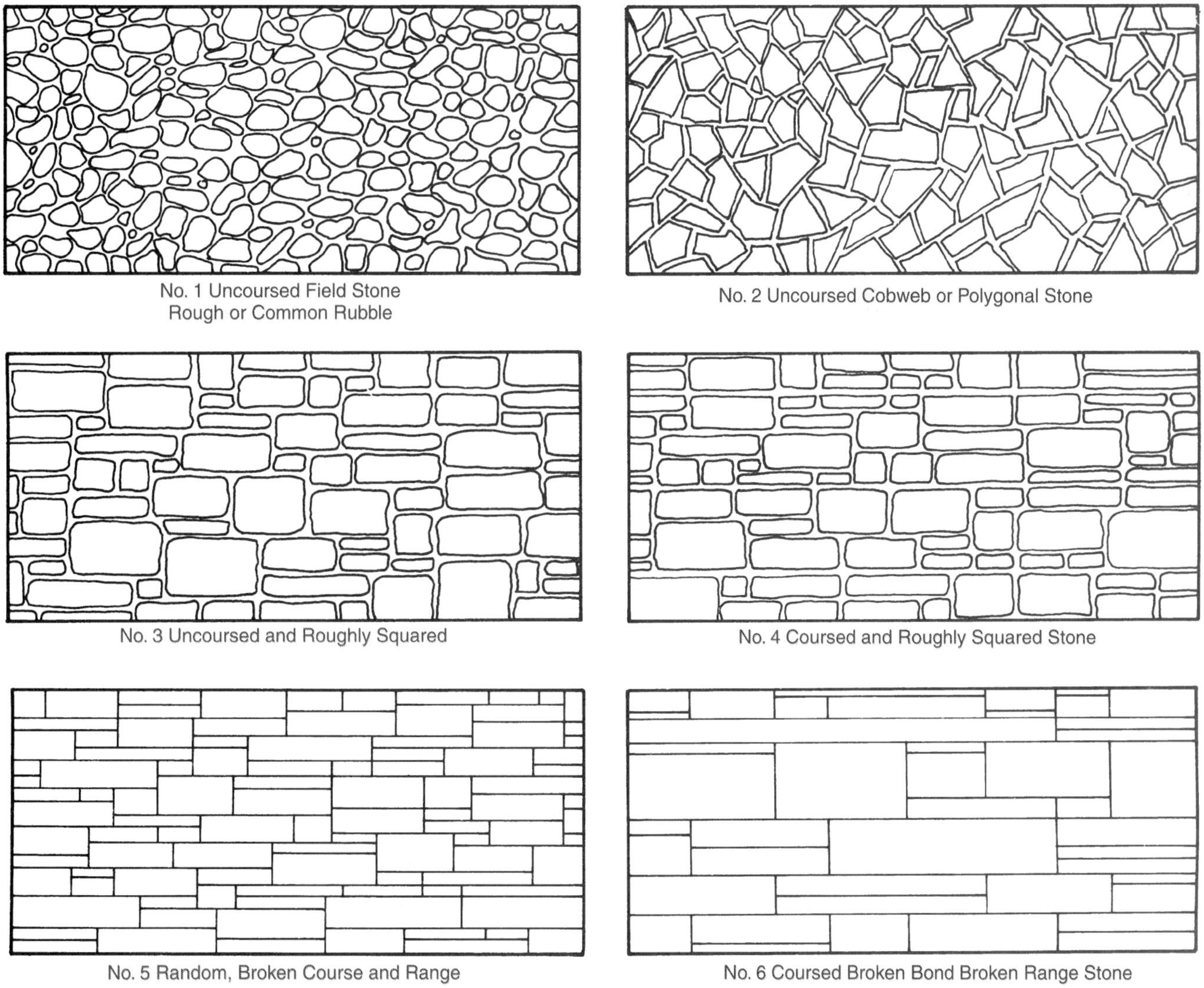

No. 1 Uncoursed Field Stone
Rough or Common Rubble

No. 2 Uncoursed Cobweb or Polygonal Stone

No. 3 Uncoursed and Roughly Squared

No. 4 Coursed and Roughly Squared Stone

No. 5 Random, Broken Course and Range

No. 6 Coursed Broken Bond Broken Range Stone

**Figure 7-14.** Representative stone patterns used in stone masonry.

### Roughly squared stone

A wall pattern made from **uncoursed and roughly squared stone** is shown as No. 3 in Figure 7-14. Even though these stones have been roughly squared, no coursing is attempted. Both large and small stones are used.

Wall pattern No. 4 in Figure 7-14, is **coursed and roughly squared.** Here the horizontal coursing begins to appear. The stones are varied in size but generally are rectangular in shape.

### Dimensioned or ashlar

**Dimensioned** or **ashlar stone** is cut stone of specific dimensions. It is cut, dressed, and finished to precise job requirements at the mill. It is transported to the site as a finished product.

**Random, broken course and range pattern** is shown in No. 5 in Figure 7-14. This pattern makes no attempt to maintain coursing or range. **Range** is squared stones laid in horizontal courses of even height.

A **coursed broken bond, broken range masonry pattern** is shown in No. 6 in Figure 7-14. This pattern keeps the basic coursing without maintaining it fully. Again, the range is broken.

# Stone Applications

Stone was once widely used for foundations, exterior walls, paving, and trim. In recent years, it has been used much less for foundations and solid exterior walls. Stone masonry is slower and more difficult due to the different sizes, and in some styles, odd shapes. Presently, stone is mostly limited to veneer, trim, and floor or paving applications.

Limestone is probably the best sill material that can be used on exterior masonry walls. See Figure 7-15. The fact that it is one piece insures better protection against water leakage.

Continuous courses of stone just above the foundation, or at other levels, can add to the appearance as well as function of the masonry wall. A grade course of stone prevents moisture from the earth entering the brickwork. It may function as a water table to divert water away from the foundation wall.

Stone is still used extensively for copings (a covering or top) on masonry walls. It prevents moisture from entering through the top. See Figure 7-16.

Overhangs and Drips Both Sides. Sloped to Back

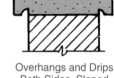
Overhangs and Drips Both Sides. Sloped Both Sides

Overhang and Drip One Side. Sloped One Side

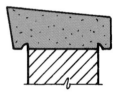
Overhangs and Drips Both Sides. Sloped to Back

**Figure 7-16.** Stone copings on top of a masonry wall prevent moisture from entering the wall. (Eldorado Stone Corporation)

**A**

**B**

**C**

**Figure 7-15.** A—Limestone slip sill in brick masonry. B—Limestone lug sill. C—Close-up detail of a limestone slip sill in a brick veneer wall.

Stone is popular in the form of quoins (large squared stones set at the corners of buildings) for buildings constructed with certain architectural styles.

Paving of patios, courts, entryways, and walks using stone is also very popular. See Figure 7-17. Slate and marble are favorites for paving. They afford a durable, hard surface which requires a minimum of maintenance and provides years of beauty.

**Figure 7-17.** Left—Roughly squared stones create an attractive formal paving. Right—Polygonal stone paving presents a more casual appearance.

## Stone characteristics summary

Several characteristics of the most popular building stones are summarized in Figure 7-18.

# Manufactured Stone ▬▬▬▬

*Manufactured stone* is a simulated stone veneer made from lightweight concrete. It is colorfast, weatherproof, and has the look and feel of natural stone, Figure 7-19. Produced in a variety of colors and textures, it simulates various types of natural stone. Figure 7-20 shows some of the popular types.

Manufactured stone may be applied directly to a base coat of stucco, concrete block, brick, concrete, or any masonry surface that has not been treated or sealed and which is rough enough to provide a good mechanical bond. Figure 7-21 shows manufactured stone being applied to concrete block. Wallboard, plaster, or wood surfaces must be covered with wire mesh or metal lath before manufactured stone is applied. Painted or sealed stucco and painted or sealed masonry surfaces require the attachment of wire mesh, sand blasting, or a coating of suitable masonry adhesive.

Mortar used for manufactured stone should be a mixture of one part cement to three parts clean sand. To this is added from 1/4 to 1/2 part lime, mortar cream, or fire clay. Packaged dry mix mortar that requires the addition of water also works well with manufactured stone.

| | Building Stone Characteristics | | | | | |
|---|---|---|---|---|---|---|
| | Igneous Rock | | Sedimentary Rock | | Metamorphic Rock | |
| | **Granite** | **Traprock** | **Limestone** | **Sandstone** | **Marble** | **Slate** |
| **Color** | Almost white to pink and white or gray and white gray, olive green to black | Dark gray to black; also dark olive green | White, light gray to light buff | Very light buff to light chocolate brown or brick red; may tarnish to brown | Highly varied: snow white to black; also blue-gray and light to dark olive green | Grayish-green, brick red or dark brown, usually gray; may be banded |
| **Texture** | Fine or coarse crystals, crystals may be varicolored | Usually coarsely crystalline but may be fine grained | Fine to crystalline; may have fossils | Granular, showing sand grains, cemented together | Finely granular to very coarsely crystalline showing flat-sided crystals | Finely crystalline; flat crystals give slaty fracture |
| **Weight** | 156 to 170 pcf | 180 to 185 pcf | 147 to 170 pcf | 135 to 155 pcf | 165 to 178 pcf | 170 to 180 pcf |
| **Hardness** | Harder than limestone and marble; keeps a good edge | About like granite; keeps a good edge | Fairly soft and easily scratched by steel | Fairly hard if well cemented | Slightly harder than limestone | Softer than granite or quartzite; scratches easily |
| **Strength** | 15,000 - 30,000 psi | 20,000 psi | 4000 - 20,000 psi | 3000 - 20,000 psi | 10,000 - 23,000 psi | 10,000 - 15,000 psi |
| **Chief Uses** | Building stone and panels | Crushed stone | All building uses | All building uses | Building stone and panels | Roofing and paving |

**Figure 7-18.** Summary of some of the most important characteristics of popular building stones.

**Figure 7-19.** These are applications of manufactured stone. (Stone Products Corporation)

**Figure 7-20.** Manufactured stone is produced in several colors and textures to simulate natural stone. (Stone Products Corporation)

In any case, the mortar should be mixed to a consistency similar to that of brick mortar. It may be applied either to the wall surface, the back of the stone, or both.

Manufactured stone should be laid out in a convenient location to the work area to ensure even distribution of sizes and colors. The stone can be cut with a brick trowel, hatchet, or similar tool to achieve sizes and shapes required for fitting stones and keeping mortar joints to a minimum.

While the mortar is still soft and pliable, the stone should be pressed into place with enough pressure so that the mortar is squeezed out around the edges of the stone. This will ensure a good bond between the stone and wall surface.

When the mortar joints become firm (from one to three hours depending on climate and suction of the base)

**Figure 7-21.** Manufactured stone being applied to a concrete block wall. (Stone Products Corporation)

they should be pointed up with a wood or metal striking tool. Excess mortar should be raked out and the stones sealed around the edges to give the finished job the appearance of a natural stone wall.

At the end of each day, the finished wall should be broomed or wire brushed to remove loose mortar and to clean the face of the stone. A wet brush should not be used to treat the mortar joints for at least 24 hours. This is likely to cause staining and discoloration that is difficult to remove.

# REVIEW QUESTIONS
# CHAPTER 7

Write all answers on a separate sheet of paper. Do not write in this book.

1. Stone is almost always used today as a nonstructural material. True or False?
2. Stone may be grouped into three categories based on the process of formation. They are igneous, sedimentary, and metamorphic. True or False?
3. Two popular types of _____ stone are granite and traprock.
4. _____ is the quarryman's term for diabase, basalt, or gabbro.
5. Two popular types of sedimentary stone are _____ and _____.
6. What type of sedimentary stone varies widely in color and texture, consists mainly of the mineral calcite, and is usually marine in origin?
7. Marble is a metamorphic stone. True or False?
8. Metamorphic stone is formed through reconstitution due to great _____ and _____.
9. Identify four popular metamorphic stones that are used in construction.
10. _____ is recrystallized limestone that may be white, yellow, brown, green, or black.

11. What metamorphic stone is frequently a blue-gray color, but may be green, red, or brown, and is used for roofing or flagstone?
12. _____ is a type of sandstone which has been recrystallized.
13. Stone is generally classified according to four grades. They are statuary, _____, _____, and rustic.
14. The **Indiana Limestone Institute** recognizes the following four grades of limestone: select, standard, rustic, and varigated. True or False?
15. What are the two colors of sandstone recognized by the **Indiana Limestone Institute**?
16. Name three surface finishes that may be applied to stone.
17. When stones are used as they are found in fields and streams, the pattern is known as uncoursed field stone or _____.
18. List the three most popular stone wall bond patterns?
19. Dimensioned or ashlar stone is cut stone of specific dimensions. True or False?
20. Large, square stones set at the corners of buildings are called _____.
21. Manufactured stone is a(n) _____ stone veneer made from lightweight concrete.
22. Manufactured stone may be applied directly to a base coat of stucco, concrete block, brick, concrete, or any masonry surface that has not been treated or sealed and which is rough enough to provide a good mechanical bond. True or False?
23. Mortar used for manufactured stone should be a mixture of one part cement to _____ parts clean sand.

Good design and the abilities of highly skilled stonemasons have combined to produce this attractive curving stone wall and stairway.

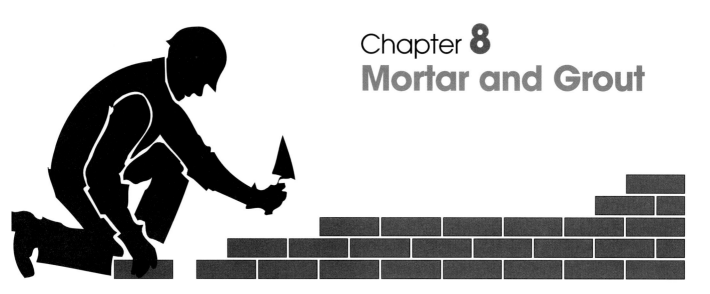

# Chapter 8
# Mortar and Grout

Mortar is the bonding agent that ties masonry units into a strong, well-knit, weathertight structure. It secures each of the units into a wall or other building element. Because mortar is the bonding agent that integrates a masonry wall, it must be strong, durable, capable of keeping the wall intact, and it must help to create a water resistant barrier. These requirements are influenced by the composition, proportions, and properties of mortar.

## Mortar Materials

Historically, mortars have been made from a variety of materials. Burned gypsum and sand were used to make mortar in ancient Egypt while lime and sand were used extensively in this country before the 1900s. Currently, the basic mortar ingredients include Portland cement, hydrated lime, and sand. Masonry cement is sometimes used to replace Portland cement and hydrated lime or combined with Portland cement to make mortar. Most of these materials make a definite contribution to mortar performance.

Mortar serves four functions:

1. It bonds the units together, sealing the spaces between them.
2. It makes up for differences in sizes of the units.
3. It provides bonding for metal ties or other types of reinforcement.
4. It provides esthetic (attractive) qualities by creating pleasing lines and color effects.

The material properties of mortar which influence the structural performance of masonry are compressive strength, bond strength, and elasticity. Because the compressive strength of masonry mortar is less important than bond strength, workability, and water retentivity, the latter properties should be given principal consideration in mortar selection.

Laboratory testing indicates that masonry constructed with Portland cement-lime mortar exhibits greater flexural bond strength than masonry constructed with masonry cement mortar or air-entrained Portland cement-lime mortar of the same type. Masonry cement mortars may not be used in seismic zones 3 and 4.

The American Society for Testing and Materials has set up standards for materials commonly used in mortars. Figure 8-1 shows the standards that apply to these materials.

| Material | ASTM Designation |
|---|---|
| Portland Cement (Types I, II, III) | C 150 |
| Air Entraining Portland Cement (Types IA, IIA, IIIA) | C 175 |
| Blended Cement (Types IS, ISA, IP, IPA, S, SA) | C 595 |
| Masonry Cement | C 91 |
| Quicklime | C 5 |
| Hydrated Lime | C 207 |
| Aggregate | C 144 |

Figure 8-1. The American Society for Testing and Materials standards for materials used in mortars.

Mortar is generally made up of cementitious (cement-like) materials together with sand and water.

## Cementitious Materials

These materials may include Portland cement (Type I, II, or III), masonry cements, and hydrated lime (Type S).

### Portland cement

Mortar is mostly Portland cement. It is a hydraulic material, which means that it hardens under water. ASTM C150, Standard Specifications for Portland Cement, covers eight types. Only the following three are recommended for use in mortar:

1. **Type I.** For general use when the special properties of Types II and III are not needed.
2. **Type II.** For use when moderate sulfate resistance or moderate heat of hydration is wanted.
3. **Type III.** For use when high early strength is desired.

Types IA, IIA, and IIIA are not recommended for masonry mortar because air entrainment reduces the bond between the mortar and masonry units or reinforcement. Types IS, ISA, IP, IPA, S, and SA are not recommended for masonry mortar unless strengths of masonry constructed with mortar containing such cements are first established by appropriate tests.

### Masonry cements

Masonry cements usually contain Portland cement, ground limestone as a filler, and additives that provide workability, water retentivity, and air entrainment. The constituents of masonry cements can vary widely among different brands. Therefore, reproducing masonry cement mortars consistently from batch to batch may not be as easy as with Portland cement-lime mortars.

### Hydrated lime

Hydrated lime is quicklime that has been slaked before packaging. **Slaked** means it has been formed into a putty by combining it with water. This converts the calcium oxide. Hydrated lime can be mixed and used immediately. For this reason, it is more convenient to use than quicklime. Quicklime, which is essentially calcium oxide, must be mixed with water and stored for as long as two weeks before using.

Hydrated lime is available in two types—S and N. Only Type S hydrated lime is recommended for masonry mortar because unhydrated oxides and plasticity are controlled. Unhydrated oxides and plasticity are not controlled in Type N.

## Sand (Aggregate)

Sand is the primary aggregate used in masonry mortar. Either natural or manufactured sand may be used. Figure 8-2 shows recommended sand gradation limits for both natural and manufactured sand. If the sand grains passing through the sieve fall within the percentage range given, it meets the ASTM gradation standards. Well-graded aggregate retards (holds back) separation of material in a plastic mortar mix. This reduces bleeding and improves workability. Sand that has too few fine particles generally produces harsh mortars. Sand with too many fine particles results in weak mortar.

Manufactured sand is made by crushing stone, gravel, or air-cooled blast furnace slag. Natural sand has rounder, smoother particles than manufactured sand. Manufactured sand, with its sharp and angular particle

**Recommended Sand Gradation Limits**

| Sieve Size | Percent Passing | |
|---|---|---|
| | Natural Sand | Manufactured Sand |
| No.    4 | 100 | 100 |
| No.    8 | 95 – 100 | 95 – 100 |
| No.   16 | 60 – 100 | 60 – 100 |
| No.   30 | 35 – 70 | 35 – 70 |
| No.   50 | 15 – 35 | 20 – 40 |
| No. 100 | 2 – 15 | 10 – 25 |
| No. 200 | – | 0 – 10 |

**Gradation limits are given in ASTM C 144**

**Figure 8-2.** Recommended sand gradation limits for both natural and manufactured sand.

shapes, can produce mortars with workability properties different than mortars made with natural sand.

## Water

Water for masonry mortar is required by ASTM C270 to be clean and free of harmful amounts of acids, alkalis, or organic materials. Whether or not the water is potable (safe for drinking) is of no concern. However, water from city mains or private wells is generally suitable for mixing mortar.

## Measuring Mortar Materials

The goal of mixing mortar to be consistent from mix to mix with respect to yield, workability, and color can be achieved when proper methods of measuring the ingredients are followed.

Even though proportions of aggregates are usually expressed in terms of loose volume, sand can vary greatly in volume due to moisture bulking. For example, fine sands can increase in volume by almost 40% when the percent of moisture is only 5% by weight. Coarse sands are not affected as much, but may experience an 18% increase in volume under the same moisture conditions. ASTM C 270 states that 1 cu. ft. of loose, damp sand contains 80 lb. of dry sand. Most sand on the construction site contains from 4% to 8% moisture content.

The common practice of measuring sand by the shovel can result in oversanding or inconsistent batches of mortar. Cubic foot measuring boxes are recommended for the job site to provide accurate mixes.

Ingredients such as Portland cement, blended cement, masonry cement, and hydrated lime are sold in bags that are labeled by weight. This presents a problem, because mortar is proportioned by volume. Add cement or lime by the bag.

Water can be measured accurately by using a pail. When the desired consistency is reached, the amount can be recorded and used each time.

# Mixing Mortar

The ingredients of mortar must be thoroughly mixed to obtain good workability and performance of plastic and hardened mortar. Mortar on the job site can be mixed by machine or by hand.

## Machine mixing

When possible, mortar should be mixed by machine. Most mortar mixers have a capacity from 4 cu. ft. to 7 cu. ft. and are of the rotating spiral or paddle-blade design with a tilting drum. Mixing of the ingredients should be from 3 to 5 minutes. Too little mixing may result in poor workability, nonuniformity, low water retention, and too little air content. Too much mixing can reduce the strength of the mortar.

There is no one correct way to load the mixer. Some prefer to add about three-fourths of the water, one-half of the sand, and all of the cementitious materials, and then mix briefly before adding the remaining sand and water. Other procedures are also used. However, mixing is most effective when the mixer is charged to its capacity. Be sure not to exceed the design specifications and never begin a new batch of mortar when some remains in the mixer from the previous batch.

## Hand mixing

Some very small jobs lend themselves to hand mixing the mortar. When hand mixing, add all the dry materials to the box, or wheelbarrow, and mix thoroughly using a hoe. Work the materials from one end of the box to the other several times before adding any water.

Add two-thirds to three-fourths of the water and mix with the hoe until the batch is uniformly wet. Carefully add additional water until the proper consistency is reached. Remember, a little water can make a big difference in the mortar consistency. Allow the batch to stand for about 5 minutes and then mix again. The mortar is now ready for use.

## Retempering mortar

Mortar should be prepared at the rate it is being used to maintain best workability. Mortar that has stood for awhile before being used may dry out and stiffen due to loss of water. The workability can be restored by adding the lost water and mixing very briefly. Mortar that has begun to harden due to hydration should be discarded. Mortar should be used within 2 1/2 hours after mixing to avoid hardening due to hydration.

# Mortar Properties

Mortars have two distinct sets of properties, which are plastic (wet) mortar properties and hardened mortar properties. Plastic (wet) mortar properties consist of workability and water retentivity. Hardened mortar properties consist of compressive strength, bond strength, and durability.

## Properties of plastic mortar

Plastic mortar is mortar that has just been mixed. It must be workable. It is uniform, cohesive, and of a consistency that makes it usable. A usable mortar is easy to spread. Yet, it clings to vertical faces of masonry units, readily extrudes from the mortar joint, but does not drop or smear. It will also support the weight of the masonry unit being used. See Figure 8-3. The water content has a great deal to do with the workability of masonry mortar.

**Figure 8-3.** A good workable mortar is soft, but has good body. It adheres readily to masonry units and will not drop away while handling. (Portland Cement Association)

Water retention is the ability of the mortar to prevent rapid loss of mixing water to an absorptive masonry unit or to the air on a hot, dry day. Water retention is also related to bleeding.  Bleeding is the ability of mortar to prevent floating of a masonry unit with low absorption. Water retentivity improves with higher lime content and the addition of fine sand. However, the mixture must be within the allowable gradation limits.

Water content is possibly the most misunderstood aspect of masonry mortar. This is probably due to the similarity between mortar and concrete materials. Many designers have mistakenly based mortar specification on the assumption that mortar requirements are similar to concrete requirements, especially with regard to the water-cement ratio. Figure 8-4 shows a comparison of the desired slump for concrete, mortar, and grout. Many specifications incorrectly require mortar to be mixed with the minimum amount of water consistent with workability. Often, retempering of the mortar is also prohibited. These provisions result in mortars that have higher compressive strengths but lower bond strengths. Mixing mortar with the maximum amount of water consistent with workability will provide maximum bond strength within the capacity of the mortar. Retempering is permitted, but only to replace water lost by evaporation. This can usually be controlled satisfactorily by requiring that all mortar be used within 2 1/2 hours after mixing.

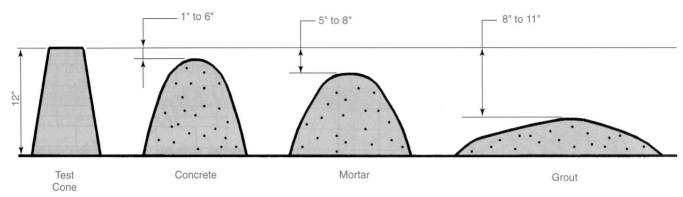

**Figure 8-4.** Slump test comparisons of concrete, mortar, and grout.

## Properties of hardened mortar

The compressive strength of mortar increases as the cement content is increased. It decreases as lime is increased. Air entrainment also reduces compressive strength. Structural failures due to compressive loading are rare, but bond strength is generally more critical.

**Bond** is that property of hardened mortar that holds the masonry units together. Bond strength is probably the most significant property of hardened mortar. It is also the most difficult to predict. Bond failure results from the masonry structure being subjected to:

1. Eccentric (not exactly vertical) gravity loads.
2. Loads from earth pressure, winds, and other lateral (sideways) loads.

Close contact between mortar and masonry unit surface is essential for a good bond. This will more likely be achieved using mortar with good workability. To insure a good bond, mortar should be placed in final position within 2 1/2 hours after the original mixing. This is required by ASTM C270.

Movement of a masonry unit after the mortar has begun to harden will destroy the bond. The mortar will not be plastic enough to reestablish the bond.

Durability of mortar is measured primarily by its ability to resist repeated cycles of freezing and thawing under natural weather conditions. Mortars of high-compressive strength have been found to have good durability.

Air-entrained mortar is highly resistant to freezing and thawing and has good durability. The tiny air bubbles absorb the expansive forces of freezing. Laboratory tests showed that mortars with adequate air entrainment withstood hundreds of freeze-thaw cycles, while other mortars soon failed.

## Mortar Proportions and Uses

Picking the right type of mortar depends upon the type of masonry and where it will be in the structure. No mortar will produce the highest rating in all properties. Adjustments in the mix to improve one property usually hurts other properties. Figure 8-5 shows proper proportioning of materials.

The five types of mortars recommended for various classes of masonry construction are designated by the letters M, S, N, O, and K. See Figure 8-6. The type of mortar designations are derived from every other letter in the word **masonwork.**

### Type M mortar

**Type M mortar** has high-compressive strength and somewhat greater durability than other mortar types. It is very good for unreinforced masonry below grade and in contact with earth. Structures would include foundations, retaining walls, walks, sewers, and manholes.

### Type S mortar

**Type S mortar** is a medium-high-strength mortar. It is intended for use where Type M is recommended but where bond and lateral strength are more important than high compressive strength. Tests indicate that the tensile bond strength between brick and Type S mortar is near the best obtainable with cement-lime mortars. It is recommended for use in:

1. Reinforced masonry.
2. In unreinforced masonry where maximum flexural (bending) strength is required.
3. Where mortar adhesion is the sole bonding agent between facing and backing as it is with ceramic veneers.

### Type N mortar

**Type N mortar** is a medium-strength mortar suitable for general use in exposed masonry above grade. It is best for parapet walls, chimneys, and exterior walls that are subjected to severe exposure. Its principal property is weather resistance. Type N mortar may not be suitable in seismic zones 3 and 4.

### Type O mortar

**Type O mortar** is a low-strength mortar. It is suitable for general interior use in nonloadbearing masonry. It can

| Mortar Proportions by Volume | | | | |
|---|---|---|---|---|
| Mortar Type | Parts by Volume of Portland Cement* or Portland Blast Furnace Slag Cement** | Parts of Volume of Masonry Cement | Parts by Volume of Hydrated Lime or Lime Putty | Aggregate, Measured in a Damp, Loose Condition |
| M | 1<br>1 | 1 (Type II)<br>– | –<br>1/4 | Not less than 2 1/2 and not more than 3 times the sum of the volumes of the cements and lime used. |
| S | 1/2<br>1 | 1 (Type II)<br>– | –<br>Over 1/4 to 1/2 | |
| N | –<br>1 | 1 (Type II)<br>– | –<br>Over 1/2 to 1 1/4 | |
| O | –<br>1 | 1 (Type I or II)<br>– | –<br>Over 1 1/4 to 1 1/2 | |
| K | 1 | – | Over 2 1/2 to 4 | |

\* Types I, II, III, IA, IIA, IIIA
\*\* Types IS, ISA
Data From ASTM C-270

**Figure 8-5.** Mortar types are shown with the proper proportions of materials by volume.

| Mortar Types for Classes of Construction | |
|---|---|
| ASTM Mortar Type Designation | Construction Suitability |
| M | Masonry subjected to high compressive loads, severe frost action, or high lateral loads from earth pressures, hurricane winds, or earthquakes. Structures below grade, manholes, and catch basins. |
| S | Structures requiring high flexural bond strength, but subject only to normal compressive loads. |
| N | General use in above grade masonry. Residential basement construction, interior walls and partitions. Concrete masonry veneers applied to frame construction. |
| O | Nonloadbearing walls and partitions. Solid load bearing masonry of allowable compressive strength not exceeding 100 psi. |
| K | Interior nonloadbearing partitions where low compressive and bond strengths are permitted by building codes. |

**Figure 8-6.** The type of mortar required is shown for a variety of classes of construction.

be used for loadbearing walls of solid masonry where compressive stresses do not exceed 100 psi. In such cases, however, exposures must not be severe. Type O mortar is not recommended by the *Masonry Standards Joint Committee* (MSJC) Code.

**Warning!** In general, do not use Type O mortar where it will be subjected to freezing.

## Type K mortar

*Type K mortar* is a very low-strength mortar. It is best used in interior nonloadbearing partition walls where high strength is not needed. Type K mortar is not recommended by the MSJC Code.

## Admixtures

No air-entraining admixtures or cementitious materials containing air-entraining admixtures or agents shall be used in the mortar.

Antifreeze compounds, accelerators, retarders, water-repellent agents, or other admixtures should not be added to mortar. Calcium chloride or admixtures containing calcium chloride should not be added to mortar. Mortar colors may be added to the mortar as specified.

## Specific mortar uses

The following specific uses are recommended based on properties of the various mortars.

### Cavity walls

For cavity walls where wind velocities will exceed 80 mph, use Type S mortar. For locations where lesser winds are expected, use Type S or Type N mortars.

## Facing tile

For facing tile, any of the first four mortar types are acceptable. Where 1/4" joints are specified, all aggregate (sand) should pass through a No. 16 sieve. For white joints, use white Portland cement and white sand in construction mortar. An alternate method is to rake all joints to a depth of 1/2" and point or grout with mortar of the desired color.

## Tuck-pointing mortar

For tuck-pointing, use only prehydrated mortars. To prehydrate mortar, all the ingredients (except water) should be mixed thoroughly, then mixed again. Use only enough water to produce a damp unworkable mix which will retain its form when pressed into a ball. After one to two hours, add enough water to bring it to the proper consistency. Prehydrated Type N mortar may be used.

## Dirt-resistant mortar

When a dirt-resistant and/or stain-resistant mortar is required, add aluminum tristearate, calcium stearate, or ammonium stearate to the mortar. The amount added should be 3% of the weight of the Portland cement.

## Grout

Grout is used in brick masonry to fill cells of hollow units or spaces between wythes of solid unit masonry. Grout increases the compressive, shear, and flexural strength of the masonry element and bonds steel reinforcement and masonry together. For compliance with the MSJC Specifications, grout which is used in brick or structural clay tile masonry should conform to the requirements of ASTM C 476, *Specification for Grout for Masonry*.

The amount of mixing water and its migration from the grout to the brick or structural clay tile will determine the compressive strength of the grout and the amount of grout shrinkage. Grouts with high initial water content exhibit more shrinkage than grouts with low initial water content. Consequently, use of a non-shrink grout admixture is recommended to minimize the number of flaws and shrinkage cracks in the grout while still producing a grout slump of 8" to 11", unless otherwise specified.

The MSJC Specifications require grout compressive strength to be at least equal to the specified compressive strength of masonry, but not less than 2000 psi (13.8 MPa) as determined by ASTM C 1019, *Method of Sampling and Testing Grout*. In general the compressive strength of ASTM C 476 grout by proportions will be greater than 2000 psi (13.8 MPa).

## Coloring Mortar

Mortar can be colored through the use of aggregates or pigments. Colored aggregates are preferred. White sand, ground granite, marble, or other stone usually have permanent color and do not weaken the bond or mortar.

For white joints, use white sand, ground limestone, or ground marble with white Portland cement and lime.

Mortar pigments must be capable of giving the desired color when used in permissible quantities, sufficiently fine to disperse throughout the mix, and non-reacting with other ingredients. These requirements are usually met by metallic oxides. Iron oxides, manganese oxides, chromium oxides, carbon black, and ultramarine blue have been used successfully as mortar colors.

**Trade Tip.** Do not use organic colors, particularly those containing Prussian blue, cadmium lithopone, and zinc and lead chromates. Paint pigments may not be suitable for mortars either.

Use the minimum quantity of pigments necessary to produce the desired results. Too much can impair strength and reduce durability. Carbon black, a popular pigment, will impair mortar strength when used in greater amounts than 2% or 3% of the cement weight.

For best results, dry mix the color with Portland cement in large controlled quantities. This will ensure a more uniform color than when mixing smaller quantities at the job site.

## Cold Weather Mortar

The temperature of the mortar should be between 70°F and 100°F when used. Higher temperatures may cause fast hardening, which will make it more difficult for the mason to produce quality work.

Heating the mixing water is one of the easiest methods of raising the temperature of the mortar. However, the water should not be heated above 160°F. This prevents the danger of *flash* set when the water comes in contact with the cement.

In freezing weather, moisture in the sand will turn to ice. It must be thawed before using. Never use sand that has been scorched from overheating.

**Warning!** Antifreeze materials should not be used to lower the freezing point of mortar. Such large quantities would be required that it would seriously impair the mortar strength and other properties.

Mortar with calcium chloride added will resist freezing. It should be used in a solution instead of in dry or flake form. To prepare the proper solution dissolve 100 lb. of flake calcium chloride in 25 gallons (gal.) of water. Not more than one quart should be used with each 94 lb. bag of masonry cement.

The length of protection against freezing should follow requirements of the *American Standard Building Code*, (A

The length of protection against freezing should follow requirements of the *American Standard Building Code*, (A 41.1). It states:

*"Masonry shall be protected against freezing for at least 48 hours after being laid. Unless adequate precautions against freezing are taken, no masonry shall be built when the temperature is below 32°F on a falling temperature, at the point where the work is in progress. No frozen materials shall be built upon."*

# Recommended Practices

The following items detail the recommended practices for mortar in the masonry trades.

❖ **Storage of Materials.** Cementitious materials and aggregates shall be stored in such a manner as to prevent deterioration or contamination by foreign materials.

❖ **Measurement of Materials.** The method of measuring materials for the mortar used in construction shall be by either volume or weight, and such that the specified proportions of the mortar materials can be controlled and accurately maintained. Measurement of sand by shovel should not be permitted.

❖ **Mixing Mortars.** All cementitious materials and aggregate shall be mixed for at least 3 minutes and not more than 5 minutes in a mechanical batch mixer, with the maximum amount of water to produce a workable consistency.

❖ **Retempering.** Mortars that have stiffened because of evaporation of water from the mortar should be retempered by adding water as frequently as needed to restore the required consistency. Mortars should be used and placed in final position within 2 1/2 hours after initial mixing.

❖ **Climatic Conditions.** Unless superseded by other contractual relationships or the requirements of local building codes, cold or hot weather masonry construction practices relating to mortar shall conform to the conditions below.

❖ **Cold Weather**—Cold weather masonry construction relating to mortar should comply with the *International Masonry Industry All-Weather Council's Guide Specification for Cold Weather Masonry Construction*, Section 04200 Unit Masonry, Article 3.

❖ **Hot Weather**—When the ambient air temperature exceeds 100°F or 90°F with a wind velocity greater than 8 mph, mortar beds shall not be spread more than 4' ahead of masonry units. Units shall be laid within one minute after spreading mortar.

# Estimating Mortar Quantities

Estimating the amount of mortar needed for a specific job is difficult for the novice. Most estimates are too low. Job delays can result from poor estimates and waste occurs when too much material is ordered. The objective is to order the proper amount of all ingredients. Most students are surprised to learn that 0.33 cu. ft. of cementitious material and 0.99 cu. ft. of sand will make 1 cu. ft. of mortar.

## Estimating mortar for concrete block wall

The typical approach to estimating mortar for concrete block construction is to calculate how much mortar there is in a typical 8" × 8" × 16" concrete block wall. It is common practice to use only face-shell bedding with 3/8" thick joints. Since a typical 8" concrete block has 1 1/4" thick face shells, the amount of mortar required for each block would be: $2 \times 0.375" \times 1.25" \times 23.62" = 22.1$ cu. in./block. There are 112.5 blocks in 100 sq. ft. of wall, so the amount would be: $22.1 \times 112.5$ divided by 1728 equals 1.44 cu. ft.

Allowance for waste is generally considered to be 10%. Therefore, the estimator would order $1.10 \times 1.44 = 1.6$ cu. ft. of mortar for each 100 sq. ft. of wall. There are practical reasons to order a greater amount than the calculation indicates. Some of these reasons include the following:

1. Shell thickness may be 50% greater at the top than at the bottom, thus requiring more mortar.
2. Some mortar falls to the ground as droppings.
3. Some masons just use more mortar.
4. Some mortar sticks to the mixer, wheelbarrow, and mortar board.
5. Some sand cannot be used because it was dumped on the ground.
6. Some cement, lime, or mortar is spilled, lost, or spoiled.
7. Mortar left at the end of the day, or at lunch time, must be discarded.
8. Weather turns bad before mortar can be used.
9. Human error causes waste.
10. The mortar is rejected—the mix isn't correct.
11. Variation in masonry units may require more mortar.
12. Cutouts for plumbing or electrical are closed with mortar.
13. The first course requires a full mortar bed.
14. Solid cores around anchor bolts require more mortar.
15. Door and window jambs are filled with mortar.
16. Change orders may require more materials.

The **National Concrete Masonry Association** has a chart that shows the following amounts of mortar needed for 100 sq. ft. of wall for various concrete masonry units. See Figure 8-7. Notice that this chart shows considerably more mortar needed than does the calculation method.

| Mortar Needed for Concrete Masonry Units | |
|---|---|
| Nominal Height and Length of Units in Inches | Cubic Feet of Mortar per 100 Sq. Ft. |
| 8 x 16 | 6.0 |
| 8 x 12 | 7.0 |
| 5 x 12 | 8.5 |
| 4 x 16 | 9.5 |
| 2 1/4 x 8 | 14.0 |
| 4 x 8 | 12.0 |
| 5 x 8 | 11.0 |
| 2 x 12 | 15.0 |
| 2 x 16 | 15.0 |

**Figure 8-7.** Mortar requirements for 100 square feet of single wythe wall area. (National Concrete Masonry Association)

## Estimating mortar for a single wythe brick wall

Estimating the mortar required for a typical single wythe brick wall is potentially more confusing than estimating for a concrete block wall because more sizes exist and mortar coverage is different. In response to the problem, the **Brick Institute of America** has developed charts for estimating mortar for 100 sq. ft. of running bond and collar joints. See Figures 8-8, 8-9, and 8-10. In addition, the Institute has also developed a manual, sliding calculator called the *Brick Masonry Estimator* for calculating brick and mortar needed for various walls.

# Grout

Grout is a primary element in reinforced brick and block masonry construction. It is generally used in rein-

| Nonmodular Brick | | | | |
|---|---|---|---|---|
| | 3/8" Mortar Joints | | 1/2" Mortar Joints | |
| Size of Brick in Inches | Number of Brick per 100 Sq. Ft. | Cubic Feet Mortar per 100 Sq. Ft. | Number of Brick per 100 Sq. Ft. | Cubic Feet Mortar per 100 Sq. Ft. |
| t       h       l | | | | |
| 2 3/4 x 2 3/4 x 9 1/4 | 455 | 3.2 | 432 | 4.5 |
| 2 5/8 x 2 3/4 x 8 3/4 | 504 | 3.4 | 470 | 4.1 |
| 3 3/4 x 2 1/4 x 8 | 655 | 5.8 | 616 | 7.2 |
| 3 3/4 x 2 3/4 x 8 | 551 | 5.0 | 522 | 6.4 |

**Figure 8-8.** Nonmodular brick and mortar required for single wythe walls in running bond with no allowances for waste or breakage. (Brick Institute of America)

| Modular Brick | | | |
|---|---|---|---|
| Nominal Size of Brick in Inches | Number of Brick per 100 Sq. Ft. | Cubic Feet of Mortar Per 100 Sq. Ft. | |
| t       h       l | | 3/8" Joints | 1/2" Joints |
| 4 x 2 2/3 x 8 | 675 | 5.5 | 7.0 |
| 4 x 3 1/5 x 8 | 563 | 4.8 | 6.1 |
| 4 x 4 x 8 | 450 | 4.2 | 5.3 |
| 4 x 5 1/3 x 8 | 338 | 3.5 | 4.4 |
| 4 x 2 x 12 | 600 | 6.5 | 8.2 |
| 4 x 2 2/3 x 12 | 450 | 5.1 | 6.5 |
| 4 x 3 1/5 x 12 | 375 | 4.4 | 5.6 |
| 4 x 4 x 12 | 300 | 3.7 | 4.8 |
| 4 x 5 1/3 x 12 | 225 | 3.0 | 3.9 |
| 6 x 2 2/3 x 12 | 450 | 7.9 | 10.2 |
| 6 x 3 1/5 x 12 | 375 | 6.8 | 8.8 |
| 6 x 4 x 12 | 300 | 5.6 | 7.4 |

**Figure 8-9.** Modular brick and mortar required for single wythe walls and running bond with no allowance for waste or breakage. (Brick Institute of America)

| Cubic Feet of Mortar Per 100 Sq. Ft. of Wall | | |
|---|---|---|
| 1/4" Joint | 3/8" Joint | 1/2" Joint |
| 2.08 | 3.13 | 4.17 |

**Figure 8-10.** Cubic feet of mortar for collar joints. (Brick Institute of America)

forced loadbearing masonry walls or in those spaces that contain steel reinforcement. The grout bonds the steel and masonry units together so that they function as a single unit to resist imposed loads. Sometimes grout is also used in some or all of the cores in nonreinforced loadbearing masonry walls to provide added strength.

Grouting brick and block walls accomplishes three main purposes:

1. Bonds the wythes of masonry together.
2. Transfers stress from the masonry to the reinforcing steel when lateral forces are present.
3. Increases the cross-sectional area of the wall to aid in supporting vertical loads and lateral shear loads.

## Codes, specifications, and requirements

Grout used in masonry wall construction should comply with the requirements of ASTM C 476; *Building Code requirements for Masonry Structures*, ACI 530/ASCES; and *Specifications for Masonry Structures*, ACI 530.1/ASCE 6.

The ingredients of grout should meet the requirements of ASTM C 150 and ASTM C 95 for cement, ASTM C 5 for lime putty, ASTM C 207 for hydrated lime, and ASTM C 404 for aggregates. All of the materials covered in ASTM C 476 may be used in grout. Most grout used in large quantities are purchased from ready mixed concrete suppliers.

## Admixtures

When concrete masonry units are highly absorbent, a grouting-aid admixture can be used to reduce early water loss to the masonry units, promote bonding, and produce a slight expansion. Chlorine admixtures should not be used in grout, because of probable corrosion of reinforcement, metal ties, or anchors.

## Fluid consistency

The fluid consistency of grout should be great enough to be pumped or poured, but not so fluid as to cause segregation. It should flow around reinforcing steel and into all cavities and joints without leaving voids. Further, plastic grout should not cause bridging or honeycombing.

Grout consistency is measured using a slump test (ASTM C 143) and should be based on the rate of absorption of the masonry units, on ambient temperature, and humidity conditions. The desired slump is between 8" for units with low absorption and 10" for units with high absorption. As with concrete and mortar, the cone height is 12".

## Strength

Grout material proportions are covered in ASTM C 476. These proportions control the strength, but grout can be proportioned to produced a compressive strength equal to or greater than the recommended strength of masonry as specified in ASTM C 1019. Grout strength should not be less than 2000 psi.

Remember that the amount of water greatly affects the strength of grout. Water that is absorbed by the masonry units before the grout has set reduces the water-cement ratio and increases the compressive strength.

## Mixing

ASTM C 94 covers ready mixed concrete. It is recommended that grout should be batched, mixed, and delivered in accordance with this standard. Ready mixed grout should be continuously agitated after mixing and until placement to prevent segregation. Its high slump causes grout to be highly susceptible to segregation. On site mixing of grout is generally not recommended. Grout should be placed within 1 1/2 hours after water is first added.

## Placing grout

It is good practice to consolidate grout by vibration or rodding to ensure that it fills all voids and adheres completely to reinforcing steel. Pours up to 12" high can be consolidated by rodding or vibration. Pours higher than 12" should be vibrated and then reconsolidated after settlement and initial water loss occurs.

The volume of grout will be slightly reduced after placement due to water absorbed by the masonry units. To remedy this condition, shrinkage-compensating admixtures or expansive cement is sometimes recommended for highlift grouting situations. The maximum height of grout lifts is usually 5'.

## Curing grout

Adequate moisture for curing grout is usually provided by the water absorbed by the masonry units due to the high water content of masonry grout. However, in dry areas with high winds, some moist curing through fogging or plastic sheet protection may be required.

Grout is especially vulnerable to freezing because of its high water content. Therefore, care should be taken in cold weather to prevent freezing. Sand and water used in the grout mixture can be heated, or heated enclosures can be provided to protect the masonry during these adverse conditions.

## Sampling and testing

ASTM Standard C 1019 should be used for quality control and as a guide in selecting grout proportions. Grout specimens to be used in compressive tests should be prepared as specified in ASTM C 1019 rather than cast in cylinder molds used for concrete samples. For grout slump and strength tests, a minimum of 1/2 cu. ft. of grout should be sampled from materials being placed in the wall. One grout sample should consist of three specimens 3 1/2" $\times$ 3 1/2" $\times$ 7". A sample should be taken whenever there is a change in mix proportions, materials used, or the method of mixing is changed. Also, a sample must be taken for each 5000 sq. ft. of masonry.

 **Warning!** Contact with wet (plastic) concrete, cement, mortar, grout, or cement mixtures can cause skin irritation, severe chemical burns, or serious eye damage. Wear waterproof gloves, a long-sleeved shirt, full-length trousers, and proper eye protection when working with these materials. If you must stand in wet concrete, wear high top waterproof boots. Wash wet concrete, mortar, grout, cement, or cement mixtures from your skin immediately. Flush eyes with clear water immediately upon contact. Seek medical attention if you experience a reaction to contact with these materials.

# REVIEW QUESTIONS
# CHAPTER 8

Write all answers on a separate sheet of paper. Do not write in this book.

1. What is the function of mortar?
2. Mortar is generally composed of _____, _____, and _____.
3. What is meant when the term hydraulic material is used?
4. What three types of Portland cement are recommended for use in mortar?
5. What is hydrated lime?
6. Either natural or manufactured sand may be used in mortar. True or False?
7. Fine sands can increase in volume by almost 40% when the percent of moisture is only increased _____% by weight.
8. What might be the result of measuring sand by the shovel?
9. Mortar is proportioned by weight. True or False?
10. What is the probable result of overmixing mortar?
11. When should mortar be discarded?
12. Mortars have two distinct sets of properties: plastic mortar properties and hardened mortar properties. The main properties of plastic mortar are _____ and _____.
13. How much time is generally allowed to use a new batch of mortar?
14. List the three properties of hardened mortar.
15. How is the durability of mortar measured?
16. What are the five types of mortars recommended for various classes of masonry construction?
17. What type of mortar is a low-strength mortar, suitable for general interior use in nonloadbearing masonry?
18. What type of mortar has high-compressive strength and somewhat greater durability than other mortar types?
19. No air-entraining admixtures or cementitious materials containing air-entraining admixtures or agents should be used in the mortar. True or False?
20. _____ is used in brick masonry to fill cells of hollow units or spaces between wythes of solid unit masonry.
21. The temperature of mortar should be between _____°F and _____°F when used.
22. What is one of the easiest methods of raising the temperature of mortar in cold weather?
23. Why is antifreeze not recommended for use in mortar mixtures?
24. Masonry should be protected from freezing for at least _____ hours after being laid.
25. Identify the three main purposes for grouting brick and block walls.
26. Why should chlorine admixtures not be used in grout?
27. The fluid consistency of grout should be great enough to be pumped or poured, but not so fluid as to cause segregation. True or False?
28. What is the desired slump of grout for units with low absorption?
29. What is the minimum recommended strength of grout?
30. What is the maximum height of pours of grout that may be consolidated by rodding or by vibration?

Mortar for all but the smallest jobs is usually mixed in a powered mixer of 4 cu. ft to 7 cu ft. capacity, like the one shown in use here. The drum tilts to permit emptying into a wheelbarrow.

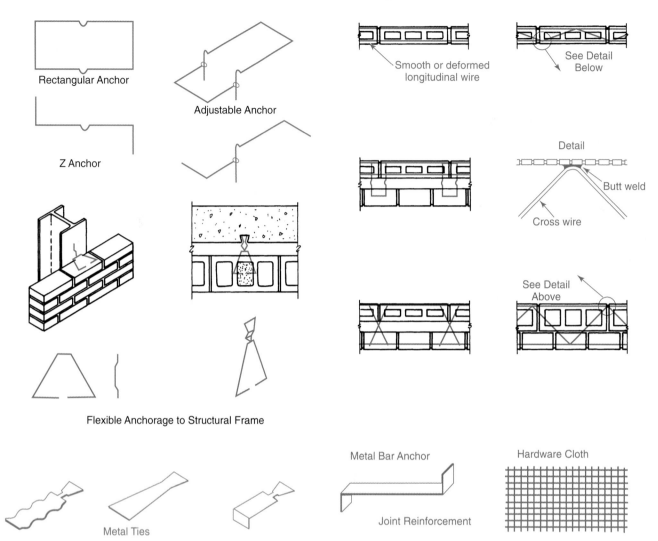

**Figure 9-1.** Common masonry anchors, ties, and joint reinforcement used in masonry wall construction.

**Figure 9-2.** Rectangular anchors used in cavity wall construction. (Portland Cement Association)

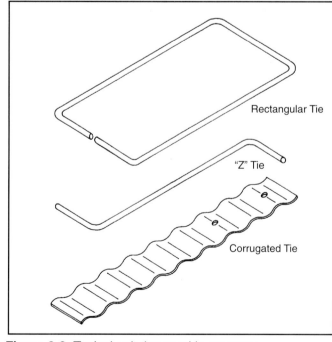

**Figure 9-3.** Typical unit ties used in masonry construction. (Brick Institute of America)

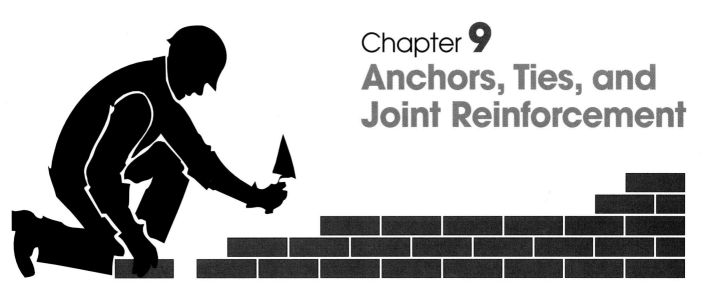

# Chapter 9
# Anchors, Ties, and Joint Reinforcement

Many types of masonry anchors, ties, and joint reinforcement are used in masonry construction. Figure 9-1 shows several of the most common. These materials are placed inside masonry walls by the mason to give it greater strength or to hold it in place. These materials include metal ties, reinforcing rods or bars, anchors, and joint reinforcement.

## Metal Ties or Wires

Wall ties normally perform three functions: they transfer lateral loads, provide a connection, and permit or restrain movement between different materials in a wall. In addition, metal wall ties can serve as horizontal structural reinforcement. However, these functions can only be achieved if the tie system is securely attached and embedded, has the required stiffness to transfer lateral loads, has little mechanical play, is corrosion-resistant, and is easily installed.

Structural bonding with metal ties or wires is widely used in three types of construction:
1. Masonry veneer to frame
2. Solid masonry
3. Cavity wall. See Figure 9-2.

### Types of ties

The basic types of metal wall ties include unit ties, continuous horizontal joint reinforcement, adjustable ties, and re-anchoring systems.

### Unit ties

Three types of unit ties are used in masonry wall construction. They are rectangular ties, Z ties, and corrugated ties. See Figure 9-3. Rectangular and Z ties are generally made from cold-drawn steel wire or copper-clad wire that meets ASTM A 82 or ASTM B 227, respectively. See Figure 9-4. Stainless steel rectangular and Z ties are also available for highly corrosive applications, as described in ASTM A 167. Corrugated ties are usually made from sheet steel that conforms to ASTM A 570, grade D. They are also available in copper and stainless steel. Figure 9-5 shows some typical applications.

**Figure 9-4.** A Z tie may be used in concrete block wall construction. (Portland Cement Association)

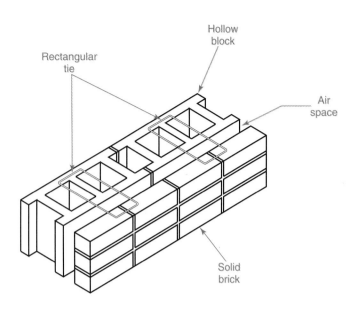

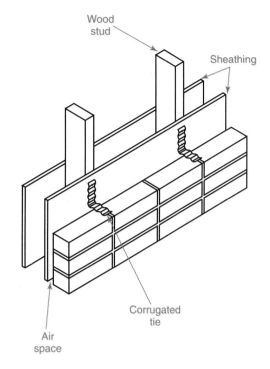

**Figure 9-5.** Typical applications of unit ties.

## Horizontal joint reinforcement

Continuous horizontal joint reinforcement is ordinarily made from #8, 9, 10, or 11 gauge wire or 3/16" diameter wire, ASTM A 82. Lengths available are 10' to 12'. Styles of horizontal joint reinforcement include ladder, truss, and tab types. See Figure 9-6. Figure 9-7 shows typical applications.

## Adjustable ties

Adjustable ties are used to attach a masonry facing to a backup system of masonry or typical stud construction. The use of adjustable ties speeds construction because they consist of two pieces. One piece is installed as the backup is constructed and the other piece is installed as the facing is constructed. One disadvantage is that the performance with respect to strength and stiffness is generally less than unit ties or joint reinforcement.

Adjustable ties are available as unit ties and adjustable assemblies. See Figure 9-8. Adjustable unit ties are intended for use with masonry backup, concrete backup, and steel studs or frames. Adjustable assemblies are available in ladder and truss styles. These are intended for use in masonry-backed cavity, veneer, and grouted walls. See Figure 9-9.

## Masonry re-anchoring systems

This class of masonry ties is the newest type. Typical approaches include mechanical expansion systems, screw systems, and epoxy adhesive systems. Masonry re-anchoring systems are used where ties were not installed during the original construction or to replace a failed system.

## Reinforcing Rods or Bars

Reinforcing rods or bars of various sizes are used in masonry construction elements such as walls, collars, pilasters, and beams. See Figure 9-10. They are placed in grouted cavities, pockets, cores, cells, or bond beams of masonry. The sizes of reinforcing bars permitted by the MSJC Code are shown in Figure 9-11.

Bar sizes are indicated by numbers. Bar sizes #3 through #8 are the number of eighths of an inch of the nominal diameter of the bars. Sizes #9 through #11 are 1 1/8", 1 1/4", and 1 3/8" in diameter, respectively. Bar sizes continue up to #18, but these larger sizes are not generally used in masonry construction.

Coated bars or fiber-reinforced plastic rebars are used when corrosion protection is needed. Coatings are generally epoxy or zinc (galvanized). Reinforcing bars have deformations on the surface that aid in bonding. See Figure 9-12. They are formed from rolled steel and are similar to those used in reinforced concrete construction.

Identification marks on ASTM standard bars are rolled into the surface of one side of the bar to denote the producer's mill designation, bar size, type of steel, and yield strength indicated by grade marks, which are used for indicating grade 60 and grade 75 bars. grade 40 and grade 50 bars show the identification marks of the producing mill (usually an initial), bar size number (#3 through #18), and type. See Figure 9-13. Grade marks are not used for grade 40 and grade 50 bars.

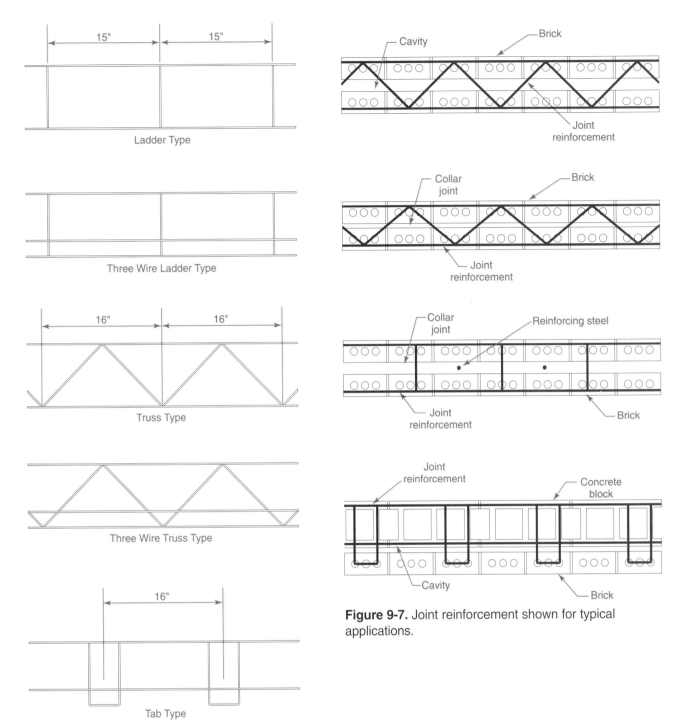

Ladder Type

Three Wire Ladder Type

Truss Type

Three Wire Truss Type

Tab Type

**Figure 9-6.** Types of continuous joint reinforcement.

**Figure 9-7.** Joint reinforcement shown for typical applications.

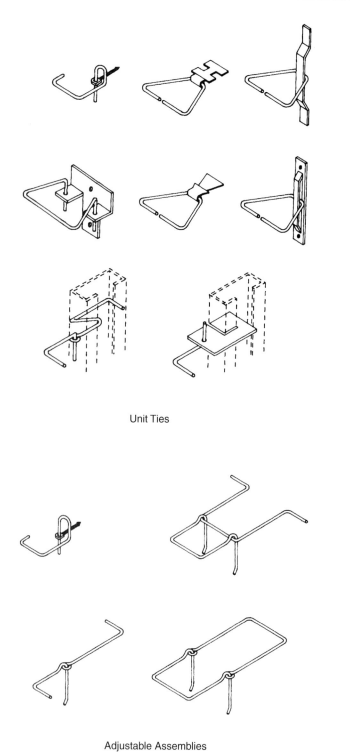

Figure 9-8. Top—These are adjustable unit ties for steel, concrete, and stud wall backup. (Brick Institute of America) Bottom—Typical adjustable unit ties for masonry backup. (Brick Institute of America)

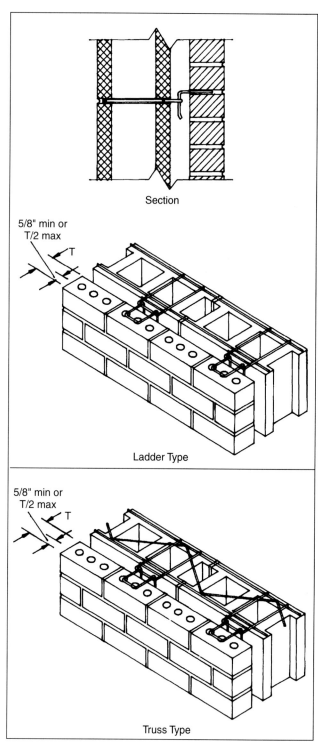

Figure 9-9. These are adjustable assemblies shown in typical applications. (Brick Institute of America)

**Figure 9-10.** Reinforcing rods are used to strengthen this wall. (Brick Institute of America)

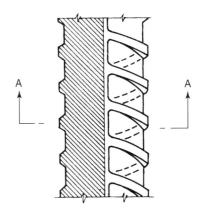

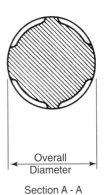

Section A - A

**Figure 9-12.** Bar diameters are nominal, with the actual diameter outside of deformations being somewhat greater. This deformation style is one typical type. (Concrete Reinforcing Steel Institute)

| Steel Reinforcement Sizes | | | |
|---|---|---|---|
| Reinforcement Type | Designation | Diameter in. (mm) | Area sq. in. (sq. mm) |
| Bars | No. 3 | 0.375 (10) | 0.110 (71) |
| | No. 4 | 0.500 (13) | 0.196 (126) |
| | No. 5 | 0.625 (16) | 0.307 (198) |
| | No. 6 | 0.750 (19) | 0.442 (285) |
| | No. 7 | 0.875 (22) | 0.601 (388) |
| | No. 8 | 1.000 (25) | 0.785 (506) |
| | No. 9 | 1.128 (29) | 1.000 (645) |
| | No. 10 | 1.270 (32) | 1.267 (817) |
| | No. 11 | 1.410 (36) | 1.561 (1007) |

**Figure 9-11.** Steel bar reinforcement sizes used in masonry construction.

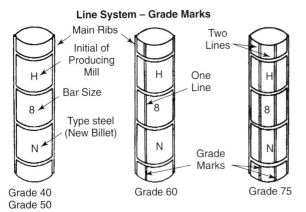

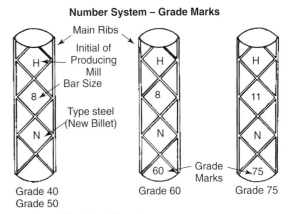

VARIATIONS: Bar identification marks may also be oriented to read horizontally (at 90° to those illustrated above.)
Grade mark lines must be continued at least 5 deformation spaces.
Grade mark numbers may be placed within separate consecutive deformation spaces to read vertically or horizontally.

**Figure 9-13.** Two systems of identification marks are used for ASTM standard bars—line system and number system. (Concrete Reinforcing Steel Institute)

# Anchors

Anchors are produced in a variety of shapes and sizes. One type is corrugated with a dovetail on one end for veneer construction. These ties are galvanized steel and the dovetail fits into a slot formed in the concrete beam or column. The other end is attached to the veneer.

Another type of anchor is the anchor bolt. Anchor bolts are used to attach roof plates, sills, etc. to masonry work. See Figure 9-14. They are available in various diameters or gauges and lengths. Some are made of band steel and have a bent end for better holding power. Others are round and may have a head or bend on the embedded end. See Figure 9-15.

**Figure 9-14.** Anchor bolts can be used to attach roof plates or sills to masonry work. (Portland Cement Association)

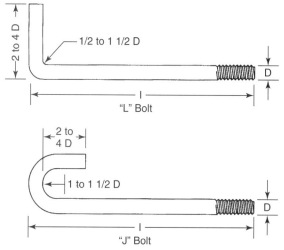

**Figure 9-15.** These are "L" and "J" bent bar anchors. (Brick Institute of America)

Bearing walls that intersect are frequently connected with a strap anchor. See Figure 9-16. Anchors are usually 1 1/4" × 1/4" × 30". Three inch, right-angle bends are made at each end for embedding in mortar-filled cores of masonry block.

Anchors used in stone panel construction include many of the types used in brick and concrete masonry construction. In addition to these, many others are used as well. Figure 9-17 shows these anchors.

Concrete anchor devices intended for installation after the concrete has hardened generally involve some type of expansion device. Figure 9-18 shows some of the more common types.

# Joint Reinforcement

Joint reinforcement is produced in many shapes and sizes for various applications. It is especially designed for placement in the horizontal joints of masonry walls to provide greater lateral strength. One popular style of joint reinforcement is made in the shape of a truss or a ladder. Refer to Figure 9-1.

Materials used in the fabrication of truss and ladder style reinforcements are smooth or deformed wire or steel rod. Joint reinforcement must be corrosion-resistant. Cross wires are welded perpendicularly or diagonally to the longitudinal wires, generally at a 16" spacing.

Joint reinforcement is available in most widths and lengths of 10' to 12'. The longitudinal and cross wires are generally #8 or #9 gauge or 3/16" diameter. Refer to Figure 9-6. The maximum size of the wire is 1/2 the mortar joint thickness.

Another type of joint reinforcing is the use of hardware cloth. See Figure 9-19. This material is usually galvanized and may be purchased in various widths and lengths. It is frequently used with a metal tie bar to tie nonloadbearing walls to intersecting walls.

Many other highly specialized anchors and ties are available for unique construction. See Figure 9-20. These products are described in the manufacturer's literature.

149

**Figure 9-16.** A strap anchor being put into place connecting a partition wall to a bearing wall.

**Figure 9-17.** Steel wire reinforcement sizes used in masonry construction.

| Steel Reinforcement Sizes | | | |
|---|---|---|---|
| Reinforcement Type | Designation | Diameter in. (mm) | Area sq. in. (sq. mm) |
| Wires | W1.1 (11 gauge) | 0.121 (3.1) | 0.011 (7) |
| | W1.7 (9 gauge) | 0.148 (3.8) | 0.017 (11) |
| | W2.1 (8 gauge) | 0.162 (4.1) | 0.021 (14) |
| | W2.8 (3/16 in.) | 0.188 (4.8) | 0.028 (18) |
| | W4.9 (1.4 in.) | 0.250 (6.4) | 0.049 (32) |

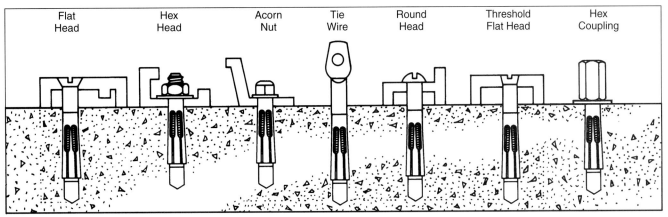

**Figure 9-18.** Available head styles of common anchor devices. (ITW Ramset/Red Head)

**Figure 9-19.** Hardware cloth can be used to tie intersecting walls together. (Portland Cement Association)

# REVIEW QUESTIONS
# CHAPTER 9

Write all answers on a separate sheet of paper. Do not write in this book.

1. Metal ties are placed inside masonry walls to give them greater strength or to hold them in place. True or False?
2. Identify the three normal functions of wall ties.
3. What three types of masonry construction use metal ties or wires for structural bonding?
4. Identify the four basic types of metal wall ties.
5. Identify three types of unit ties used in masonry wall construction.
6. What type of metal tie is generally used to attach masonry veneer to a frame structure or masonry backup?
7. What type of anchor is usually used to attach roof plates and sills to masonry work?
8. In what lengths are horizontal joint reinforcement available?
9. Name three applications where adjustable ties are used.
10. Corrugated ties are used to attach masonry veneer to a frame structure or other masonry veneer. There should be one tie for every _____ sq. ft. of wall area.
11. Reinforcing rods are used in solid reinforced masonry walls, masonry lintels, sills, and pilasters. True or False?
12. When are coded bars or fiber-reinforced plastic rebars used?
13. What three identification marks are shown on grade 40 and grade 50 bars?
14. Why do anchor bolts usually have a bent end?
15. Hardware cloth is used in masonry construction to tie _____ walls to _____ walls.

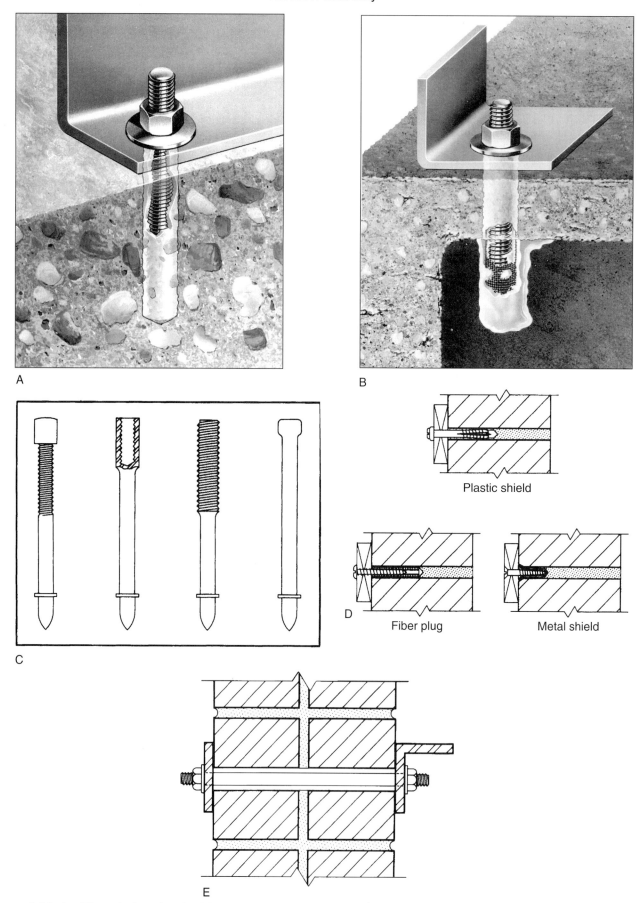

**Figure 9-20.** A—Threaded stud embedded in a concrete slab using epoxy adhesive. (ITW Ramset/Red Head) B—Threaded stud attached to a hollow concrete masonry unit using epoxy adhesive. (ITW Ramset/Red Head) C—Power-driven pins. (Brick Institute of America) D— Screw shields and plugs. (Brick Institute of America) E—Through bolt. (Brick Institute of America)

This room utilizes glass block to lighten the area.

# SECTION **III**
# TECHNIQUES

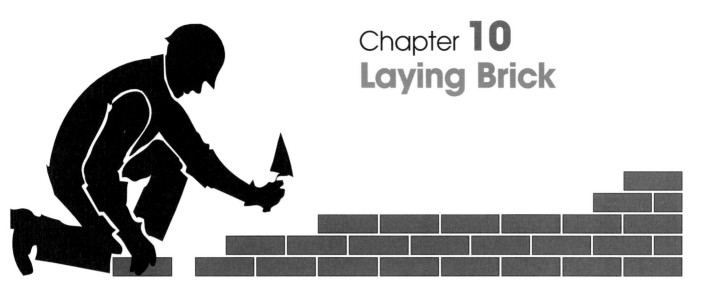

# Chapter **10**
# **Laying Brick**

Laying brick masonry units involves planning, knowledge of the materials, and ability to use the tools properly. Laying masonry is an art and requires a great deal of practice. However, the sense of satisfaction gained from doing a job well is worth the effort. This chapter is designed to help you master the proper techniques used in laying several common types of brick walls.

## Basic Operations

A number of basic operations must be mastered before a masonry structure can be built. Operations such as spreading mortar, holding a brick, using both hands at the same time, forming a head joint, cutting brick, and using a mason's line are basic operations that must be mastered. Most will need lots of practice.

### *Spreading mortar*

Spreading mortar is part of all masonry construction. It is a simple process, but requires mastery if the mason is to be an expert at it. With practice you will be able to determine just how much mortar to use for a given joint. Placing the unit will be easier if the right amount of mortar has been laid down.

Select a trowel that is best for the job being performed and which "feels good" to you. If it is too small, effort will be wasted in needless movements. If the trowel is too large, you will tire easily. You may want to begin with a standard mason's trowel about 10" long.

Grasp the trowel in the right hand, if you are right handed. Fingers should be under the handle and the thumb on top of the ferrule as shown in Figure 10-1.

### Loading the trowel

Mortar may be loaded on the trowel in several ways:
1. From the side of the board
2. From the middle
3. From the top of the pile

**Figure 10-1.** Hold the trowel with the fingers under the handle and the thumb on top of the ferrule.

Figure 10-2 shows the steps used by one mason to load the trowel. Develop the method which works best for you and perfect it.

Spread the mortar with a quick turn of the wrist toward the body and a backward movement of the arm. As the trowel is near empty, tip the trowel blade even more to help the remaining mortar slide off. See Figure 10-3. The kind of mortar and its consistency most often determines the speed of the stroke.

When solid masonry units are used, or the first course of concrete block is being laid, the mortar is spread and furrowed with the point of the trowel as in Figure 10-4. This helps to form a uniform bed to lay the masonry on. When mortar is laid down it should be slightly heavier on the outside (face) edge than on the inside. This will force out enough mortar for the next cross joint. Mortar hanging over the sides should be cut off using the front half of the trowel.

### Mortar consistency

Mortar on the board should be kept well tempered (sprinkled lightly) with water until it is used. Avoid constant working. This causes the mortar to dry out and become stiff. Stiff mortar does not spread well. Use mortar before it begins to set up.

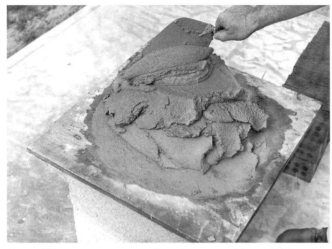

Step 1: Work the mortar into a pile in the center of the mortar board.

Step 2: Smooth off a place with a backhand stroke.

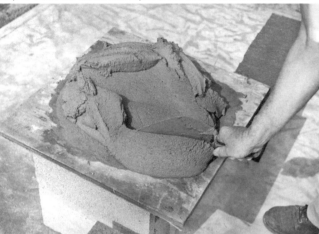

Step 3: Cut a small amount from the larger pile with a pulling action.

Step 4: Scoop up the small pile with a quick movement of the trowel.

**Figure 10-2.** Steps in loading the trowel.

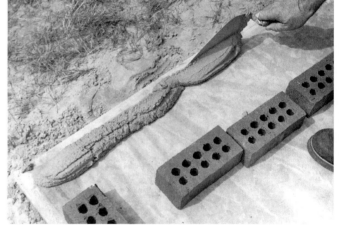

**Figure 10-3.** Steps in emptying the trowel. As the trowel is unloaded, the outside edge is tilted upward with a swift wrist motion (left) while the arm is snapped backward in the direction of the mortar line. As the dump is completed (right) the trowel is nearly vertical.

**Figure 10-4.** Furrowing the first course to form a uniform bed.

**Figure 10-5.** Proper way to hold a brick when laying to the line.

 **Warning!** Contact with wet (plastic) concrete, cement, mortar, grout, or cement mixtures can cause skin irritation, severe chemical burns, or serious eye damage. Wear waterproof gloves, a long-sleeved shirt, full-length trousers, and proper eye protection when working with these materials. If you must stand in wet concrete, wear high top waterproof boots. Wash wet concrete, mortar, grout, cement, or cement mixtures from your skin immediately. Flush eyes with clear water immediately upon contact. Seek medical attention if you experience a reaction to contact with these materials.

## Holding a brick

The proper way to hold a brick to apply mortar and to lay it on the mortar bed is to let the thumb curl down over the top edge and be slightly away from the face of the brick. The fingers should be extended outward. This hand position will allow the line to fit between the thumb and the top edge of the brick. When laying a veneer facing this position will allow the line to fit between the fingers and top edge of the brick. See Figure 10-5. It is important not to touch the line when laying units to the line. This hand position will also not interfere with cutting off the excess mortar.

## Using both hands at once

Learn to use both hands at the same time when laying brick. For example, if you are right handed, pick up the brick with your left hand and load the trowel with the right when laying a veneer facing. Both hands should be used to pick up bricks and stack them on the finished back wythe of a solid 8" wall. If the bricks are stacked before any mortar is spread, less bending and reaching will be necessary. Using both hands at the same time will increase efficiency and reduce fatigue.

## Forming a head joint

Mortar for the head joint should be applied to the brick before it is placed on the mortar bed. The first brick is placed on the mortar bed with no mortar on it. Mortar for the head joint is applied to each brick thereafter using a swiping or throwing action across the end of the brick. This motion will attach the mortar to the brick and form it into a wedge shape. See Figure 10-6. The brick is then pressed into place, squeezing out some of the mortar. Getting just the right amount of mortar on the brick will take some practice.

## Cutting brick

Most every bricklaying job requires some brick to be cut to size for a particular application. Figure 10-7 shows the names used to describe cut or broken brick pieces.

**Figure 10-6.** Applying mortar to a brick to form a head joint.

They can be cut several different ways—brick hammer, brick set chisel, trowel, or masonry saw.

 **Warning!** When doing any cutting procedure, be sure to wear safety glasses.

## Cutting brick with the brick hammer

The method of cutting brick most frequently used by masons is the brick hammer. The recommended procedure is to hold the brick in one hand, keeping the fingers away from the side where the cutting will take place. With the other hand, strike the brick with light blows with the blade edge of the brick hammer along the line where the cut is planned. Turn the brick over to the adjacent edge and continue the cut along that side. Repeat the process until all sides have been scored. Then strike the face of the brick with a sharp blow with the hammer. The brick should break along the scored lines. This procedure will produce a rough cut, but is acceptable for applications where the cut edge will be hidden by mortar.

## Cutting brick with the brick set chisel

When a more accurate, straight cut is needed, the brick set chisel is used. First, mark the brick with a pencil or other marking device where the cut is to be made. Place the brick on a soft surface such as soil or board. Hold the

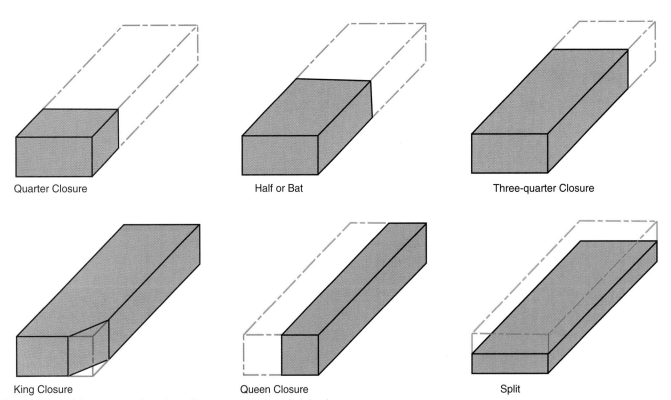

| Quarter Closure | Half or Bat | Three-quarter Closure |

| King Closure | Queen Closure | Split |

**Figure 10-7.** Names used to describe cut or broken brick pieces.

chisel of the brick set vertically with the flat side of the blade facing the direction of the finished cut. Strike the brick set sharply with the brick hammer. The resulting cut should be relatively smooth and not require additional chipping before use in the wall.

## Cutting brick with the trowel

The brick trowel can be used to cut softer brick, but is not recommended for most cutting, especially hard brick. Even though it is common practice for most masons to cut brick with the trowel when the brick set chisel and hammer are not within reach, the trowel was not intended to be a brick cutting tool. However, when cutting a brick with the trowel, hold the brick in one hand, keeping the fingers away from the side where the cutting is to be done. Hold the brick down and away from the face. Strike the brick with the edge of the trowel using a quick, sharp blow at the spot where the cut (break) is intended.

## Cutting brick with the masonry saw

When an exact, smooth cut is required, a power masonry saw is used. This process is slower, but produces the highest quality cut. Be sure you have had instruction in the proper use of the masonry saw before using it. Hold the brick against the fence firmly in one hand and move the saw slowly through the cut. Move the saw back to its original position before removing the pieces.

### *Using a mason's line*

Any straight wall longer than the plumb rule (4') is usually laid to a line. The line is a guide that helps the bricklayer build a straight, level and plumb wall. Mason's line is a light, strong cord that can be stretched taut with little or no sag. Pull the line to the same degree of tautness for each course. This is necessary so that all bed joints will be uniform and parallel.

Leads are usually built first at each corner to establish proper height and provide a place to attach the mason's line. See Figure 10-8. The height of each course is determined by a mason's rule or a story pole with height marks on it.

On long walls, it is impossible to eliminate sagging in the line. The proper course height should be checked with the story pole to ensure accuracy.

Laying brick too close to the line (crowding) should be avoided. Stay a line width away so that you do not risk moving the line destroying its usefulness as a reference.

### *Related tasks*

There are a few tasks or procedures that the mason may or may not be involved with that need to be considered. They include locating the building line, erecting batter boards, and the function of the story pole and gauge stick.

**Figure 10-8.** Leads are the built-up sections of the wall at each corner. This is done to establish proper height for each course and provide a place to attach the mason's line. It is important that these corners be as near perfect as possible.

## Locating the building line

The building line is generally the face or side of the building being constructed. It is an important line, because all crafts involved in the construction of the building refer to it. Frequently, the line will be referenced on the drawings. In residential construction, the contractor usually establishes the building line for the foundation walls. In commercial construction, a surveying crew will most likely locate the building line as defined on the plot plan.

In some towns and cities, the zoning commission has already determined the distance the building line is to be from the property line, center of the street, or some other place. This is to make the front of all buildings on the street conform to a straight line. Therefore, it is very important that the building line be established exactly to prevent a code violation or other complications.

## Erecting batter boards

*Batter boards* are boards that are used to preserve the building lines during excavation and construction. Right angle boards are formed by nailing two boards to three stakes to form the angle. Straight batter boards are made with one board and two stakes. Material used for batter boards is usually 1 × 4 or 1 × 6 boards and the stakes are 2 × 4s.

The stakes are driven into the ground beyond the corners of the foundation walls and the excavation. Lines are secured to the boards to locate the building line. The boards are marked in some way (usually with a saw kerf) to locate the position of the cord so that it can be located in exactly the same place each day. Masons and other workers can refer to the line whenever they need to locate a corner or face of the building.

## Story pole and gauge stick

The **story pole** is a narrow board, generally 1" × 2" × 10' long (about one story in height), that is used to measure the mortar joint heights of each course of masonry. Each joint is marked on three sides of the board (pole) as well as window heights, door heights, sill heights, or any other feature that needs reference.

Some jobs require more than one story pole because different materials are used, or windows, doors, etc. are at different elevations. A bench mark (nail or other permanent object) is generally used to ensure the proper placement of the story pole with respect to the building elevation. It is always positioned at that point for measurements.

A gauge stick is shorter than a story pole—about 4' long and 3/4" square. The purpose of a gauge stick is to be able to lay all brick courses in uniform bed joints. Gauges are generally marked in terms of 4 brick courses. For example, an 11" gauge would require 1/2" bed joints for brick 2 1/4" high (4 brick × 2 1/4" + 4 mortar joints 1/2" each = 11"). Masons also use the 6' mason's folding rule for this purpose.

# Laying Common Brick Walls

The procedure for laying several common brick walls is presented to give the student practice in learning the proper construction and procedure in building these walls. The types of walls covered in this chapter include: 4" common bond wall with leads, 8" common bond, double wythe wall with leads, 12" common bond solid wall with leads, corner layout in Flemish bond, corner layout in English bond, cavity wall, and single wythe brick bearing wall.

## Laying a four course, single wythe, running bond lead

Corners (leads) must be constructed very carefully, because they are generally used as a guide for the entire wall. Laying the corners is called **laying the leads.** Two leads are used, one at each end of the wall. A mason's line is stretched between the leads to serve as a guide for laying the rest of the brick between the corners. Sometimes an intermediate lead, called a *rack-back lead*, is constructed about mid-point between the corner leads.

The following procedure is suggested for laying up a four course, single wythe, running bond lead:

1. The first step is to snap a chalk line where the face of the masonry wall is to be located along adjacent sides. See Figure 10-9. Mark the actual head joints along the chalk line for proper placement of the masonry units.

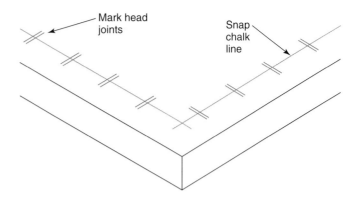

**Figure 10-9.** Step One. Building a four course corner lead.

2. Start the first course by laying the first unit from the corner. Note, the placement marks for the head joints are visible even when the unit is in place. See Figure 10-10.

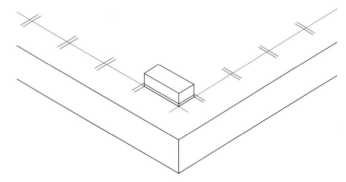

**Figure 10-10.** Step Two. Building a four course corner lead.

3. Complete the first course along both legs of the lead. Each leg should be about the same length. Leads are generally about seven or nine courses high, but a shorter lead may be used for practice, as in this case. See Figure 10-11.

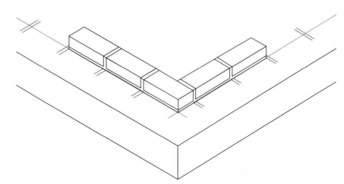

**Figure 10-11.** Step Three. Building a four course corner lead.

4. Lay up the second and third courses alternating the pattern of brick placement. Lay up the third course similar to the first course. See Figure 10-12.

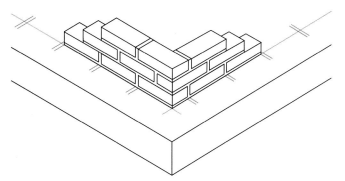

**Figure 10-12.** Step Four. Building a four course corner lead.

5. Lay up the fourth course to complete the lead. Two brick on the highest course of the lead will provide enough resistance to maintain a tight line. See Figure 10-13.

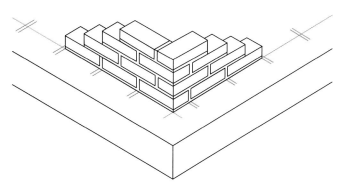

**Figure 10-13.** Step Five. Building a four course corner lead.

## Laying a 4" running bond wall with leads

Place materials and arrange your work space for efficient work. The mortarboard should be located in the center of the work space about 24" from the wall. Brick should be stacked on both sides of the board as in Figure 10-14.

The following procedure is recommended for laying a 4" running bond wall with leads:

1. Establish a wall line. See Figure 10-15. This line can be easily made using a chalk line. The line is coated with colored chalk. When held taut it may be snapped to produce a chalked line on the surface you wish to mark.

**Figure 10-14.** Materials arranged for efficient work.

**Figure 10-15.** Establishing a wall line with a chalk line.

Check the slab or footing to be sure that it is square before snapping the wall line. The diagonal distance from corner to corner should be the same.

Another method which may be used to check for squareness is to measure 6' along one side and 8' along the intersecting side. The diagonal distance between these points should be 10'. This procedure is known as the **6-8-10 rule**.

2. After the wall line has been drawn or chalked a dry course of brick should be laid out from corner to corner as in Figure 10-16. Arrange them to eliminate as much cutting of units as possible. Be sure to allow for a mortar joint between each brick. Mark the joints and then move the bricks aside. See Figure 10-17.

3. Spread the mortar on one side of the corner and furrow it as in Figure 10-18.

4. Lay the corner brick exactly on the point where the corner is located. See Figure 10-19. It must be set level and square with the wall line.

**Figure 10-16.** The first course is laid out dry to eliminate as much cutting as possible.

**Figure 10-17.** Marking location of each brick.

**Figure 10-18.** Mortar in place and furrowed for first side of lead corner.

**Figure 10-19.** The corner brick is laid at the exact corner. Use the trowel to tap it into position.

5. Lay the remaining four or five bricks of the lead corner. This is called *tailing out* the lead of the corner. After the bricks have been laid, level them with the bricklayer's plumb rule. See Figure 10-20.

**Figure 10-20.** Leveling the first five bricks of the corner with the mason's level (plumb rule).

6. Plumb the corner brick. See Figure 10-21. Then plumb the tail end. See Figure 10-22. Leveling is done on the outside and top edge of the brick.

**Figure 10-21.** After leveling, plumb the corner brick to make its edge exactly vertical.

**Figure 10-22.** Next plumb the tail end brick. Note how the mason braces his arm against his knee to steady the plumb rule. Brick is tapped with the trowel to adjust it to proper position.

**Figure 10-23.** First five bricks are straightedged with the plumb rule.

**Figure 10-24.** The other side of the corner is begun. Be careful not to move corner brick. See how the mason supports the corner brick with fingers.

7. Line up the bricks between the two plumb points. See Figure 10-23. Here the level is used as a straightedge and the bubbles are disregarded. Always follow this sequence when building a corner: level the unit, then plumb, and then line up.

8. After one side of the corner has been laid and trued, start the other side of the corner as in Figure 10-24. Again, spread the mortar and furrow it. Lay three or four bricks. Level, plumb, and line them up. Do not tap the level with the trowel or hammer. Use your hand. This time the corner brick does not require plumbing because it has already been plumbed. But, if you think that you might have moved it, check it again.

9. Lay the second course following the same sequence used for the first course. Check for proper height just after the course has been laid, but before the brick have been leveled. See Figure 10-25. If the brick are too high, tap them down as they are leveled. If they are too low, remove them and add more mortar before leveling.

10. Repeat the sequence until the corner is built. See Figure 10-26.

11. Straightedge the rack of the lead, as shown in Figure 10-27. This will eliminate any wind, belly, or cave-in in the wall.

12. Lay the second corner of the wall following the same procedure used for the first. See Figure 10-28.

**Figure 10-25.** Checking for proper height. Notice that these bricks are being laid on 5s. Hold the rule straight up and down for accurate measurement.

**Figure 10-26.** Completed lead corner will be used as a guide for the rest of the wall.

**Figure 10-27.** Mason straightedges the rack of the lead to check for proper spacing of brick, wind, belly, or cave-in in the wall. The rack refers to the "steps" formed by the last brick in each course. If the rule rests on each of them, the spacing is proper.

**Figure 10-28.** Top—Beginning the second lead corner using the same procedure as for the first corner. Center— Laying up the corner. Bottom—Completed corner.

13. Stretch the mason's line between the corners at the top of the first course. See Figure 10-29. The line may be secured with line pins or corner blocks. Pull the line taut enough that it is straight and level. Be sure to use the same amount of tension each time.

**Figure 10-29.** Laying the first course to the line. The closer brick is being placed. Note that both ends have been buttered. It must be fully pressed into position.

14. Begin laying the wall from the lead toward the center, one course at a time. See Figure 10-30. Each brick on the second course should be centered over the cross joint of the first course. The cross joints must be of uniform width or the closer brick (the last brick) will not fit correctly. The line should be worked from both leads toward the center. Level, plumb, and line up this course as you lay it.

15. Tool or strike the joints when the mortar can be indented with a thumbprint without too much pressure. The brick will be smeared if the mortar is too wet. If too dry, the mortar will turn black from metal worn off the jointer. Experience will help determine the proper time to tool the joints.

**Figure 10-30.** Laying the second course from both ends toward the middle. Top—Mason's line in position. Bottom—Next to last brick being placed. Note how the fingers extend out over the line without touching it.

In striking a wall use the long jointer for the bed joints. See Figure 10-31. Next, use the short jointer for vertical or head joints. See Figure 10-32. With the edge of the trowel, cut off the tags of mortar that have been forced out by the jointer. See Figure 10-33.

The proper procedure for removing the tags is to move the trowel horizontally, along the joint. After the head joints have been struck, use the long jointer again to strike the bed joints for best appearance.

16. Brush the wall when the mortar is stiff enough. See Figure 10-34. This may be done at the end of the day or more frequently, if desired. Brushing reduces the amount of cleaning required later.

**Figure 10-31.** Bed joints are struck with the long jointer when the mortar is ready.

Figure 10-32. Short jointer is used to strike the head joints.

Figure 10-33. Remove the tags by moving the blade of the trowel forward along the bed joint.

Figure 10-34. The wall is brushed when the mortar is stiff enough.

17. Lay successive courses in the same manner until the wall is the required height. See Figure 10-35.

Figure 10-35. The completed wall.

## Laying an 8" common bond, double wythe wall with leads

The common bond is strong and can be laid up fast. It consists of a course of headers between every five to seven courses of stretchers. The heading course serves as a bond between the inside and outside 4" tiers. The mortar joint is 3/8" or 1/2" thick.

The recommended procedure for constructing an 8" American bond wall with leads follows:

1. Lay out the wall location as with the 4" wall. In addition, locate a second line (about 8") inside the first wall line. This distance is equal to the average length of the brick being used.
2. Lay out the bond to eliminate excessive cutting for windows and doors. This is an important step!
3. Lay the corner bricks as shown in Figure 10-36. The header course is shown here starting at the second course, but it could start wherever specified.
4. Lay the first course of stretchers for the corner. Four or five stretchers should be sufficient. Lay the header course as shown in Figure 10-36.
5. Lay successive courses of stretchers until reaching the next header course. Level, plumb, and line up each course as it is laid.
6. Lay the header course breaking the bond in the usual manner.
7. Finish laying the first corner and complete the second in the same manner.
8. Construct the first stretcher course of the wall by laying the outside first and then the inside. These should be laid from the leads to the center. Level, plumb, and line up each course before beginning the header course. See Figure 10-37. A line should be used on all backing courses (the inside tier) to insure a good face.
9. Lay the header course from each lead toward the center. Level, plumb, and line up the header course.
10. Lay the outside tier up to the next header course keeping it straight, level, and plumb. This is called **header high**.

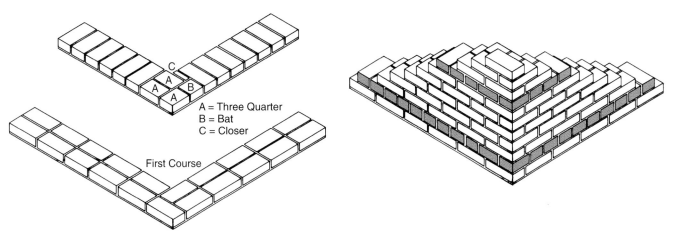

A = Three Quarter
B = Bat
C = Closer

First Course

**Figure 10-36.** An 8" common bond wall lead with first and second courses shown in detail.

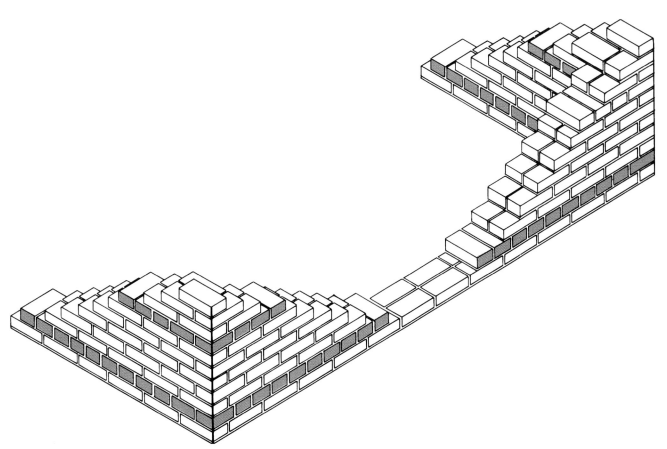

**Figure 10-37.** The first stretcher course and leads for the corners are in place in this 8" common bond wall. The wall will now be constructed between the leads.

11. Lay the inside tier up to the same height. Be careful to keep it level with the outside tier as it is laid up. This is important as you must have a level surface for the header course.

12. Continue as before until the desired height is reached. Strike the joints and brush the wall at the proper time.

## Laying an 8", two wythe intersecting brick wall

Intersecting brick walls are often necessary in the construction of most brick structures. They are laid out so that the courses are tied together to form an integrated unit. The individual units are placed so that they interlock the wall segments together. Reinforcement ties (Z-shaped ties) are generally used.

The following procedure may be used to construct an 8", two wythe intersecting brick wall in standard running bond with joints broken so that they do not line up vertically. See Figure 10-38.

1. Locate the face of the walls on the floor and snap a chalk line to preserve the location.

2. Lay out the first course using no mortar (dry bond) to maintain proper spacing and identify any difficulties or problems.

3. Spread the mortar bed and lay out the face wythe beginning with brick #1 through brick #2. See Figure 10-38. Level, align, and plumb the course of brick.

4. Lay the backing wythe beginning with brick #3 and continuing with brick #4. Do not forget to butter the back side of each brick where the backing wythe sits against the face course. This joint (collar joint) should be completely filled. Level, align, and plumb the course and check the wall thickness.

5. Lay the intersecting wall brick beginning with brick #5 and then brick #6, running each course to the outside wall. Level, align, and plumb each course and check the wall thickness. This will complete the first course plan.

6. Lay the second course following the layout shown in Figure 10-38. Level, align, and plumb the course. Notice how the intersecting wall is interlocked with the outside wall.

7. Locate the metal Z-shaped ties as shown in the second course layout plan. Embed the metal ties in a mortar bed and lay the third course identical to the first. Metal ties will be added again on the eighth course in the wall.

8. Continue laying courses alternating the pattern used in the first and second courses until the wall

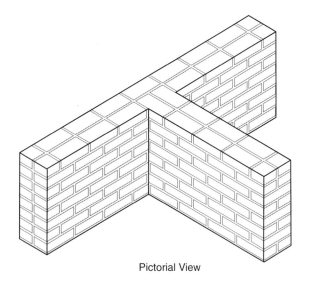

Pictorial View

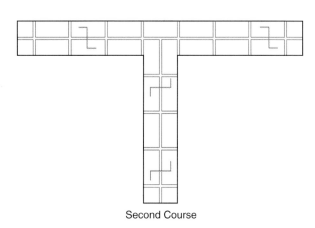

Second Course

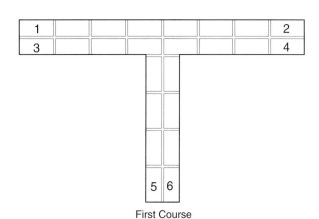

First Course

**Figure 10-38.** Procedure for laying an 8", two wythe intersecting brick wall.

has reached eight courses high. You may desire to build end wall leads to serve as guides for the wall.

9. Level and plumb the completed wall and clean off any mortar splatter.

10. Tool the joints with a concave jointer when the mortar is thumbprint hard. Tool the head joints first, then the bed joints. Use a sled jointer for the bed joints.

11. Remove any mortar tailings remaining after finishing the joints. Clean the wall by brushing with a bricklayer's brush.

## Laying a 12" common bond solid wall with leads

The 12" solid wall is essentially an 8" solid wall with a third wythe added. Figure 10-39 shows the first course which is repeated every sixth or seventh course. It contains three-quarter closures, quarter closures, stretchers, and headers arranged in a particular configuration. Headers are visible on the outer face of the wall.

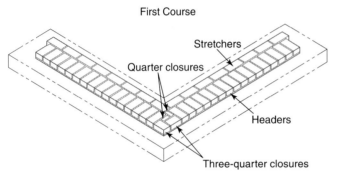

**Figure 10-39.** First course of 12" common bond solid wall.

The second course is composed of stretchers, headers, and quarter closures. See Figure 10-40. The stretchers are visible on the outer face and headers show on the inside. The quarter closures are generally placed after the headers and stretchers are laid.

The third course is composed of all stretchers. See Figure 10-41. This pattern is continued for the fourth and fifth courses in typical running bond. The sixth course is identical to the first as the process is repeated.

Twelve-inch brick masonry walls are rare today, but may be used as thermal storage units. Solid, dark colored brick should be used for this application as solid brick provide more mass than hollow or cored brick and dark brick absorb heat more efficiently than lighter colored units.

## Corner layout in various bonds

Laying up a corner is very similar for most bonds. After a beginner has built several of the more common ones

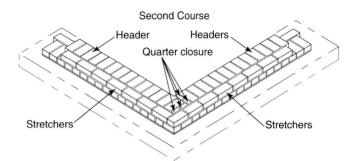

**Figure 10-40.** Second course of 12" common bond solid wall.

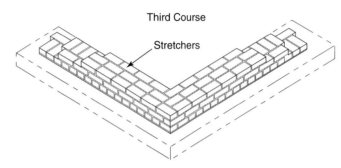

**Figure 10-41.** Third course of 12" common bond solid wall.

there should be no difficulty raising any corner. In addition to the common or American bond, the bonds described in the following sections represent some of the most frequently used patterns.

### Flemish bond

The Flemish bond is very popular. It is easy to lay producing an artistic and pleasing wall. It is more costly than the common bond and requires greater care. The bond consists of alternate headers and stretchers in each course. The headers are centered on the stretchers between each course. Bonds may be started at the corner with either a 3/4 bat or a 1/4 closure. See Figure 10-42.

### English bond

English bond has alternate courses of stretchers and headers. The headers center on the stretchers and on the joints between the stretchers. The stretchers all line up vertically, one over the other.

Two methods may be used to start the English bond at a corner. Figure 10-43 shows the English bond with the 1/4 and the 3/4 closure at the corner.

## Constructing a 10" brick masonry cavity wall

No changes are required in basic bricklaying techniques in the construction of a cavity wall. One basic

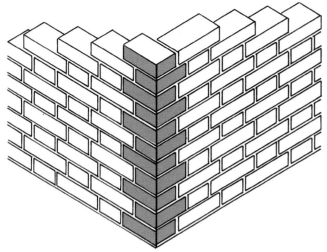

Flemish bond with 3/4 Closure at Corners

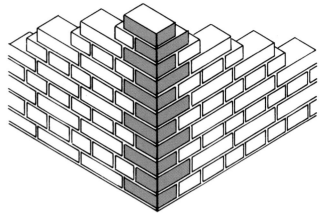

English Bond with 3/4 Closure at Corners

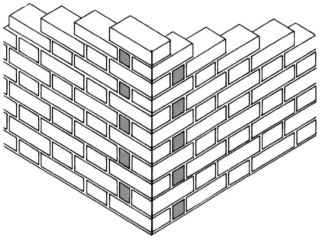

Flemish Bond with 1/4 Closure

**Figure 10-42.** Corner layout in Flemish bond with 3/4 and 1/4 closure.

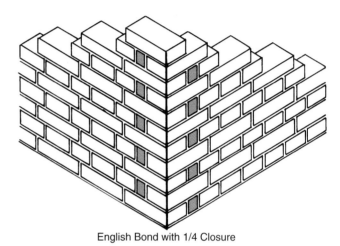

English Bond with 1/4 Closure

**Figure 10-43.** Corner layout in English bond with 3/4 and 1/4 closure.

principle, however, is that no bridge of solid material capable of carrying water across the minimum 2" cavity space should be permitted. The construction of two separate wythes, with a clean cavity, is the objective. See Figure 10-44.

## Workmanship

Extensive tests by a variety of groups and observations of masonry structures show that to obtain good masonry performance, there is no substitute for the complete filling of all mortar joints (as the units are laid) that are intended to receive mortar. Leaky walls result from partially filled mortar joints that in turn can contribute to spalling due to freezing and thawing when excessive moisture is present.

## Keeping the cavity clean

The cavity must be kept clean of mortar droppings and other foreign materials during the construction process. If mortar falls into the cavity, it may form *bridges*

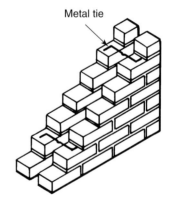

Metal tie

**Figure 10-44.** Metal ties used in cavity wall construction.

for moisture passage, or it may block the weep holes. See Figure 10-45. One method that has been used to keep the cavity clean is to place a wood or metal strip, slightly narrower than the cavity width, in the cavity or air space. The strip rests on the wall ties as the wall is built. It is pulled out when the next row of ties is placed. In this way, any mortar that has fallen into the cavity can be removed. See Figure 10-46.

**Figure 10-45.** Weep holes must be kept clear of mortar droppings to function properly.

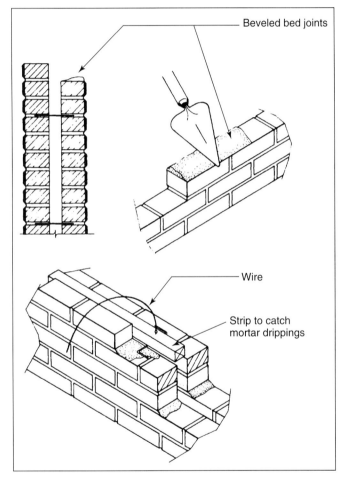

**Figure 10-46.** Procedures designed to keep the cavity clean in cavity wall construction. (Brick Institute of America)

A considerable amount of mortar should be eliminated from the cavity if the mason follows these steps:

1. After spreading the mortar bed, bevel the cavity edge with the flat of the trowel. Therefore, very little mortar will be squeezed out of the bed joints into the cavity when the units are laid.
2. Roll the units into place, keeping most of the mortar on the outside.

3. After the unit is placed on the bed joint, any protruding mortar fins should be flattened over the backs of the unit; not cut off. This procedure prevents mortar from falling into the cavity and forms a smooth surface that will not interfere with the insulation.

## Tooling

Weather tightness and textural effect are the basic considerations of mortar joint finish selection and execution. Properly *striking* or *tooling* the joint helps the mortar and masonry units bond together and seal the wall against moisture. The concave and V joints provide the most weather tight joints. Joints should be tooled when the mortar is *thumbprint hard*.

Weep holes should be placed at the base of the cavity and at all other flashing levels. They provide drainage for any moisture that may have found its way into the cavity. Weep holes must be clear of obstructions and placed directly on the flashing to function properly. They are formed or created by using any one of the following techniques:

1. Eliminating each second or third head joint.
2. Inserting oiled rods, rope, or pins in the head joint at a maximum of 16" o.c. and removing before the mortar has set.
3. Placing metal or plastic tubing in the head joint at a maximum of 16" o.c.
4. Placing sash cord or other suitable wicking material in the head joint at a maximum of 16" o.c.

## Tie placement/joint reinforcement

Both wythes of masonry must be properly tied together in a quality constructed cavity wall. The primary concern is that all of the ties are in place, remain operative, and are firmly embedded in and bonded to the mortar. In order to achieve this, the two wythes must be laid with completely filled bed joints and the ties must be in the correct position so that later disturbance of the wall assembly is unnecessary. Wall ties will do their job if properly embedded.

Ties should be 3/16" in diameter or #9 gauge wire and placed in every 4 1/2 sq. ft. of wall space when the cavity is no more than 4". The most common type of wall tie for brick masonry construction is the Z tie. There are also rectangular and U-shaped ties that may be used when the backup units are hollow masonry units with cells laid vertically. See Figure 10-47. The most important factors in terms of the performance of wall ties are:

1. Being corrosion resistant.
2. Placing ties at proper spacing. Twice as many ties should be used if they have drips. Crimping the ties reduces the strength.
3. Full bedding of the bed joint and placing the wall tie in the mortar 5/8" from either edge of the masonry unit.

Prefabricated horizontal joint reinforcement may be used to tie the interior and exterior wythes, but truss type

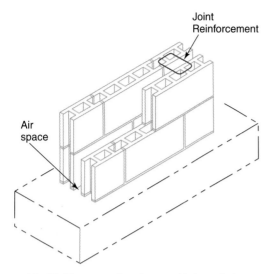

**Figure 10-47.** Rectangular ties and joint reinforcement used in cavity wall construction.

joint reinforcement should not be used to tie the wythes of a brick and block cavity wall together. Rather, ladder type reinforcement, which allows for the in-plane movement between the wythes, is recommended. Horizontal joint reinforcement is generally not required in brick masonry walls, because they are not subject to shrinkage stresses.

## Storage of materials

The proper storage of materials at the construction site may have an important influence on their future performance. Care should be taken to avoid wetting by rain or snow and contamination by salts or other chemicals that may contribute to efflorescence and staining.

- **Masonry Units.** Masonry units should be stored off the ground to avoid contamination by dirt and groundwater that may contain soluble salts. Also, masonry units should be kept dry from rain or snow.
- **Cementitious Materials.** Cementitious materials should be protected from the elements and stored off the ground.
- **Sand.** Sand for mortar should be placed on high ground or off the ground to prevent contamination. A protective cover is advisable in cold weather to prevent saturation and freezing.
- **Flashing.** Flashing materials should be stored where they will not be punctured or damaged. Plastic and asphalt coated flashing materials should not be stored in areas exposed to sunlight. Sunlight causes them to become brittle.

## Protection of walls

Masonry walls that are exposed to heavy rain during construction can become so saturated with water that they may require weeks or even months to dry out. This increases the likelihood of soluble salts going into solution which may contribute to efflorescence or contaminate other elements of the structure—concrete, plaster, trim, etc.

All walls should be kept dry by covering the top of the wall with a strong, water-resistant membrane at the end of each working day. The membrane should overhang the wall by at least 24" on each side and be secured against wind. The cavity wall should remain covered until it is completed and protected by other materials.

Leaky walls may be attributed to freezing of the mortar before it has set. It is very important to protect walls against freezing during the construction process to prevent damage.

## *Laying a 10" brick cavity wall with metal ties*

A typical 10" brick cavity wall with a corner is shown in Figure 10-48. Use the standard planning and layout procedure in laying out this practice wall. Flashing has purposely been omitted in this exercise. Be especially careful not to drop mortar into the cavity.

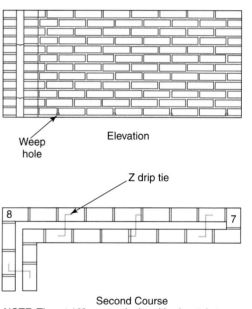

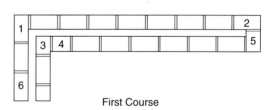

**Figure 10-48.** Procedure for laying a 10" cavity wall with metal ties.

The following procedure is suggested for constructing this wall:

1. Locate the outside wall edges and snap a chalk line to preserve the location.
2. Layout the brick dry to determine proper spacing and any layout problems. Locate the inside line for the backing wythe such that it is 10" away from the outside wall edge. This should form a 2" cavity between the wythes.
3. Lay the inside bed course beginning at brick #1 and continuing to brick #2. Work from both ends toward the middle. Use a full bed joint checking the bricks to be sure they are plumb and level. Use the chalk line to ensure a straight wall.
4. Continue the outside wythe by laying brick #6 and the closure between brick #6 and brick #1. Level, align, and plumb this part of the wall. Check to be sure the corner is square.
5. Brick #3, #4, and #5 are three-quarter brick. Cut them.
6. Lay bed brick #3, #4, and #5 in the position shown in Figure 10-48. Level and plumb the brick. Be sure the outside edge of brick #3 and #4 are 10" from the inside face. Check the position of brick #2 and #5 to be sure they are square.
7. Lay the brick between bed brick #4 and #5 and finish the leg from bed brick #3. Level, align, and plumb these sections of the wall. The first course is now complete and should be square at the corners and 10" wide throughout.
8. If you did not use empty head joints for the four weep holes shown in the elevation, open those joints or insert tubing or wicking at the weep hole locations.
9. Cut 3 three-quarter brick for the second course and lay the mortar bed for the second course. Bevel the mortar bed to avoid dropping mortar into the cavity.
10. Lay three-quarter brick #7 at the right end of the wall over bed bricks #2 and #5.
11. Lay second course brick #8 at the other end of the wall and complete the wythe between them.
12. Complete the second course following the procedure established for the first course. Plaster back any mortar that is squeezed out of the inside bed joints. Keep the cavity clean.
13. Locate the four metal Z-shaped ties as shown in the second course plan. Embed the ties in the mortar bed being sure that the drip on the metal ties points downward if your ties are of this design.
14. Build corner leads six courses high on each end of the wall so that a mason's line can be used to aid in the construction of the wall.
15. Lay the wall to a height of eight courses and then embed metal Z-shaped ties as previously

instructed. Note: the ties should be offset from the ties on the second course.
16. Attach wires at each end of a piece of wood 1" × 2" × 4' long. The board will be used to catch any mortar droppings and to flatten any mortar squeezed out inside the cavity.
17. Complete the cavity wall to a height of twelve courses using the wood piece to catch mortar. Check the wall to be sure that it is square and plumb. Remove any mortar splatter on the outside wall and clean with a bricklayer's brush.
18. When the mortar has reached thumbprint hard, finish the joints with a convex jointer. Remove any tailings that are produced from tooling.

## *Constructing a single wythe brick bearing wall*

Brick masonry bearing wall systems have been used for years for their strength, durability, and other inherent values. Once widely used in single family residential construction, this application is experiencing a resurgence in interest. The rising cost of wood framing members has created a renewed interest in alternative building systems. The use of light-gauge steel framing is one alternative. Another is the use of single wythe brick bearing walls. The use of brick masonry as the load-carrying element of the structure provides several benefits over other alternate systems. It capitalizes on brick masonry's strength, the look of permanence and beauty, lower maintenance costs, often lower insurance costs because of the fire resistance characteristics, energy efficiency, and high resale value.

In a single wythe brick bearing wall system, the brick masonry serves as both the structural system and the exterior facing. See Figure 10-49. A backing system of wood, steel, or masonry is not required. The interior living space of a brick bearing wall home is the same as a framed home. Roof, floor, and interior partitions are constructed with the same materials used in frame homes. One- and two-story home designs are typical and all of the popular amenities are easily incorporated into a single wythe brick bearing wall system. This system is also equally appropriate in commercial construction.

## Design considerations

Single wythe brick loadbearing walls include the same design considerations as other types of wall systems. Minimum loads to be resisted by the structural system, minimum thermal performance requirements, necessary fire resistance of the wall system, and resistance to moisture penetration, are covered by the model building codes.

## Structural considerations

Determining the design loads is generally the first step in the design of any structural system. These loads include vertical loads from the weight of the building materials and occupants and lateral (horizontal) loads from wind, soil, and seismic forces. In most regions of the U.S., only verti-

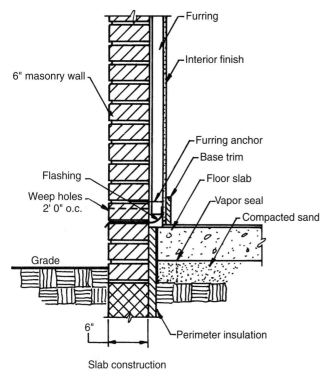

Slab construction

**Figure 10-49.** This is a detail of a 6" exterior loadbearing single wythe wall.

cal loads and wind loads are a major concern, but in some areas, seismic loads may control the structural design. Additionally, severe wind loads, as in hurricane prone areas, may also dictate the structural design.

The model building codes and the associated structural loads will dictate the size of the building's structural members. In the case of loadbearing masonry, all model building codes specify the minimum wall thicknesses and maximum wall height or number of stories for empirical designs.

The most widely used residential building code is the CABO *One- and Two-Family Dwelling Code.* The CABO specifies an 8" minimum nominal wall thickness for load-bearing masonry structures over one story in height. A nominal 6" wall is permitted for a single-story home so long as the wall height does not exceed 9' and the gable height does not exceed 15' In those cases where these limits are applied, the minimum wall thickness requirements will influence the type and size of brick unit used. Other model building codes have similar proscriptive restrictions.

## Energy considerations

Requirements for the thermal performance of a building are contained in the model building codes. Minimum levels of insulation are required, and sometimes, air leakage is covered. The method of installation and type of insulation is somewhat different in a single wythe brick bearing wall from other types of residential construction. In wood or metal frame construction, batt insulation is typically used between the studs. For brick bearing wall structures, rigid board insulation is generally placed on the interior face of the brick wythe, Figure 10-50. This type of insulation is

easily installed between furring strips and provides high insulation values. It can also be attached using mechanical fasteners and adhesives or placed in the cells of hollow brick units. Insulation placed in hollow brick is generally not as effective as surface mounted rigid insulation.

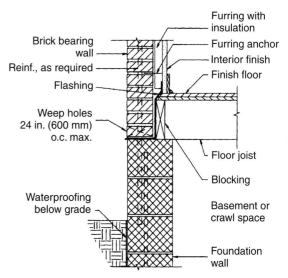

**Figure 10-50.** Typical method of insulating a single wythe exterior brick wall. (Brick Institute of America)

Air leakage is a concern in single wythe exterior brick walls. There will be some leakage through the weep holes and at the top of the brickwork. Materials such as *house wraps* are not appropriate for this type of construction. Limiting air leakage must be accomplished with papers or films on the insulation or gypsum board. All joints between panels and around doors and windows must be taped or sealed.

## Water penetration resistance

One of the chief performance concerns of any exterior wall is its resistance to water penetration. This concern is of major importance in a single wythe wall. Full mortar joints and a good bond between units is necessary to reduce water penetration. Head joints must be full, not face-shell mortar bedded only. Even with good workmanship, a single wythe wall may not prevent water penetration completely. Therefore, a drainage cavity with flashing and weep holes should be provided. See Figure 10-51. Another alternative is to use a bituminous, dampproof coating on the inside face of the brick wall that is applied before the insulation and finishes are installed.

## Location of interior work

Since there is no cavity within the exterior wall to place plumbing, heating, and electrical systems, it must be placed between the furring strips, in the floor or ceiling, or in interior frame walls. Naturally, the type of foundation/floor system used will influence the location of interior systems. In structures with basements or crawl

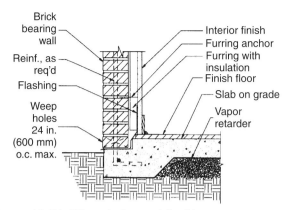

**Figure 10-51.** Weep holes and flashing are necessary in exterior single wythe masonry walls to reduce moisture penetration. (Brick Institute of America)

spaces, the mechanical systems can be located between the floor joists. In slab type structures, it is easier to route the mechanical systems through the ceiling or attic space.

## Masonry material selection

When selecting the masonry materials for a single wythe brick bearing wall system, consider the structural, energy, and reinforcement performance requirements.

- **Brick.** The brick masonry units used in a single wythe bearing wall are solid (ASTM C 216) or hollow (ASTM C 652). Solid bricks are commonly manufactured in nominal widths of 3", 4", and 6". Hollow bricks are available in nominal widths of 4", 5", 6", and 8". Nominal 5" and 6" wide hollow bricks are the most common units used to build this type of wall for residential purposes. The cells in most hollow bricks are made to accommodate vertical reinforcement and grout.

- **Mortar.** Portland cement-lime mortars with an air content less than 12% are recommended for their superior bond strength and resistance to water penetration. Masonry cements are prohibited by some codes and all type N mortars are prohibited in seismic performance categories D and E (formerly zones 3 and 4). Type S, M, or N mortar may be used, generally, in loadbearing brick masonry. Type S is recommended for use in reinforced brick bearing walls.

- **Grout.** Steel reinforcement is bonded to the surrounding brick masonry using grout. Grout may be made using fine or coarse aggregate, but fine aggregate is generally used, because it is easier to fill smaller cells. Grout should meet ASTM C 476. The water/cement ratio of grout should have a slump of 8" to 11". It must be fluid enough to fill voids, but not separate into its constituent parts.

- **Reinforcement.** Lateral loads must be resisted through the use of vertical steel reinforcement or other means. See Figure 10-52. The MSJC Code limits the maximum size of reinforcement used in masonry to a No. 11 reinforcing bar. As a

rule-of-thumb, the maximum bar size should not exceed the nominal thickness of the wall in inches to ensure proper development of the reinforcement. For example, a maximum reinforcing bar size No. 6 is recommended for nominal 6" walls.

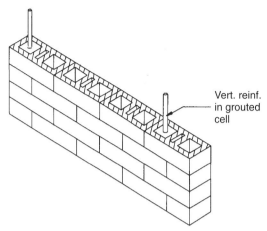

**Figure 10-52.** Vertical reinforcing used to resist lateral loads in a single wythe wall. (Brick Institute of America)

- **Joint reinforcement.** If used, joint reinforcement should comply with ASTM A 82 and be hot-dipped galvanized steel or made from stainless steel to reduce corrosion.

# Advanced and Specialized Brickwork

Other more advanced and specialized forms of brickwork are covered in Chapter 13, *Construction Details*. Specific applications include stepped footings, columns, piers, pilasters, solid masonry walls, hollow masonry bonded walls, veneered walls, reinforced masonry walls, retaining walls, wall openings, lintels, sills, arches, window details, door details, steps, paving, fireplaces, chimneys, stone quoins, garden walls, racking and corbels, and caps and copings. Many of these applications involve the integration of multiple materials—concrete block, concrete, tile, stone, and brick.

### Corbelling a 12" wall

*Corbelling* is a method used by masons to widen a wall by projecting out masonry units to form a ledge or shelf. When building a corbel, each brick course extends out further than the one below it. As a general rule, a masonry unit should not extend out more than one-third the width or one-half the height, whichever is less.

In view of the fact that corbels normally support a load, they must be carefully constructed. Headers are generally used to tie the corbel into the base. Building codes generally require the top course to be a full header course. All joints must be completely filled with mortar.

The following procedure is suggested for corbelling a 12" wall. See Figure 10-53.

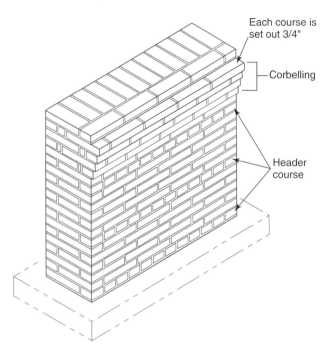

Each course is set out 3/4"

Corbelling

Header course

**Figure 10-53.** Procedure for corbelling a 12" brick wall.

1. Snap a line on the foundation or base where the front edge of the wall is to be constructed.
2. Lay out the first course as a dry course to check spacing. Running bond will be used with a header course every seventh course.
3. Lay the first course with headers on the front wythe and stretchers on the back. Use a full mortar bed. Level, plumb, and square the brick. Check the spacing.
4. Lay the second through sixth courses. Check the height of each course as it is laid and be sure bed joints are uniform and level.
5. Lay the seventh course as a header course with full mortar joints.
6. Continue to lay the wall until you reach the height where the corbel course is to begin. Check your progress often to be sure it is plumb and straight.
7. Begin the corbel course by projecting headers out 3/4" beyond the course below. Fill the extra wide head joint with mortar.
8. Lay the second corbel course using stretchers along the front and headers along the back wythe. Fill in the space between with bats. Be sure the course is level and straight.
9. Lay the third corbel course using stretchers on the front and back wythes and three-quarter headers between the two wythes. Each corbel should project out 3/4" beyond the course below.

10. Lay the next course the same as the first course of the wall—headers on the front wythe and stretchers on the back wythe. Continue the wall to the desired height.
11. Finish all joints when the mortar has set to thumbprint hard.
12. Clean the wall with the trowel and bricklayer's brush.

### Hollow brick pier

Piers are similar to columns except they are shorter and generally do not support a load. They are commonly used as gate posts at corners or openings or ends of a wall. For example, a garden wall with piers is generally referred to as a pier and panel wall because the piers are used to brace the wall panels.

Piers can be constructed of a one-wythe wall 4" thick. The bond pattern is usually staggered so that the wall is tied together from a different side in each course in an interlocking fashion. Piers that are exposed to the weather must be capped to prevent water from entering at the top. Weep holes may be required. Piers are generally built on a square footing.

### Laying a 16" by 20" hollow brick pier

The following procedure shown in Figure 10-54 is suggested for laying a 16" by 20" hollow brick pier:

1. Locate the exact position of the pier and snap a chalk line to preserve the location of the outside faces of the brick.
2. Lay the bed course on a generous mortar bed. Check the brick to be sure that they are level and straight and that the corners are square. You may use either 1/2" or 3/8" joints. Refer to the first course plan in Figure 10-54.
3. Lay the second course as shown in the second course plan. Brick should be positioned so that head joints are offset to provide an interlocking connection. Level, plumb, and square the course.
4. Lay successive courses alternating the patterns used in the first and second courses until the pier has reached the desired height. Twelve courses should be adequate for practice. Avoid dropping any mortar inside the pier.
5. When the mortar is thumbprint hard, tool the joints and clean off any fins with the trowel. Clean the surface with a bricklayer's brush.

## Cleaning New Masonry

The finished appearance of a masonry wall depends not only on the skill used in laying the units but on the cleaning procedure as well. The appearance of a masonry structure can be ruined by improper cleaning. In many instances, the damage caused by faulty cleaning

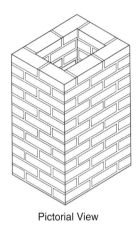

Pictorial View

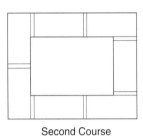

Second Course

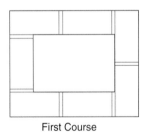

First Course

**Figure 10-54.** Procedure for laying a 16" by 20" hollow brick pier.

techniques or the use of the wrong cleaning agent cannot be repaired. All cleaning should be applied to a sample test area of approximately 20 sq. ft.

Some general precautions that can be taken to promote a cleaner wall during construction are as follows:

1. Protect the base of the wall from rain-splashed mud and mortar splatter.
2. Scaffold boards near the wall should be turned on edge at the end of the day to prevent possible rainfall from splashing mortar and dirt directly on the completed masonry.
3. Cover walls with a waterproof membrane at the end of the workday to prevent mortar joint wash out and entry of water into the completed masonry.
4. Protect site stored brick from mud. Store brick off the ground under protective covering.
5. Careful workmanship should be practiced to prevent excessive mortar droppings. Excess mortar should be cut off with the trowel as the brick are

laid. Joints should be tooled when *thumbprint hard*. After tooling, excess mortar and dust should be brushed from the surface. Avoid any motion that will result in *rubbing* or *pressing* mortar particles into the brick faces. A medium soft bristle brush is preferable.

 **Warning!** When cleaning bricks with chemicals and acids, always take proper safety precautions, wear the proper safety clothing and eye protection.

## Cleaning brick

Cleaning new brick masonry is mainly concerned with removing mortar, mortar stains and any other materials, such as dirt, deposited on the masonry during construction.

Small quantities of various minerals found in some burned clay masonry units will react with some cleaning agents to cause staining. These reactions cannot be predicted in advance and it is therefore recommended that before applying any cleaning agent to a masonry wall, it be applied to a sample section of 10 sq. ft. to 20 sq. ft. to judge its effectiveness. Wait a minimum of one week before inspection.

In the construction of masonry walls, some mortar stains will be present even if the mason is very skilled and careful. Therefore, most specifications require a final washing down of all masonry work.

## Acid solutions

A solution of hydrochloric acid is used extensively as a cleaning agent for new masonry. The following procedures are recommended as good practice when using an acid or base solution.

### Cleaning dark brick

Dark colored bricks include red, red flash, brown, and black. They are most likely to show light gray, brown, or yellow discoloration from failure to rinse off dissolved mortar or dirt. The first procedure requires no acid.

1. When the mortar is thoroughly set and cured, begin the cleaning operation.
2. Remove large particles of mortar with wooden paddles or scrapers before wetting the wall. A chisel or wire brush might be necessary.
3. Saturate the wall with clean water and flush away all loose mortar and dirt.
4. Scrub down the wall with a solution of 1/2 cup of trisodium phosphate and 1/2 cup of household detergent dissolved in one gallon of clean water. Use a stiff fiber brush.
5. Rinse off all cleaning solution and mortar particles using clean water under pressure.

When acid cleaning becomes necessary, the following procedure is recommended:

1. When the mortar is thoroughly set and cured, begin the cleaning operation.
2. Remove large particles of mortar with wooden paddles or scrapers before wetting the wall. A chisel or wire brush might be necessary.
3. Saturate the wall with clean water and flush away all loose mortar and dirt.
4. Use a clean, stain-free commercial grade of hydrochloric (muriatic) acid mixed one part of acid to nine parts water. Mix in a nonmetallic container. Pour the acid into the water, not the water into the acid! Use a long handled fiber brush to scrub the waft. Be careful when using this type of chemical.
5. Keep the area not being cleaned flushed free of acid and dissolved mortar. This scum, if allowed to dry, may be impossible to remove later.
6. Scrub the brick, not the mortar joints. Do not use metal tools. Clean only a small area at a time.
7. Rinse the wall thoroughly with plenty of clean water while it is still wet from scrubbing with the acid.

## Cleaning light colored brick

Light colored bricks include buff, gray, speck, and pink. They are more likely to be burned by acid than darker brick. Therefore, do not use acid except in extreme cases.

The first procedure for light colored brick is exactly the same as the first procedure given for dark brick—no acid!

The second procedure for light colored brick is the same as the second procedure for dark brick except for the following:

1. Use the highest grade acid available. It should be free of any yellow or brown discoloration.
2. Mix 1 part acid with 15 parts water and scrub the wall with a fiber brush. Rinse the wall well with clear water. The acid wash may be neutralized with a solution of potassium or sodium hydroxide, consisting of 1/2 lb. hydroxide to 1 qt. of water (2 lb. per gal.). Allow this to remain on the wall for two or three days before washing again with clear water.

Cleaning failures generally fall into one of three following categories:

- Failure to thoroughly saturate the brick masonry surface with water before and after application of chemical or detergent cleaning solutions. Saturation of the surface prior to cleaning reduces the absorption rate, permitting the cleaning solution to stay on the surface rather than be absorbed.
- Failure to properly use chemical cleaning solutions. Improperly mixed or overly concentrated acid solutions can etch or wash out cementitious materials from the mortar joints.
- Failure to protect windows, doors, and trim. Many cleaning agents, particularly acid solutions, have a corrosive effect on metal. If permitted to come in contact with metal frames, the solutions may cause pitting of the metal or staining of the masonry surface and trim materials, such as limestone and cast stone.

## Sandblasting

Sandblasting is used extensively in some areas to clean new masonry. It costs about the same as acid cleaning. With an experienced sandblasting operator, there is virtually no change in texture of hard brick. However, care must be exercised when sandblasting sand finished brick. This method eliminates the possibility of mortar smear, acid burn, and efflorescence which are present in acid cleaning.

For best results with sandblasting, use a very low pressure (from 60 lb. to 120 lb.) and a 1/4" sandblast nozzle with white urn sand. The secret to successful sandblast cleaning is the distance the cleaner stands from the wall and the manner in which the blast is directed at the brick. Concentrate on hitting the brick rather than the mortar joints.

# REVIEW QUESTIONS CHAPTER 10

Write all answers on a separate sheet of paper. Do not write in this book.

1. What three things does laying brick masonry units involve?
2. Identify five basic operations that must be mastered before a masonry structure can be built.
3. What is the likely outcome of using a trowel that is too large?
4. Identify three standard methods of loading mortar on the trowel.
5. Mortar is usually furrowed with the point of a _____.
6. Mortar on the board should be kept well tempered with water until it is used. True or False?
7. What is the correct way to hold a brick to apply mortar and to lay it on the mortar bed?
8. When should the head joint be applied to the brick?
9. List four ways (tools) to cut brick.
10. Any straight wall longer than 4' is usually laid to a line. True or False?
11. Why is it important to place the same amount of tension on a mason's line each time it is moved?
12. What is the purpose of batter boards?
13. What is the purpose of a story pole?
14. The corner of a masonry wall that is used to establish proper height for each course is called a(n) _____.

15. Height marks are made on a(n) _____.
16. Bricks are laid up tight to the line to ensure accuracy. True or False?
17. A dry course of brick is laid out along the wall line when starting a wall to eliminate as much _____ of units as possible.
18. A mason's _____ (level) is used to check whether the wall is level and plumb.
19. A long jointer is used to _____ or _____ the bed joints of brick masonry.
20. What tool is used to strike the vertical or head joints?
21. Tags of mortar are removed from the brick with a(n) _____.
22. A brick wall is usually laid one course at a time from the leads toward the center. True or False?
23. In an 8" American bond brick wall, a(n) _____ course is laid every five to seven courses.
24. Mortar may be tooled when it is _____ hard.
25. When a brick wall is completed, all joints are struck, and mortar tailings have been removed, the wall may be brushed with a(n) _____.
26. How is a 12" solid wall similar to an 8" solid wall?
27. The _____ bond is composed of alternate courses of stretchers and headers. The headers center on the stretchers and on the joints between the stretchers.
28. What is the objective of a 10" brick masonry cavity wall?
29. Why is the mortar bed beveled along the cavity edge in cavity wall construction?

30. Which of the following is an acceptable method of creating weep holes at the base of the cavity wall?
    A. Eliminating each second or third head joint.
    B. Inserting oiled rods, rope, or pins in the head joint and removing before the mortar has set.
    C. Placing metal or plastic tubing in the head joint.
    D. Placing sash cord or other suitable wicking material in the head joint.
    E. All of the above are acceptable methods.
31. Masonry walls that are exposed to heavy rain during construction can be become so saturated with water that they may require weeks or even months to dry out. True or False?
32. Which way should the drip on metal ties point in cavity wall construction?
33. List four benefits of single wythe brick bearing walls over alternative systems.
34. In a(n) _____ wythe brick bearing wall system, the brick masonry serves as both the structural system and the exterior facing.
35. What is the most widely used residential building code?
36. Air leakage is not a concern in single wythe exterior brick walls. True or False?
37. What is the purpose of corbelling?
38. A solution of _____ acid is used extensively as a cleaning agent for new masonry.
39. When cleaning new masonry by sandblasting, a very low pressure from _____ to 120 psi should be used.

This stone is both beautiful and functional as a structural column.

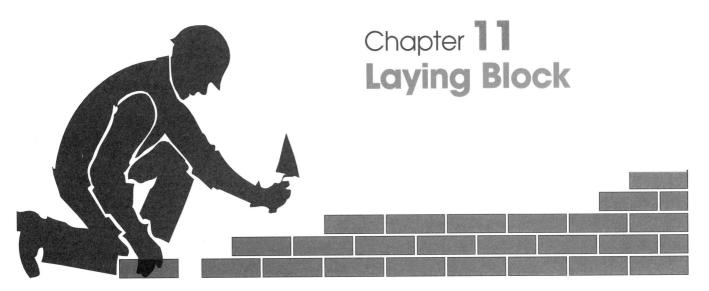

# Chapter 11
# Laying Block

Most concrete blocks are used in the construction of a wall (basement or foundation), exterior above grade wall, interior partition, or retaining wall. The design and construction of the wall will depend on its desired appearance, fire resistance, economy, strength, insulation, acoustics, etc. The layout of the wall and use of specific masonry units will take into by necessity, consideration such things as modular planning, internal arrangement of components, provision for expansion and contraction, and weather resistance. All of these factors will, at one time or another, affect the success of a concrete masonry wall assembly. Figure 11-1 shows a variety of applications and designs of concrete masonry construction.

Laying concrete masonry units involves planning, knowledge of the materials, and ability to use the tools properly. Many of the procedures for laying concrete blocks are identical to those used in laying clay brick, but some differences exist. Procedures that are unique to concrete masonry units will be covered in this chapter, but the preceding chapters should be studied prior to this one.

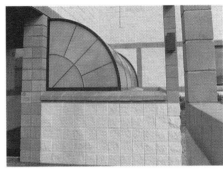

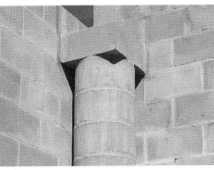

**Figure 11-1.** A variety of uses of concrete block construction. (Besser Company; Trenwyth Industries, Inc.)

# Types of Concrete Masonry Walls

The types of walls possible using concrete masonry units are essentially the same as those using clay masonry units. Concrete masonry walls are classified as solid, hollow, cavity, composite, veneered, reinforced, and grouted (nonreinforced) walls.

## Solid masonry walls

*Solid masonry walls* are walls built with solid masonry units. All joints are completely filled with mortar or grout, at least 75% of the net area of an ASTM C 145 solid unit is solid concrete. Facing units are generally either brick or solid architectural units that are laid with full head and bed joints. See Figure 11-2. Backup units are also solid masonry units laid with full head and bed joints.

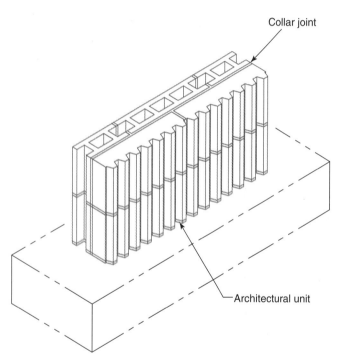

**Figure 11-2.** An 8" composite block wall with 4" × 8" × 16" ribbed face lightweight block with 4" × 8" × 16" lightweight concrete block backup. All joints are filled with mortar.

If flanged end units are used, the end cavity must be filled with grout. The spaces between units are also filled with mortar.

Masonry headers, metal ties, and grout are used between wythes to provide a solid structural bond. Some codes require no less than 4% of the wall area to be headers. Headers should extend at least 3" or 4" over the adjacent wythe and should be between 24" and 36" apart vertically, depending on the code.

**Trade Tip.** The type of wall just described is not popular today. ASTM considers it to be obsolete.

## Hollow masonry walls

*Hollow masonry walls* are built with hollow masonry units, but solid units can be combined with hollow units laid in face-shell mortar bedding. See Figure 11-3.

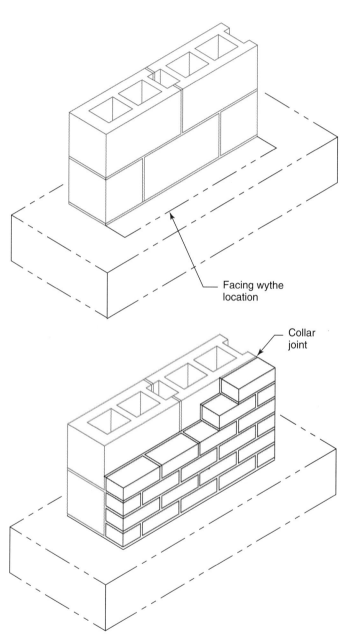

**Figure 11-3.** Top—An 8" × 8" × 16" lightweight concrete block partition wall. Bottom—A 12" composite wall with 4" × 2 2/3" × 8" face brick with 8" × 8" × 16" lightweight concrete block backup.

Hollow masonry walls are constructed in any standard thickness of 4", 6", 8", 10", 12" etc., using a single or multiple wythe where required. Multiple wythes generally consist of a facing wythe and a backup wythe. These walls may be classified as composite walls. Bonding between wythes may be masonry headers, metal ties, or grout. Mortar is used to fill all collar joints.

## Cavity walls

A *cavity wall* is a two-wythe wall that allows each wythe to react independently to stress. It usually consists of two walls separated by a continuous airspace that is 2" or more wide. The 2" cavity is most common. The walls (wythes) are tied together by rigid metal ties embedded in the mortar joints of both walls. See Figure 11-4. The facing wall generally is made from one wythe of solid or hollow masonry units that are 3 1/2" to 4" thick. The backing can be a single or multiple wythe wall of solid or hollow masonry units. The backing thickness may be equal to, or greater than, the facing depending on the design requirements of the wall. Most of the weight is usually carried by the backing while the facing resists wind loads.

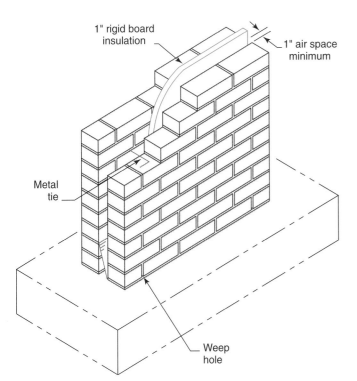

**Figure 11-4.** A 10" concrete block cavity wall under construction. The airspace between the wythes is 2".

Insulation may be placed within the space between the walls. A material that is non-water-absorbent is required such as foamed plastics, foamed glass, or glass fiber. Mats or rigid boards are recommended.

## Composite walls

A *composite wall* is a multi-wythe wall that is designed to act as a single member when subjected to loads. Stress is shared by the wythes through the collar joint and metal ties or headers. See Figure 11-5.

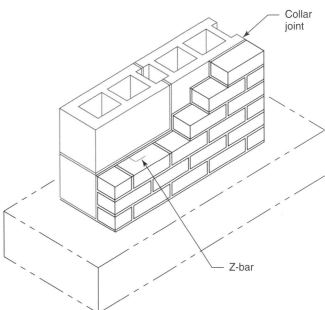

**Figure 11-5.** This is a composite wall using 8" x 8" x 16" concrete blocks with brick facing. Note the Z-bar and mortar collar joint.

When ties are used, at least one tie for every 2 2/3 sq. ft. of wall is required for No. 9 gauge ties. The area expands to 4 1/2 sq. ft. when 3/16" diameter ties are used. Regular ties should be spaced no more than 36" horizontally and 24" vertically. Adjustable ties should not exceed 1.77 sq. ft. of wall area or be spaced more than 16" apart.

Composite walls that use header-bonded wythes of solid masonry should have headers that compose at least 4 % of the wall area. Distance between headers should not exceed 24" in solid masonry and 34" in hollow masonry.

## Veneered walls

Masonry veneer is commonly used as a nonload-bearing facing material in residential and light commercial construction. See Figure 11-6. Veneer carries its own weight only and is anchored, but not bonded, to the backing. In some commercial construction, veneer can be bonded to the concrete masonry backing through the use of architectural units and joint reinforcement. By definition, if the cavity is less than 2", it is a veneered wall. The space is 2" or greater in a cavity wall.

The primary objective of veneer is to provide an attractive, durable exterior surface that will prevent (or reduce) the entrance of moisture into the building. A

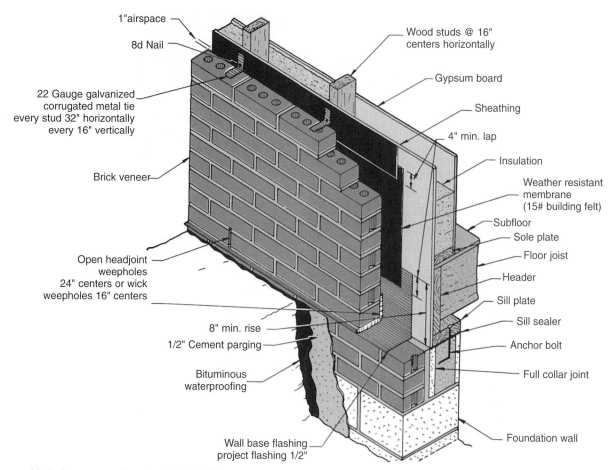

1"airspace

8d Nail

22 Gauge galvanized
corrugated metal tie
every stud 32" horizontally
every 16" vertically

Brick veneer

Open headjoint
weepholes
24" centers or wick
weepholes 16" centers

8" min. rise

1/2" Cement parging

Bituminous
waterproofing

Wall base flashing
project flashing 1/2"

Wood studs @ 16"
centers horizontally

Gypsum board

Sheathing

4" min. lap

Insulation

Weather resistant
membrane
(15# building felt)

Subfloor

Sole plate

Floor joist

Header

Sill plate

Sill sealer

Anchor bolt

Full collar joint

Foundation wall

**Figure 11-6.** A construction detail of a brick veneer wall section.

system of flashing and weep holes provides for the exit of collected moisture.

Metal ties are used in residential veneer construction to anchor the veneer to frame or masonry backing. The ties are generally 22 gauge corrugated, galvanized steel strips 7/8" wide. Spacing varies by code, but is generally about 16" vertically and 32" horizontally. Adjustable ties are preferred for commercial construction.

## Reinforced concrete masonry walls

**Reinforced concrete masonry walls** are walls designed for use in applications that experience high stress, high wind loads, or severe earthquakes. The strength of the wall is greatly increased through the use of steel that is embedded both horizontally and vertically. This type of wall may be either a single or multiple wythe wall. See Figure 11-7.

Hollow masonry units can be used for single wythe walls laid with face-shell mortar bedding. Here the vertical cores are lined up to form continuous vertical spaces that provide space for grouted reinforcement. Two-core blocks are preferred because they provide large spaces to locate the reinforcement.

Two wythes are most common in multi-wythe wall construction. The wythes have a continuous airspace from

1" to 6" between them. The masonry units are hollow, solid, or a combination of unit types. Reinforcement is placed in the space between the wythes and grouted solid. More complicated reinforced masonry walls are common in commercial construction that involve three-wythe walls with No. 9 rectangular ties that are 4" wide and 2" narrower than the wall thickness.

Higher strength concrete masonry units are sometimes available for use in reinforced concrete masonry walls. They have compressive strength ranging from 3000 to 5000 psi (net area).

## Grouted masonry walls

The primary difference between grouted masonry walls and reinforced masonry walls is that grouted masonry walls do not include reinforcement. Grout is sometimes added to loadbearing walls to provide additional strength. It is also used to fill bond beams and the collar joint in two-wythe wall construction.

## Basic Operations

There are a number of basic operations in laying concrete block that must be mastered before quality wall construction can be accomplished. These basic operations

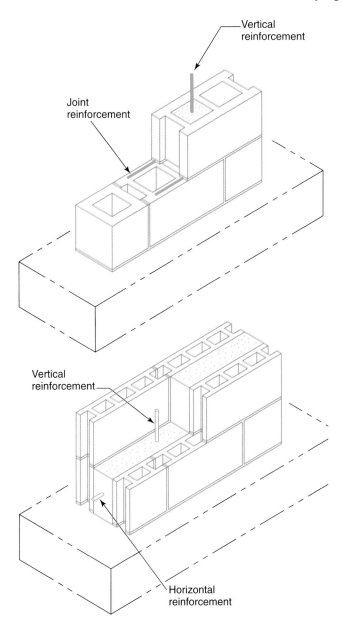

**Figure 11-7.** Top—A 6" reinforced wall using 6" × 4" × 12" concrete masonry face brick with vertical reinforcing #4 rebar at 48" o.c. and horizontal reinforcing at 24" o.c. Bottom—A 10" reinforced cavity wall with 4" × 4" × 12" concrete masonry face brick on the interior and exterior wythes. The 2" grouted cavity has vertical reinforcing #4 rebar at 48" o.c. and horizontal reinforcing at 24" o.c.

include spreading mortar, handling concrete blocks, applying head joints, cutting blocks, and using a mason's line.

## *Spreading mortar*

Spreading mortar is part of all masonry construction. It is a simple process, but requires mastery if the mason is to be an expert. With practice you will be able to determine just how much mortar to use for a given joint. Placing the unit will be easier if the right amount of mortar has been laid down.

Select a trowel that is best for the job being performed and one that "feels good" to you. If it is too small, effort will be wasted in needless movements. If the trowel is too large, you will tire easily. You may want to begin with a standard mason's trowel about 10" long.

Grasp the trowel in the right hand, if you are right-handed. Fingers should be under the handle and the thumb on top of the ferrule as shown in Figure 11-8.

**Figure 11-8.** Hold the trowel with the fingers under the handle and the thumb on top of the ferrule as shown.

## Loading the trowel

Mortar can be loaded on the trowel from the side of the board, from the middle, or from the top of the pile. Figure 11-9 shows the steps used by one mason to load the trowel. Develop the method that works best for you and perfect it.

Spread the mortar with a quick turn of the wrist toward the body and a backward movement of the arm. As the trowel is nearing empty, tip the trowel blade even more to help the remaining mortar slide off. The kind of mortar and its consistency most often determines the speed of the stroke.

When solid masonry units are used or the first course of concrete block is being laid, the mortar is spread and furrowed with the point of the trowel as in Figure 11-10. This helps to form a uniform bed on which to lay the masonry.

Generally, mortar is bedded only on the outside edges of concrete block. Very seldom is a solid bed joint used. ***Face shell bedding*** is bedding on the outside edges of blocks. See Figure 11-11. A swiping motion is used to apply the mortar to the shells. Be careful to fill all joints and holes when the blocks are laid so that all mortar dries at the same time.

## Mortar consistency

Mortar should be the proper consistency. If it is too runny, it will not support the weight of the block. If it is too stiff (dry), it will make setting the block difficult. Blocks should never be moved after they have been laid.

Step 1: Work the mortar into a pile in the center of the mortar board.

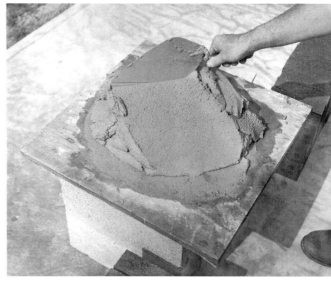

Step 2: Smooth off a place with a backhand stroke.

Step 3: Cut a small amount from the larger pile with a pulling action.

Step 4: Scoop up the small pile with a quick movement of the trowel.

**Figure 11-9.** Steps in loading trowel.

**Figure 11-10.** A full mortar bed is being furrowed with the trowel. It is important that ample mortar be used to ensure a good bond between the footing and the bed course of masonry. (Portland Cement Association)

**Figure 11-11.** Face shell bedding is being used on this concrete block wall. (Portland Cement Association)

Mortar on the board should be kept well tempered (sprinkled lightly) with water until it is used. Avoid constant working. This causes the mortar to dry out and become stiff. Stiff mortar does not spread well. Use mortar before it begins to set up.

## Handling concrete blocks

Concrete blocks are large units that generally require both hands for placement. For example, a typical 8" × 8" × 16" concrete block made with sand and gravel aggregates weighs about 40 lb. A comparable lightweight block weighs between 22 lb. and 28 lb. depending on the specific aggregate used.

Use both hands to lift the block. See Figure 11-12. Grasp the web at each end of the block to lay it on the mortar bed. Even movements are preferred to jerking motions. This will prevent dislodging the mortar from the block. The trowel should remain in your hand while placing the block to save time when only one or two blocks are being set. If

**Figure 11-13.** The proper procedure for placing a concrete block in the wall. (Portland Cement Association)

**Figure 11-12.** Use both hands to lift and place a regular concrete block on the mortar bed. (Portland Cement Association)

several blocks have been prepared, then the trowel can be laid aside while the blocks are placed on the mortar bed.

By tipping the block slightly forward toward you when you place it on the mortar bed and looking down the face of the block, you can position the block in proper position with respect to the top edge of the course below. See Figure 11-13. Then roll the block back slightly to correctly align the top of the block with the line. During this movement, the block should be pressed toward the last block to form a good head joint. Mortar should squeeze out slightly.

Blocks are positioned in a wall with the wide flange on the top. This provides a wider space for the bed joint. As each block is laid, the excess mortar should be cut off with the trowel held at a slight angle to the block.

## Applying the head joint

The objective of making head joints on concrete blocks is to form full head joints on both ears (end edges) of each block to be laid. First, set the block on end so the ears face up that are to receive the mortar. Pick up enough mortar with the trowel to form the head joints. This could be one or more blocks depending on the situation, but the trowel should not be fully loaded as this makes it difficult to apply the right amount of mortar to each ear.

Use a downward swiping action to apply the mortar to the ears. Next, press down the mortar on the inside of each ear of the block to attach it to the block. If the mortar is not attached sufficiently, it will fall off when the block is lifted and turned for placement in the wall. If the mortar falls off during placement, start again with fresh mortar. Enough mortar should be applied so that a full head joint is formed that will be watertight. Too much mortar is wasteful and creates extra work and cost. Just the right amount of mortar will be evident after some practice.

## Cutting block

Even though concrete blocks are available in half-length units as well as full-length units, it is sometimes necessary to cut block to fit. Blocks can be cut with a brick hammer and blocking chisel. See Figure 11-14. Another method is to use a masonry saw. See Figure 11-15. When using the chisel, hold the beveled edge toward you. The piece of block to be cut off should be facing away from you. Score the block on both sides to get a cleaner break.

If a neater, cleaner cut is desired, a masonry saw should be used. It is fast and accurate. The block should be dry when cut with the masonry saw. Be sure to wear safety glasses when cutting blocks.

**Figure 11-14.** A brick hammer and blocking chisel can be used to cut concrete block.
(Portland Cement Association)

**Figure 11-15.** A masonry saw being used to cut a concrete block.

### Using a mason's line

Any straight wall longer than about 4' is usually laid to a line. Strong, thin mason's line is used for this purpose. It is used as a guide that helps the blockmason build a straight, level, and plumb wall. mason's line is a light, strong cord that can be stretched taut with little or no sag. It should be pulled to the same degree of tautness for each course. This is necessary so that all bed joints will be uniform and parallel. See Figure 11-16.

**Figure 11-16.** A mason's line is generally used to ensure a straight and level course of block.
(Portland Cement Association)

Leads are usually built first at each corner to establish the proper height and provide a place to attach the mason's line. The height of each course is determined using a mason's rule or story pole with the height marks on it.

On long walls, it is impossible to eliminate sagging in the line. The proper course height should be checked with the story pole to insure accuracy. A trig block can be set about midpoint to support the line in a very long wall. This block is set with the proper mortar joint thickness and checked with the mason's rule to maintain the proper height. A metal trig is placed on top of the block and held in place with a smaller piece of block or brick. A *metal trig* is a thin piece of metal designed to hold the line between fingers.

Laying block too close to the line (crowding) should be avoided. Stay a line width (1/16") away from the line so that you do not risk moving the line that would destroy its usefulness as a reference guide.

## Laying Concrete Block Walls

A well-planned concrete block structure will involve mainly stretcher and corner blocks. See Figure 11-17.

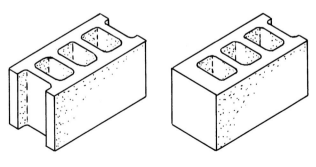

Stretcher                          Corner

**Figure 11-17.** These two concrete blocks are the primary units used to construct most concrete masonry walls. Nominal size is 8" × 8" × 16".

These blocks are nominally 8" × 8" × 16". Actual size is 7 5/8" × 7 5/8" × 15 5/8". This allows for a 3/8" mortar joint that is standard. Blocks of other sizes are sometimes used, but they also use a 3/8" mortar joint.

Concrete blocks must be protected from excess moisture before use. If they are wet when placed, they will shrink when dry and cause cracks. They should be stacked on platforms and covered with plastic or a tarpaulin to protect them from rain.

Mortar for concrete block masonry should be mixed in accordance with specifications in Chapter 8. The consistency of mortar for concrete block masonry is just as critical as it is for brick or stone.

## Laying an 8" running bond concrete block wall

Good construction requires that adequate reinforcement be used or that the joints be staggered. Therefore, advance planning is necessary for a strong wall with a good appearance. The length of the blocks must be considered as they relate to window and door openings in the wall. It is a good idea to check the building dimensions carefully before beginning the work. Problems can be identified before the wall is begun.

The recommended procedure for laying an 8" concrete block wall follows:

1. The outside wall line should be established. If desired, a chalk line is used to provide a straight line for the first course of block. The wall line should be checked for squareness and proper length before proceeding to the next step.
2. String out the blocks for the first course without mortar to check the layout. See Figure 11-18.

**Figure 11-18.** When the corners have been located, the first course of blocks are strung out to check the spacing. Note that the space between the blocks is equal to the desired mortar joint. (Portland Cement Association)

Allow 3/8" for each mortar joint. When you are satisfied with the layout, the blocks are set aside.

**Figure 11-19.** A full mortar bed is being furrowed with the trowel. It is important that ample mortar be used to ensure a good bond between the footing and bed course of masonry. (Portland Cement Association)

3. A full mortar bed is spread and furrowed with the trowel. See Figure 11-19. Provide plenty of mortar on which to set the block.

**Figure 11-20.** Corner blocks are laid carefully and accurately. (Portland Cement Association)

4. Lay the corner block. See Figure 11-20. Position it carefully and accurately. Concrete blocks should be laid with the thicker edge of the face shell up to provide a wider mortar bed.
5. Lay several stretcher blocks along the wall line. Several blocks can be buttered on the end of the face shells if they are stood on end. This speeds the operation. To place them, push them

downward into the mortar bed and sideways against the previously laid block.

6. After three or four blocks have been placed into position, they may be aligned, leveled, and plumbed with the mason's level. See Figures 11-21 and 11-22. Tap on the block rather than on the level.

**Figure 11-21.** The blocks are leveled and settled to the proper elevation. (Portland Cement Association)

**Figure 11-22.** The plumb rule (mason's level) is used to plumb the block. (Portland Cement Association)

7. After the first course has been laid, the corner lead is built up as in Figure 11-23. This corner is very important since the remainder of the wall is dependent upon its accuracy. The lead corner is usually laid up four or five courses high above the center of the wall. Each course is checked to be sure it is aligned level and plumb as it is laid. See Figures 11-24 and 11-25. The faces of the blocks should be in the same plane. Each block is stepped back half a block. The spacing is checked by placing the level diagonally across the corners of the block as shown in Figure 11-26. All corners should be lined up on the edge of the level.

**Figure 11-23.** Alignment is being checked with the level (used as a straightedge) to ensure accuracy before completing the corner. (Portland Cement Association)

**Figure 11-24.** The last block is laid and leveled on the lead. (Portland Cement Association)

**Figure 11-25.** The corner is checked to be sure that it is plumb.

**Figure 11-26.** Spacing is being checked by lining up the corners with a straightedge. (Portland Cement Association)

**Figure 11-28.** All edges of the opening and the four vertical edges of the closer block are buttered with mortar before it is placed. (Portland Cement Association)

8. After the corner leads have been constructed, the blocks are laid between the corners. A mason's line should be stretched from corner to corner at the proper height for each course. See Figure 11-27. Line blocks or line pins may be used to secure the line. Tension should be the same each time to ensure a uniform wall. Work from the corners toward the center of the wall. Do not allow a block to touch the line as this will push the line beyond its normal position causing a curve in the wall. Keep the block a line width away from the line and even with the bottom of the line. If the blocks have been positioned accurately, the closer block will fit properly. See Figure 11-28.

9. Tool the mortar joints. The most effective joint is one that has been compacted or pressed into place. For this reason, the concave or V-joint is best for exterior work. Joints may be tooled when the mortar has become thumbprint hard. The tool should be slightly larger than the width of the joint so that it will make contact along the edges of the blocks. A 5/8" diameter bar is usually used for a 3/8" concave mortar joint. A 1/2" square bar is used for making a 3/8" V-joint. Tools for tooling horizontal joints in concrete block construction should be at least 22" long. See Figure 11-29. After the horizontal joints are shaped the vertical joints can be tooled with an S-shaped jointer. See Figure 11-30. A trowel should be used to remove any mortar burrs left from the tooling. See Figure 11-31.

**Figure 11-27.** The wall is being filled in using a mason's line to ensure proper height and a straight wall. (Portland Cement Association)

**Figure 11-29.** Horizontal joints are tooled before the vertical joints. (Portland Cement Association)

**Figure 11-30.** Vertical joints are tooled with an S-shaped jointer. (Portland Cement Association)

**Figure 11-31.** The mason's trowel is used to remove any mortar burrs after tooling joints.

10. When the mortar is sufficiently dry, the wall can be brushed or rubbed with a stiff fiber brush or a burlap bag to remove dried particles. See Figure 11-32.

**Figure 11-32.** Use a stiff fiber brush to clean the wall of dried particles of mortar when it is sufficiently dry.

## Control joints in a concrete block wall

**Control joints** are continuous vertical joints built into concrete masonry walls. They are placed where forces might cause cracks. Instead, movement of the wall occurs along the control joint and is not very visible. Such joints should be the same width as other joints. They should be sealed with caulking compound after the mortar has been raked out to a depth of 3/4". See Figure 11-33.

**Figure 11-33.** Control joint at end of lintel is sealed with caulking. (Portland Cement Association)

Other methods of constructing control joints in concrete block walls include:

1. Place a Z-bar in the horizontal mortar joint above jamb blocks. See Figure 11-34. This provides lateral (sideways) support for the wall above windows and doors.

**Figure 11-34.** Control joint using jamb block and Z-bar.

2. Use building paper or roofing felt in the end core of the block to break the bond. See Figure 11-35. The felt extends the entire height of the control joint.

**Figure 11-35.** Building paper is used to break the bond. This method is known as the Michigan control joint. (Portland Cement Association)

3. Use special tongue-and-groove block which allows movement. See Figure 11-36.

**Figure 11-36.** These are special tongue-and-groove control joint blocks. They allow up and down movement but stop sideways (lateral) motion.

## Wall intersections

Intersecting walls should not be joined together in a masonry bond as they are at corners. Instead, one wall should end at the face of the other wall with a control joint.

One method that provides for movement and also lateral support is to use a metal tiebar. Figure 11-37 shows such a tie set into a wall. The tiebar is 1/4" thick, 1 1/4" wide, 28" long with 2" right angle bends on each end. These tiebars are usually placed 4' apart vertically. The bent ends are placed in cores filled with mortar or concrete. Pieces of metal lath are placed under the cores, to prevent the mortar from falling through. See Figure 11-38.

**Figure 11-37.** The bent ends of the metal tiebar are embedded in the cores of the block. The cores are filled with mortar or concrete. (Portland Cement Association)

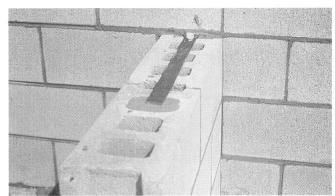

**Figure 11-38.** Cores of the next course will be filled with mortar over this metal tiebar.

Nonloadbearing walls can be tied to other walls using strips of metal lath or hardware cloth. **Nonloadbearing walls** are walls that do not carry a structural load. This material is placed across the joint between the two walls. See Figure 11-39. The metal strips are put in every other course.

## Anchorage to masonry walls

Very often a wood plate must be anchored to the top of a concrete block wall. Anchor bolts are generally used for this purpose. They are usually 1/2" in diameter, 18" long, and are spaced about 4' apart. See Figure 11-40.

The bolts are placed in the cores of the top two courses of block. The cores should have metal lath under them, so that they can be filled with mortar or concrete. The

**Figure 11-39.** Metal lath or 1/4" mesh galvanized hardware cloth can be used to tie nonloadbearing walls together. (Portland Cement Association)

**Figure 11-40.** Anchor bolts placed in the cores of the top two courses of block will be used to attach the plate to the wall.

threaded end of the anchor bolts must extend above the top of the wall so that they pass through the sill and provide enough thread for the washer and nut.

### Laying a 10" concrete block cavity wall

A *cavity wall* is a wall that consists of two walls (wythes) separated by a continuous air space 2" to 4 1/2" wide. The wythes are tied securely together with noncorroding metal ties that are embedded in the mortar joints. See Figure 11-41. Unit ties are generally placed at every other block horizontally and in every other horizontal joint. Continuous metal joint reinforcement serves the same purpose.

**Figure 11-41.** Rectangular anchors or ties are used in cavity wall construction to bond the wythes together. (Portland Cement Association)

Cavity walls are popular because they provide good moisture resistance if they are constructed properly. See Figure 11-42. Three areas need particular attention: weep holes at the bottom of the wall, flashing to direct water to the weep holes, and a clean cavity. Weep holes should be located in the bottom course at about every second or third head joint in the outside wythe. The cavity must be kept clear of mortar droppings that could form a bridge from one wythe to the other. Such a condition would transmit moisture across to the interior.

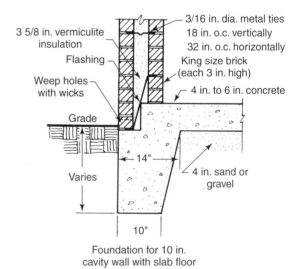

Foundation for 10 in.
cavity wall with slab floor

**Figure 11-42.** A typical cavity wall foundation detail for slab floor construction.

The following procedure is used when building a 10" concrete block cavity wall using two wythes of 4" × 8" × 16" concrete blocks:

1. Lay down a full bed of mortar for the inside wythe. This bed joint must be watertight.
2. Place the blocks for the first course of the inside wythe. Be sure all head joints are solid and watertight.
3. Position the flashing over the top edge of the first inside course and rest it on the foundation (under the outside wythe). Embed the flashing with mortar between the first and second course of the inside course. See Figure 11-43.

**Figure 11-43.** Flashing should be used to prevent moisture from entering the inner wall from the cavity. Note how the flashing is embedded in the mortar joint. (Portland Cement Association)

4. Lay the first course of the outside wythe allowing a 2" continuous airspace between the wythes. The flashing will be beneath this course. Maintain weep holes between every second or third block. Insert wick or other acceptable material. Keep the cavity clean as the work proceeds up to the second course.
5. Lay the second outer course with full bed and head joints. Be sure all joints and holes are filled with mortar. Do not allow any mortar to fall into the cavity. Mortar can be spread about 1/2" back from the edge of the cavity to reduce the chance of falling into the cavity. If rigid insulation is to be used in the wall, now is the time to install it. It should be placed against the inside wythe and should not prevent the escape of moisture from the cavity. A 1" airspace should always be maintained as the minimum.
6. Position unit ties or continuous joint reinforcement over the second course of the two wythes. Embed it within the mortar joint under the third course on the inside wythe.
7. Lay the third outer course with the reinforcement embedded in the mortar joint.

8. Place a board on the ties or reinforcement to catch mortar droppings as the next course is laid. It can be removed when the next reinforcement is placed. See Figure 11-44.

**Figure 11-44.** A recommended method of keeping the cavity free of mortar droppings is to use a board resting on the joint reinforcement. The board is moved up when the next reinforcement level is reached. (Portland Cement Association)

9. Repeat the procedure until the wall has reached the proper height.

## Laying an 8" composite wall with concrete block backup

A **composite wall** is two wythes bonded together with masonry, metal ties, or joint reinforcement. See Figure 11-45. The two wythes are joined together in a continuous mass using a vertical collar joint that prevents the passage of water through the wall. See Figure 11-46. A **collar joint** is the narrow space between the facing units and the backup units in a wall.

Concrete blocks are often used as backup for bricks to make a composite wall. The basic construction of a composite wall can be similar to a cavity wall, except the cavity is reduced to the width of the collar joint and filled with mortar. However, the composite wall can be quite different when masonry headers are used to bond the two wythes together, as in Figure 11-47.

Modern Masonry

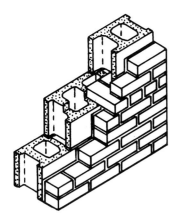

Masonry bond used to
tie wythes together

**Figure 11-45.** Composite walls can be bonded together using a masonry bond to tie the wythes together.

**Figure 11-46.** Bonding a wall across a collar joint by parging the backside of the facing. (Portland Cement Association)

**Figure 11-47.** Header course being laid in composite wall.

Use the following procedure when building an 8" composite wall with brick facing and concrete block backup:

1. Lay up the inner wythe of 4" concrete block two courses high. See Figure 11-48. Parge the side adjacent to the outer wythe being careful not to upset the bond. **Parging** is the application of a coating of mortar. The use of a plasterer's trowel is more efficient than a brick trowel for this operation. Any mortar that has hardened should be cut off with the trowel before parging.

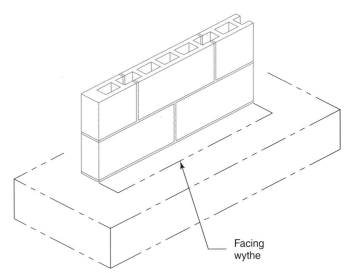

Facing wythe

**Figure 11-48.** Step 1 in laying up an 8" composite wall with brick facing and concrete block backup.

2. Next, lay six courses of bricks in running bond to bring the outer wythe to 16" high. See Figure 11-49. Be sure the vertical collar joint is filled with mortar and the brick are laid with full bed joints.

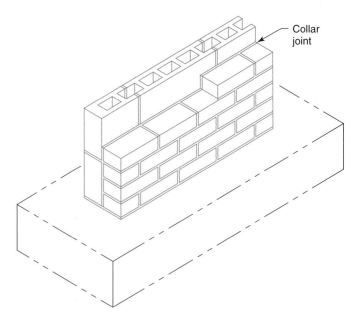

Collar joint

**Figure 11-49.** Step 2 in laying up an 8" composite wall with brick facing and concrete block backup.

3. Lay a header course of brick across the two wythes to tie the courses together. See Figure 11-50. In header courses, the cross joints should be completely filled with mortar spread over the entire side of each brick before it is placed in the wall. Composite walls with header courses are susceptible to water leakage if all joints are not solid and all holes filled with mortar.

**Figure 11-50.** Laying a header course to tie the two wythes together.

4. Repeat the previous steps until the wall has reached the desired height.

Composite walls greater than 10" in thickness generally use a specially shaped header block to bond the facing headers and backup units with 6th-course bonding. The special header block can be laid with the recessed notch up or down and with block backup laid in face-shell mortar bedding.

Concrete block is often used as a backup for a stone facing. See Figure 11-51. Since the mortar joints in the block and stone do not line up, corrugated metal ties can be used to tie the wythes together.

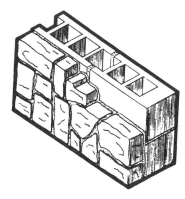

**Figure 11-51.** A 12" composite stone wall facing on 8" × 8" × 16" lightweight concrete block backup. (Colorado Masonry Institute)

# Cleaning Concrete Block Masonry

Concrete block walls are not cleaned with acid to remove mortar smears or droppings. Therefore, care must be taken to keep the wall surface clean during the construction process.

 **Trade Tip.** Any mortar droppings that stick to the wall should be allowed to dry and harden. Large particles of mortar can be removed with a trowel, chisel, or putty knife. If the mortar is removed while it is wet, it will most likely smear into the surface of the block.

Rubbing the wall with a small piece of block will remove practically all of the mortar. In some instances, a commercial cleaning agent such as a detergent can be used. Be sure to follow the manufacturer's directions and try out the product on a small section of the wall to check the results.

# REVIEW QUESTIONS CHAPTER 11

Write all answers on a separate sheet of paper. Do not write in this book.
1. What four building elements are concrete blocks usually used for?
2. A(n) _____ wall is a two-wythe wall that allows each wythe to react independently to stress.
3. Which of the following materials is generally *not* used as insulation in a cavity wall?
   A. Foamed plastics
   B. Foamed glass
   C. Glass fiber
   D. Fiberglass batts
4. Masonry _____ is commonly used as a non-loadbearing facing material in residential and light commercial construction.
5. A cavity wall may have a cavity less than 2" in width. True or False?
6. Reinforced concrete masonry walls are frequently used in conditions that experience high stress. Name two other applications that might require reinforced concrete masonry walls.
7. What size trowel is generally used for common masonry work?
8. Describe face shell bedding.
9. What problem is caused by mortar that is too runny?
10. A typical 8" × 8" × 12  16" concrete block made with sand and gravel aggregates weighs about _____ lb.

11. What is the proper procedure in lifting a typical concrete block to lay it on the mortar bed?
12. The objective of making head joints on concrete blocks is to form full head joints on both _____ of the block.
13. Name two methods (tools) for cutting concrete block.
14. What two devices are used to determine the height of each course of masonry?
15. The actual size of an 8" × 8" × 16" concrete block is _____.
16. The proper mortar joint thickness for concrete block masonry is _____.
17. To avoid laying blocks too close to the line, stay _____" away from the line.
18. How is a corner concrete block different from a stringer?
19. Concrete blocks should be laid with the thicker edge of the face shell _____ to provide a wider mortar bed. Up or Down?
20. Describe the motion required to place a concrete block in the wall.
21. Every block should be aligned, leveled, and plumbed with a mason's level before laying the next block. True or False?

22. The _____ is usually laid up four or five courses high above the center of the wall.
23. How is the mason's line generally attached to the lead corner?
24. What diameter bar is generally used to form a 3/8" concave mortar joint?
25. Vertical joints which are placed in a masonry wall to prevent cracking due to movement are called _____ joints.
26. Intersecting walls should not be joined together in a masonry bond as they are at corners. True or False?
27. _____ bolts are used to attach sills to a masonry wall.
28. When building a 10" concrete block cavity wall, which wythe is generally laid first?
29. A(n) _____ wall is two wythes bonded together with masonry, metal ties, or joint reinforcement to form a continuous mass.
30. Concrete block walls are not cleaned with acid to remove mortar smears or droppings. True or False?
31. Any mortar droppings that stick to a concrete block wall should be allowed to dry and harden. True or False?

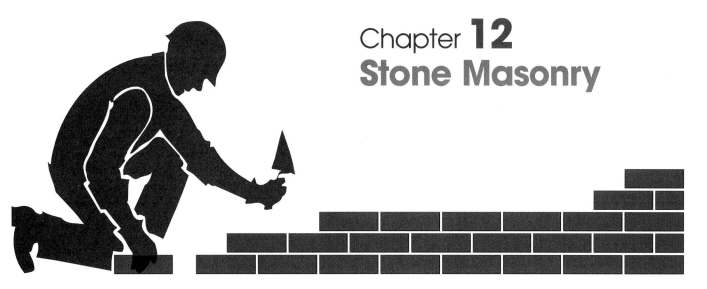

# Chapter 12
# Stone Masonry

Stone masonry is the oldest, most artistic, and most expensive of all types of masonry construction. See Figure 12-1. Over the years it has declined in importance as a solid wall building material. However, it is still being used extensively as veneer over other masonry or frame construction. Stone is also used a great deal as trim on buildings and paving for foyers, patios, drives, and walks.

Setting stone rubble, cut stone, or panels involves planning, knowledge of the materials, and ability to use the tools properly. Stone masonry is truly an art and requires a great deal of practice and creativity. However, the sense of satisfaction gained from doing a job well is worth the effort to learn the proper skills. Many of the procedures for laying brick and concrete masonry units are applicable to setting stone, but some differences exist. Procedures that are unique to setting stone and placing panels will be covered in this chapter, but the preceding chapters should be studied prior to this one.

**Figure 12-1.** A—National Home Builder Association's New American Home uses stone masonry to draw attention to one section of this striking home. B—This magnificent 1800s place of business uses stone to accent the design in a very effective manner. C—Close-up detail of precise stone carving. D—This series of carved sandstone panels produces an interesting pattern in the wall. E—Stone is the primary exterior material of this period home.
F —This Cotswold cottage is an excellent example of stone used as the only building material for the structure.
(John R. Bare & Associates; Stone Products Association)

# Basic Operations

There is a set of basic operations that is utilized when setting stone. These must be mastered before quality stone work is possible. The basic operations include spreading mortar, handling stone, forming joints, and cutting stone.

## *Spreading mortar*

Spreading mortar is part of all masonry construction. It is a simple process, but requires mastery if the mason is to be an expert. With practice you will be able to determine just how much mortar to use for a given joint. Placing the unit will be easier if the right amount of mortar has been laid down or applied to the masonry unit.

Select a trowel that is best for the job being performed and one that "feels good" to you. If it is too small, effort will be wasted in needless movements. If the trowel is too large, you will tire easily and not be able to put the mortar where you want it. You may want to select several trowels for the various tasks in setting stone. See Figure 12-2. For example, a regular mason's 10" trowel, 7" buttering trowel, 5" margin trowel, and narrow caulking trowel would be valuable tools in setting a rubble stone wall.

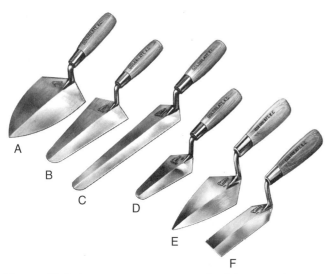

**Figure 12-2.** Stone masons use each of the trowels shown above. A—Buttering B—Gauging C —Duck bill D—Cross joint E—Margin F—Pointing.

Grasp the trowel in the right hand, if you are right handed. Fingers should be under the handle and the thumb on top of the ferrule as shown in Figure 12-3.

## Loading the trowel

Mortar can be loaded on the trowel from the side of the board, the middle, or the top of the pile. Figure 12-4 shows the steps used by one mason to load the trowel. Develop the method that works best for you and perfect it.

Spread the mortar with a quick turn of the wrist toward the body and a backward movement of the arm. As the

**Figure 12-3.** The mason is holding the trowel in the proper manner with the thumb on the top of the ferrule and the fingers under the handle.

trowel is nearly empty, tip the trowel blade even more to help the remaining mortar slide off. The kind of mortar and its consistency most often determines the speed of the stroke. Mortar used for setting stone may be slightly stiffer (dryer) than for typical masonry units due to the weight of the pieces. Stiffer mortar will be a little harder to spread than thinner mortar.

## Mortar consistency

Mortar on the board should be kept well tempered (sprinkled lightly) with water until it is used. Avoid constant working. This causes the mortar to dry out and become too stiff. Stiff mortar does not spread well. Use mortar before it begins to set up.

---

**Warning!** Contact with wet (plastic) concrete, cement, mortar, grout, or cement mixtures can cause skin irritation, severe chemical burns, or serious eye damage. Wear waterproof gloves, a long-sleeved shirt, full-length trousers, and proper eye protection when working with these materials. If you must stand in wet concrete, wear high-top waterproof boots. Wash wet concrete, mortar, grout, cement, or cement mixtures from your skin immediately. Flush eyes with clear water immediately upon contact. Seek medical attention if you experience a reaction to contact with these materials.

---

## *Handling stone*

Stone used for rubble or roughly squared stonework will most likely be fieldstone, in the sizes and shapes as they are found in fields and streams. The stone will be varied also, as to type of stone in the load. A skilled stonemason will know which ones can be split easily and which shapes will look best in various locations in the wall. But, a novice will not have that experience and will make many

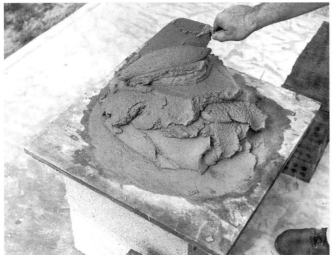

Step 1: Work the mortar into a pile in the center of the mortar board.

Step 2: Smooth off a place with a backhand stroke.

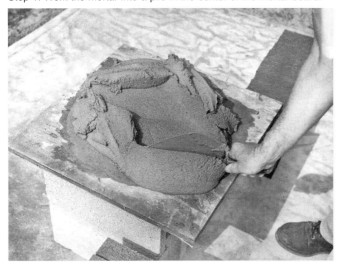

Step 3: Cut a small amount from the larger pile with a pulling action.

Step 4: Scoop up the small pile with a quick movement of the trowel.

**Figure 12-4.** Steps in loading the trowel.

mistakes. Try to learn from your experience and work with an experienced person to discover the "tricks of the trade". For example, bedding is visible in most sedimentary stones and an experienced mason will know just where to strike the stone to split it.

**Trade Tip.** Keep the pile of stone close to where you are working, but clear of the work area. If the source of stone is too far away, you will spend most of your time running back and forth. If it is too close, you won't have any room to work and may fall over the stones.

Form a particular pattern in your mind before you begin setting stone. The pattern should take into consideration the size, shape, and type of stone you have to work with. See Figure 12-5. If all of the stones are round and smooth, don't plan a pattern that requires cutting and trimming every stone. If you plan to build a polygonal stone pat-

tern, then order broken stone that already has the basic shapes needed.

Use larger stones near the bottom of a wall if some are much larger than average. A very large stone will look out of place near the top. You may decide to break larger stones into smaller pieces to produce a more uniform appearance if not enough large stones are available to develop a theme.

Do not try to lift large stones by yourself. Remember that a cubic foot of granite weighs about 170 lb. Use pry bars to move large stones. Get help to lift large stones into place. Use the proper tool for each task. For example, use a chisel to chip off an unwanted protuberance on a flat side of a stone. A mason's hammer or brick hammer can be used to powder a point on a stone. Use the trowel to spread mortar, not to cut stone.

**Figure 12-5.** The mix of large and small stones is pleasing to the eye and in balance.

 **Warning!** Wear protective clothing. Heavy-duty steel-toed work shoes or boots are required when working with stone. A pair of sturdy leather work gloves will prevent cuts on your hands. Always wear safety glasses or goggles with plastic lenses when splitting or chipping stone. Follow safe work practices. Don't take a chance with a shortcut!

Cut stone in the form of trimmings, ashlar, or panels must be handled carefully to prevent breakage, staining, or chipping. This stone is delivered to the job site already cut, dressed, and finished to precise specifications for the particular job. The following suggestions should be helpful in handling this type of stone:

1. Upon delivery of the stone, check for damage, verify that it is the proper stone (color, finish, and grade), and verify dimensions or other particular requirements.
2. As a general rule, handle stone with the same care you would use to handle any material you don't want to chip or scratch. Store it on sturdy skids or timbers in a well-drained place, protected against water and mud splatters. Plan storage so the stones will be easily available when they are needed.
3. Lean smooth-finish stones face-to-face and back-to-back. See Figure 12-6. Textured finishes should be separated with spacers. When lifting, avoid sliding one face surface over another. Make sure lifting holes are easily reached from the stacked position.
4. Be careful not to bump stones. Position stones so as to avoid damage from another stone or equipment.
5. When lifting stones, make certain that all personnel are clear. Never hoist or suspend stones over workmen or bystanders. If stones must be hoisted over scaffolds or stages, workers must

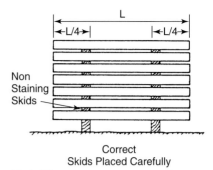

Correct
Skids Placed Carefully

**Figure 12-6.** The proper way to stack stone panels is illustrated in this drawing. (Indiana Limestone Institute)

be clear. Workers on scaffolds must wear safety lines.
6. Make certain that all lifting devices, and all parts involved with hoisting machines, are of sufficient capacity for the heaviest stone to be lifted.
7. Observe special caution in rigging, handling, and lifting large panels. Windy conditions require precautions as well. Use tag lines, and make certain that operators are instructed in their proper use.
8. Always use safety slings under stones as you lift them. Keep safety slings in place until the stones are within a few inches of their final location.
9. Protect finished work against construction traffic. Drape with tarp or pile sand on projecting courses to protect them against mortar droppings.

## Forming joints

Joints in stone masonry are generally set with mortar. Joints in stone panel construction may be mortar or other materials. Whenever mortar is used, it should be a non-staining type designed for the specific application. A mix of 1 part nonstaining cement, 1 part hydrated lime, and 6 parts clean, sharp, washed sand is recommended. This 1/1/6 mixture will provide sufficient compressive strength, good bond strength, and good weather resistance. In any case, the mortar should be appropriate for the job.

The width of the mortar joints will have a profound effect on the finished appearance of the stone construction. The most frequent error is to allow the mortar joints to become too wide, especially with the use of rubble or polygonal stones. See Figure 12-7. Stones should be cut to fit with mortar joints not exceeding 1" for rough work and 3/4" for ashlar. Joints that are 1/2" for rough work and 3/8" for ashlar are preferred. A narrow caulking trowel works well for filling narrow spaces and working the mortar into crevices.

Bed joints should be generous. In rubble, polygonal, and roughly squared stonework, the stone should be set with wedges placed beneath the stone to maintain the proper mortar joint. Heavy stones will squeeze out the mortar if not supported in some manner. Excess mortar should

**Figure 12-7.** Notice how some mortar joints in this polygonal stone wall are too wide.

be removed from the stone with the trowel. The joint can be struck once the initial set of the mortar has occurred. Joints can be raked out to about 1" deep and pointed later with mortar if preferred.

Portland cement used in preparing cement/lime mortar for Indiana limestone should conform to the requirements of ASTM C 150. If masonry cement is used, it should conform to the requirements of ASTM C 91. Either material will produce a suitable mortar for stone panel construction. Cements with low-alkali content are called **nonstaining cement**. They are best to use when setting limestone. White cements are not necessarily either low-alkali or nonstaining. To avoid staining due to alkali from mortar, wet the joint surfaces of the stone before applying mortar, and do not use too much water when mixing the mortar. To avoid major staining, keep moisture out of the wall.

## Splitting, shaping, and cutting stone

Common fieldstone and river rock can be used in their natural state, split, or shaped to fit a particular space in a stone wall, chimney, pavement, or other application. See Figure 12-8. Splitting stones takes some practice and sharp observation to achieve the desired results in a con-

**Figure 12-8.** These common fieldstones have been split and shaped to fit the contour of surrounding stones.

sharp observation to achieve the desired results in a consistent manner.

To split a stone that has a stratified (layered) structure, mark a line along the grain, then chip on the line with the chisel end of your mason's hammer until a crack begins to develop. Widen the crack slowly by driving wedges into the crack at several points. If it is a very large stone, a pry bar may be needed to finish prying it apart. Stones such as granite that are not stratified are difficult to split, but not impossible. Begin by drilling holes with the narrow bladed chisel or power drill about 6" apart along the line where you wish the stone to separate. Then drive thin wedges into the holes. Keep up the process until the stone splits.

Split stones can be shaped using the mason's hammer or stone mason's chisel. The trowel should not be used to cut natural stone. Manufactured stone that is a Portland cement product can be cut with the mason's trowel. When a large stone is to be cut into two pieces, a line should be marked on the face of the stone where the cut is to be made. Use the chisel or mason's hammer to chip along the line to form the break line, much the same as cutting brick. When a cut has been made the length of the line, strike the stone with a sharp blow away from the cut. The stone should break along the line. If the break is rough or not exactly where it should have been, chip away the stone with the chisel or hammer until the stone is the proper shape. Be patient, and do not try to remove too much material at a time.

Cutting ashlar veneer, trimmings, and panel stone is accomplished using power tools. Smaller pieces can be cut on an ordinary masonry saw. Larger pieces such as panels should be cut to size by the supplier. They have the proper equipment to do this kind of work.

# Types of Stone Masonry Construction

Natural stone is available in a wide variety of sizes, shapes, colors, and textures. Refer to Chapter 7. These characteristics make possible many wall patterns. See Figure 12-9. Stone masonry is more time-consuming, because the pieces frequently have to be selected individually and trimming is required when building some patterns. Due to the different sizes and uneven surfaces of stone units, they are more difficult to lay and can require several trial placements before the fit is perfect.

The three main classes of stone masonry construction are rubble, ashlar, and trimmings. Panel construction is not included in this classification.

## Rubble masonry

**Rubble** is uncut stone or stone that has not been cut to a rectangular shape. It is generally used in walls where a rustic appearance is desired and precise lines are not so important. See Figure 12-10. Many sizes and shapes of

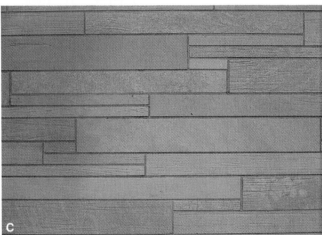

**Figure 12-9.** A—Coursed ashlar B—Cobbed or polygonal stone C—Ashlar D—Rubble. (Eldorado Stone Corporation)

stone have been used to create the pattern. A liberal amount of mortar must be used because of irregularities in the stone. Rubble stonework is not as strong as other bonds.

**Figure 12-10.** Random rubble stonework with uniform mortar joints.

### Ashlar veneer

When stone is precut with enough uniformity to allow some regularity in assembly, the wall is generally called **ashlar**. See Figure 12-11. Ashlar stone is easier to lay because the individual units are precut to fit. In this respect, it is much like brick or block.

**Figure 12-11.** Ashlar stonework.

Produced as a standard product, Indiana limestone ashlar veneer is broadly available from a network of dealers throughout the United States. Most of these dealers also stock standard trim items that can be cut to length with ordinary job site masonry saws or hand tools. Window and door sills, coping, and trim courses are produced in standard heights and thicknesses. They require no additional work other than cutting to proper length on the job. Ashlar veneer and its matching trim are set in mortar like any unit masonry product, and may be applied to any type of back-up.

When the stone is laid in regular courses, it is referred to as coursed ashlar. When it is laid in broken courses without regard to the arrangement of the joints, it is referred to as broken, random, or ranged ashlar. See Figure 12-12. Stones for ashlar work are generally produced with thicknesses of 2" to 8", heights of 1" to 48", and lengths of 1' to 8'.

Figure 12-13 shows the typical method of anchoring ashlar veneer to concrete block backing and wood frame backing.

## Trimmings

**Trimmings** are stones that are cut on all sides to specific dimensions. They are used for moldings, sills, lintels, and ornamental purposes. See Figure 12-14. They are either set in mortar or anchored using a variety of mechanical fasteners.

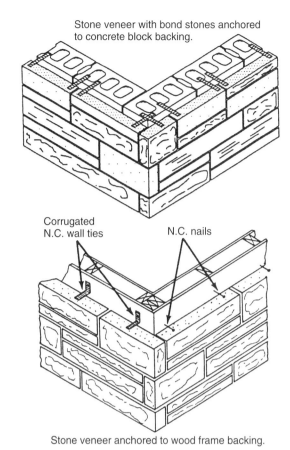

Stone veneer with bond stones anchored to concrete block backing.

Corrugated N.C. wall ties — N.C. nails

Stone veneer anchored to wood frame backing.

**Figure 12-13.** Methods of anchoring random ashlar stone veneer. (Indiana Limestone Institute)

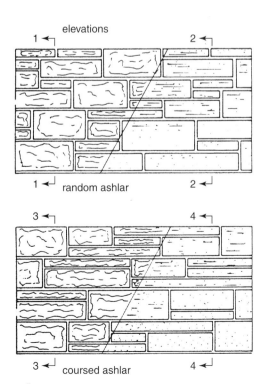

elevations

1 — random ashlar — 2

3 — coursed ashlar — 4

Random ashlar and coursed ashlar are both made from strips of limestone with lengths cut as desired at the job site. The only real difference between them is that the mason lays to a line with one (coursed) and does not with the other (random).

Standard course heights are 2 1/4", 5", 7 3/4" (and sometimes 10 1/2" & 13 1/4") based upon 1/2" beds and joints.

The most common and economical finish for this stone is split face. However, other standard industry finishes (chat or sand sawn, and shot sawn) are also available.

Either pattern may be varied further by the use of more than one thickness as shown in Section at right, thereby creating a three dimensional effect (staccato pattern).

Splitface strip ashlar may have either sawed or split backs. Sawed backs allow a close tolerance in setting. Split backs allow the mason to select a face for greater boldness.

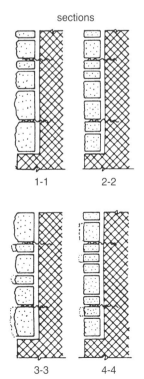

sections

1-1      2-2

3-3      4-4

**Figure 12-12.** Random ashlar and coursed ashlar in elevation and section. (Indiana Limestone Institute)

**Figure 12-14.** Stone trimmings to be used as sills, lintels, and for other purposes. (Georgia Marble Company)

## Setting a rubble stone veneer wall

To lay a rough stone masonry veneer wall using rubble use the following procedure:

1. Spread out the stone so that each one can be examined for shape, texture, or other characteristics. See Figure 12-15. This stone will most likely vary in size from 6" to 18".

**Figure 12-15.** Rubble stone to be used for a stone veneer wall.

2. Mix enough mortar to last for 1 hour of work. See Figure 12-16. Use it up before it begins to set. Be sure to use a nonstaining mortar. Mix 1 part nonstaining cement, 1 part hydrated lime, and 6 parts clean, sharp, washed sand. A comparable manufactured mortar cement can also be used.

3. Select several larger stones for the bed course and place them into position dry. With a piece of chalk, mark them for trimming.

Stones should be placed in as natural a position as possible. Have a pattern or special effect in

**Figure 12-16.** Mortar should be mixed in small amounts so that it will not begin to set before being used.

mind as you select and place stones together. Do not use too many varieties and keep textures relatively uniform. Stone veneer is generally 4" to 8" thick depending on the method of bonding and construction specifications.

Large stones must be split so that they will meet these requirements. A sledge hammer can be used for splitting large stones.

4. Trim each stone as you are ready to place it in the wall. See Figure 12-17. Each stone should be thoroughly cleaned on all exposed surfaces by washing with a brush and soap powder, followed by a thorough drenching with clear water.

⚠ **Warning!** Be careful of flying chips while trimming stone. Wear protective eye wear when trimming stone.

**Figure 12-17.** A stone mason's hammer and chisel are used to trim stone to desired shape.

5. Place each trimmed stone in its proper location in the wall and check for proper fit. When you are satisfied with the result, proceed to the next step.
6. Using the trowel, lay a full bed of mortar for the trimmed stones and place them into position. See Figure 12-18.

**Figure 12-18.** Trimmed stones are placed in a full bed of mortar.

7. Fill in the spaces between the stones with mortar. See Figure 12-19. A narrow caulking trowel works well for filling narrow spaces and working the mortar into crevices. See Figure 12-20. Joints should be 1/2" to 1" wide for rough work and 3/8" to 3/4" for ashlar. Sponge the stone free of mortar along the joints as the work progresses.

**Figure 12-19.** Spaces between stones are filled with mortar.

**Figure 12-20.** Narrow caulking trowel works well for filling spaces between stones with mortar.

8. Remove excess mortar from the stone with the trowel and strike the joint. See Figure 12-21. Joints can be tooled when the initial set has occurred. If desired they can be raked out 1" deeper and pointed later with mortar.

**Figure 12-21.** Excess mortar may be removed with the trowel.

9. Trim other stones for the deeper bed course and fit them into place using the same procedure.
10. Bend corrugated metal ties into place for successive courses of stone. See Figure 12-22. All ties must be noncorrosive. Use extra ties at all corners and large stones when possible.

Lead, plastic, or wood pads the thickness of the mortar joint should be placed under heavy stones to avoid squeezing mortar out. See Figure 12-23. Remove the pads after the mortar has been set and fill the holes. Heavy stones or projecting courses should not be set until mortar in the courses below has hardened sufficiently to avoid squeezing.

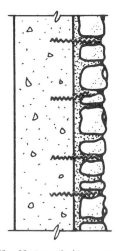
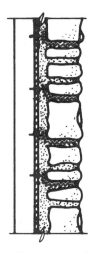
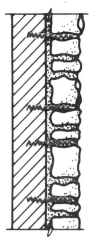

4" – 8" stone tied to concrete backing using wall ties.

4" – 8" stone tied to frame construction. Use wood sheathing and W.P. felt or W.P. sheathing board.

4" – 8" stone tied to brick or block masonry. Provide 1" – 2" air space or slush fill voids.

**Figure 12-22.** Stone veneer sections showing noncorrosive metal ties.

11. Using the above procedure, lay the remaining stones until the wall is the desired height. See Figure 12-24. Work from the corners toward the middle as in laying any masonry wall.

**Figure 12-23.** Wooden pads support heavy stones to avoid squeezing mortar from the joints.

A mason's line and level can be used to keep the wall straight and plumb. If the structure is to be inclined or tapered as the chimney in Figure 12-25, then a mason's line should be used to keep the edge straight and moving in the proper direction.

 **Warning!** The masonry should be protected at all times from rain and masonry droppings. Adequate protection must be provided during cold weather construction. See Chapter 8.

**Figure 12-24.** Stone wall being finished off at the proper height.

12. When the wall is completed, it can be scrubbed with a fiber brush and clear water. The stone should be clean and free of mortar.

Strong acid compounds generally should not be used, because they will burn and discolor certain types of stone. Limestone is one of these.

Waterproofing can be used. Use a nonstaining asphalt emulsion, vinyl lacquer, cement base masonry waterproofing, stearate, or other approved material.

Figure 12-26 shows a close-up of a rubble stone wall with thin mortar joints of uniform width. Here the stones have been trimmed with expert accuracy to fit into a specific place in the wall. Practice and patience is required to develop this talent.

## Building a solid stone wall

The procedure for building a solid stone wall using rubble or ashlar is similar to laying a stone veneer wall with a few exceptions:

2. A solid stone wall is usually thicker than a veneer wall. The backing in a veneer wall provides the structural support and the stone is a protective

**Figure 12-25.** A beautiful example of stonework in a residential structure.

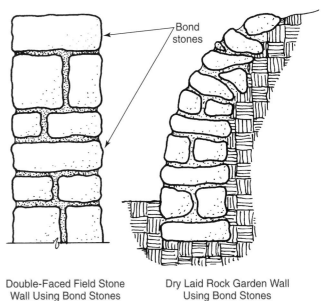

Double-Faced Field Stone Wall Using Bond Stones

Dry Laid Rock Garden Wall Using Bond Stones

**Figure 12-27.** Sections of freestanding and retaining stone walls. Note how the bond stones tie the wall together and increase stability.

covering. In a solid stone wall, the stone provides structural support as well as covering. The usual thickness is a minimum of 18". Larger stones may be used and they should be placed in their natural position for best appearance.

Reinforcing may be used in a solid stone wall to give added strength and support. The material used should be noncorrosive.

**Figure 12-26.** Uniform mortar joints are a mark of good craftsmanship.

1. The wall will depend on bond stones for strength and stability and tying the wall together. A bond stone is one that passes through the wall from side to side. See Figure 12-27. Bond stones should be placed as frequently as possible. Every 5 sq. ft. to 8 sq. ft. of wall should have one.

# Limestone Panels

Indiana limestone has been the material of choice among architects, builders, and clients in America since well before 1900. The first organized quarrying effort of record was established in 1827 in southern Indiana near Stinesville. Since that early beginning, Indiana limestone has remained popular due to its qualities of durability, beauty, designability, and good thermal performance.

## Design and construction criteria

Designing and building with Indiana limestone panels requires attention to numerous details to ensure quality appearance and performance. Following are some of the criteria that should be considered.

**Trade Tip.** Check with the Indiana Limestone Institute, Stone City Bank Building, Bedford, Indiana 47421 for more detailed information.

1. Pitch coping stones toward the roof to avoid discoloration on the exposed face and walls. See Figure 12-28 for joint treatment in copings.

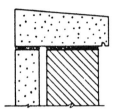 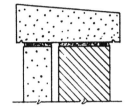

**Figure 12-28.** Stone coping details. Stones are pitched toward the roof to protect the face of the wall. (Indiana Limestone Institute)

2. It is not necessary for backs of stone above a roof line to be finished.
3. Vent attics over porticos to prevent condensation entering the wall and filtering through the joints to the face of the stone.
4. Provide washes and drips on projecting stones.
5. Butt joints are usually less expensive on smooth finishes; quirk joints on textured finishes. See Figure 12-29.

 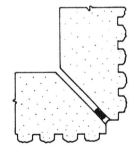

**Figure 12-29.** Stone panel joints recommended for smooth and textured finishes. (Indiana Limestone Institute)

6. Prevent physical contact between the back face of stone walls and columns, floor slabs, and the end beams; expansion can push the stone out.
7. If exposed stone patios occur over a heated room, insulate the ceiling of the room. Warm stones absorb moisture, and a sudden drop in outside air temperature may freeze and crack the stone.
8. Where design allows, step the foundation in sloping terrain to avoid placing limestone below grade. If such installation is unavoidable, utilize either the trench or the coating method shown in Figure 12-30.

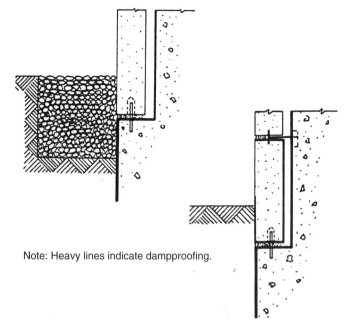

Note: Heavy lines indicate dampproofing.

**Figure 12-30.** Protect limestone panels from contact with soil. (Indiana Limestone Institute)

9. Specify mortar joints to be tooled for proper moisture resistance. Where sealants are used, follow manufacturer's specifications to obtain the desired joint profile and adherence to inner joint surfaces.
10. If false joints are used, they may be pointed to match other joints.
11. Where masonry butts against a stone pilaster, or at building corners, specify a sealant joint. Long vertical mortar joints are difficult to compress, and may tend to fail.
12. Steel lintels carrying stonework should have ample stiffness to carry the superimposed load with a deflection of less than 1/360 of the clear span.
13. Incorporate a 1" minimum clearance between stone and all structural members.
14. While jointing is a part of the design and should be placed as desired, avoid conditions that place heavy loads on long, narrow stones. Where design requires such conditions, use steel angles to relieve the weight and to avoid cracking the smaller stones.
15. Dowel pins *must not* be embedded solidly when installed through a relieving angle at an expansion joint or in horizontal tie-backs.
16. When possible, avoid setting stone with mortar in extreme cold. Stonework set in cold weather can expand and crack the mortar bond in warm temperatures.
17. A lean pointing mortar will usually perform better than a strong one. While pointing reduces leak potential whether the joint is thoroughly filled or not, a strong mix may cause spalling at joints.

18. Cavity walls should be kept clear of mortar droppings during construction.

**Warning!** Do not pour the cavity full of grout. See Figure 12-31.

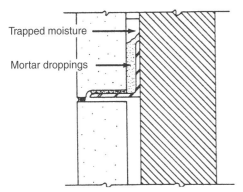

**Figure 12-31.** Avoid the condition shown above to prevent trapped moisture. (Indiana Limestone Institute)

19. In mortar systems, all mortar joints between stones should be thoroughly filled. The front 1/2" to 1 1/2" should be filled with pointing mortar after the setting bed has set. See Figure 12-32.

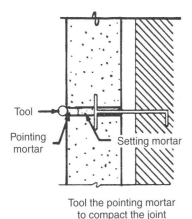

Tool the pointing mortar to compact the joint

**Figure 12-32.** Recommended method for tooling joints that have been pointed. (Indiana Limestone Institute)

20. The use of calcium chloride or other salts that act as accelerators or retarders in mortar are not recommended in setting stone.
21. Do not pour concrete against unprotected stonework. Alkali from the concrete will stain stone.
22. Store stone clear of the ground and protect adequately from the elements and construction traffic.

23. Place an adequate number of setting pads in the horizontal joint under heavy stones in order to sustain the weight until the mortar has set. See Figure 12-33.

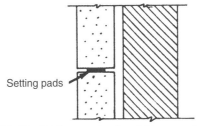

**Figure 12-33.** Setting pads should be used under heavy stones to support the weight until the mortar sets. (Indiana Limestone Institute)

24. Be sure that all projecting courses, sills, entrances, columns, column covers, and other stone elements exposed to traffic of the trades and mortar droppings are properly protected during construction. Use galvanized nails in mortar joints for support, or galvanized steel straps.
25. Do not permit wash from concrete floors or scaffolding to run down onto or behind stone walls.

 **Trade Tip.** Always cover walls and openings adequately at night and during rains. This will help prevent staining and efflorescence. Failure to do this is a source of trouble, dissatisfaction, and expense.

## Panel sizes

The panel sizes commonly produced are the maximum recommended *for efficient fabrication and handling.* Larger panels are available in either monolithic stones or epoxy assemblies. For information on larger units, consult Indiana Limestone Institute or member companies. Standard panels are available in sizes of 3' × 5' × 2", 4' × 9' × 3", 5' × 11' × 4", 5' × 14' × 5", and 5' × 18' × 6".

 **Trade Tip.** Two-inch thicknesses in Indiana limestone wall panel work should be considered only in those applications where grasping systems specially designed for thin stone can be used. The anchors and supports ordinarily used in 3" and thicker stones can often be used for thinner materials, but wind loads and other building dynamics become more critical as stone thickness is reduced. Fabrication, handling, and erection procedures associated with conventional erection methods often produce greater costs when applied to thin stones. For most purposes, the conventional support systems should be limited to 3" and thicker stones.

## Anchors, supports, and embeds

Stone fabricators are usually suppliers of limestone only. They provide the stone with the required holes, slots, chases, and sinkages for the anchoring system but seldom provide the anchors themselves. This responsibility falls usually on the mason, the erector, or can be assigned by the general contractor. It is standard practice for the architect to indicate generally the anchor system for each typical condition. Architects judge and then approve or change the extension of that system shown on shop drawings provided by the stone fabricator.

*Anchor* refers generally to the straps, rods, dovetails, and other connections between stone and structure. Refer to Chapter 9 of this text. While most anchors are intended to maintain stones in their vertical positions rather than bear weight, certain anchors are structural and their use eliminates the need for other support.

All anchors in Indiana limestone should be stainless steel, or other noncorrosive metal. In practice, anchors are embedded in the stone with mortar, sealant, or other non-expansive, stable material. Supports by definition are not embedded in the stone, but support its weight. They may touch or be adjacent to the stone. Supports are typically A36 steel, but may be any metal of adequate strength. Supports should have at minimum a shop-coat of rust protection, applied after forming and de-greasing, and a job-coat of compatible rust protection applied after supports are installed on the structure. Hot-dip galvanized supports are acceptable; stainless steel supports are not required except under special circumstances. Any damage to these coatings as a result of installation work, cutting, or drilling, must be repaired with a compatible rust protection coating.

## Mortars

Portland cement used in preparing cement/lime mortar for setting Indiana limestone should conform to the requirements of ASTM C 150. If masonry cement is used, it should conform to the requirements of ASTM C 91. Either material will produce a suitable mortar.

Mortars should be mixed to the proportion requirements of ASTM C 270, type N. Its compression resistance will approximate 750 psi when cured. This moderate strength is sufficient for most limestone installations.

Portland cement/lime mortars are mixed with one part cement, one part lime, and six parts sand, all by volume. Lime improves the workability of mortar, and helps to reduce shrinkage. This 1/1/6 mixture provides sufficient compressive strength, good bond strength, and good weather resistance.

Low-alkali cement mortars will tend to be nonstaining, although low-alkali content in mortar used to set stone is not in and of itself a guarantee of such performance. Good masonry practice, especially the protection of stone backs from moisture sources, is the best prevention of stain. It should be noted that white cement is not necessarily nonstaining. ASTM C 91 describes nonstaining cement; this type should be used with Indiana limestone.

Sand should be clean and sharp and washed free of loam, silt, and vegetable matter. Grading should be from fine to coarse complying with ASTM C 144. If the setting mortar is to be the pointing mortar, white sand should be used. Mixing water should be of potable quality.

As setting proceeds, the mason should remove mortar tags from joint edges after they have taken their initial set. Avoid smearing pointing mortar—smeared mortar can be difficult to remove. It may be desirable to use a mortar-bag in pointing vertical joints particularly, where pointing mortar cannot be easily scraped off a mortarboard. Final cleaning should be done with fiber brushes and detergents or mild soap powder.

 **Warning!** Do not use acids when doing the final cleaning.

Either mortar or joint sealant is used to fill anchor holes. Lead wool is another alternative. Shim stock of a noncorrosive material may be used provided it is not built up beyond a *snug thickness*.

Never use expanding type grout or mortar. Certain types of packaged grouts expand during the hardening period. When used in anchor slots and holes, this expansion can create sufficient pressure to fracture the stone at the anchor locations.

## Pointing

Pointing cut stone after setting, rather than full bed setting and finishing in one operation, reduces a condition that tends to produce spalling and leakage. Shrinkage of the mortar bed will allow some settling since the mortar bed hardens from the face in. If set and pointed in one operation, the settling, combined with the hardened mortar at the face, can set up stresses on the edge of the stone. For this reason, it is best to set the stone and rake out the mortar to a depth of 1/2" to 1 1/2" for pointing with mortar or sealant application at a later date.

By raking the joints to a depth of about 1/2" to 1 1/2" the pointing can be done in one, two, or three stages. This allows each stage to seal shrinkage cracks in the preceding stage and finally the concave tooled joint provides the maximum of protection against leakage.

## Cold weather setting

The bond strength of mortar is considerably reduced when mortar is frozen prior to hardening. Hydration progresses very slowly below 40°F (12° C). *Hydration* is the chemical reaction between water and cement. Protection is necessary if the outside air temperature is 40°F (12° C) and falling.

Admixtures or antifreezes should not be used to lower the freezing point of mortar. The effectiveness of most of these compounds is due to the calcium chloride they contain acting as accelerators.

Calcium chloride cannot be used on limestone. Salts cause efflorescence and can cause spalling or flaking through recrystallization (crystal growth).

Never set stone on a snow or ice-covered bed. A bond cannot develop between the mortar bed and frozen supporting surfaces.

 **Trade Tip.** If stone is to be set during cold weather, the cold weather masonry construction recommendations of the **International Masonry Industry All-Weather Council** should be closely followed.

## Sealant systems

Sealants must serve the same purpose as mortar and pointing in the exclusion of moisture. The most commonly used one-part systems are called *moisture-cure* or *air-cure*. Two-part systems depend on a catalyst or chemical agent to cure. Curing may be considered analogous to the *setting* of mortar, but sealants are not intended to bear weight and thus must be handled differently in the construction process.

The materials of a sealant system consist of a sealant backer or stop, usually a foam rope or rod placed in the joint at a predetermined depth, and the sealant itself, which is gunned in against the stop. Some systems require a primer which must be applied to the joint's inner surfaces in advance of the sealant to assure its adhesion.

In general, the sealant should not adhere to the backer rod. Joints utilizing a sealant system work best when the sealant is required to adhere to parallel surfaces only; especially in the wider joints omission of a bond-breaking stop may contribute to failure. See Figure 12-34.

Joint Sealant Design

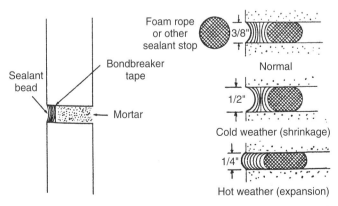

Note: Shims may be substituted in sufficient area to support the load.

**Figure 12-34.** Sealant used between stone panels must be elastic and installed properly as described by the manufacturer. (Indiana Limestone Institute)

 **Trade Tip.** Manufacturers of sealant systems are the best source of information on system types and accessories.

## Joint movement

Indiana limestone has a low coefficient of expansion, however, a 100' long wall exposed to a temperature variance of 130° will move 7/16". Seven 1/4" expansion joints can be incorporated into this wall, allowing 1/16" movement of each 1/4" joint. If the percent of elongation and compression capabilities of the sealant selected is 25%, the joint design and percent of movement are within the 25% maximum movement recommendations for a joint. The size of the joint should be adjusted to the elongation limits of the sealant. A sealant with a 15% joint movement capability would require a 7/16" joint size to withstand a 1/16" joint movement.

These examples assume that movement will be equally distributed over all joints. By sizing the joints larger than minimum design requirements, a safety factor can be obtained to accommodate atypical movement.

## Expansion joints

In exterior limestone walls, relief joints should be provided to reduce the damaging effect of thermal expansion of the building frame.

The coefficient of expansion of concrete and steel supporting structures is much greater than that of limestone. Differential movement between the building frame and walls due to thermal changes causes the pushing out of header stones at corners, the fracturing or spalling of stone at joints, and the pulling out of rigid anchors. Stonework erected in cold weather can develop temperature stress in joints in warm weather.

Expansion joints normally consist of a premolded filler and sealant compound. The premolded filler should be compressible to the amount of movement calculated with resiliency to return to its original shape. The sealant compound should be completely elastic and provide lasting adhesion to the surfaces separated.

The best location for an expansion joint is in an offset of a building when one occurs. An alternate location is at the junctions of the sections comprising a U-shaped, T-shaped, or L-shaped building.

## Typical cornice detail

The typical traditional cornice detail in Figure 12-35 shows the most commonly used method of anchoring a cornice with a projection large enough to be unbalanced in the wall.

The bed joint immediately below the heavy cornice is left open back far enough to remove any compressive stress that would have a tendency to break off stone below.

Typical Cornice Detail

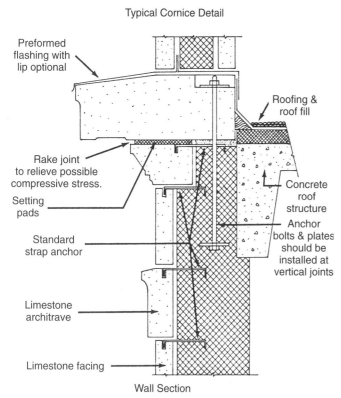

Preformed
flashing with
lip optional

Roofing &
roof fill

Rake joint
to relieve possible
compressive stress.

Setting
pads

Concrete
roof
structure

Anchor
bolts & plates
should be
installed at
vertical joints

Standard
strap anchor

Limestone
architrave

Limestone facing

Wall Section

**Figure 12-35.** A typical traditional stone cornice detail. (Indiana Limestone Institute)

The optional rod-and-plate tie-down stabilizes unbalanced, projecting courses.

# Cleaning New Masonry

The finished appearance of a masonry wall depends not only on the skill used in laying the units, but on the cleaning procedure as well. The appearance of a masonry structure can be ruined by improper cleaning. In many instances, the damage caused by faulty cleaning techniques or the use of the wrong cleaning agent cannot be repaired. All cleaning should be applied to a sample test area of approximately 20 sq. ft.

Some general precautions that can be taken to promote a cleaner wall during construction are as follows:

1. Protect the base of the wall from rain-splashed mud and mortar splatter.
2. Scaffold boards near the wall should be turned on edge at the end of the day to prevent possible rainfall from splashing mortar and dirt directly on the completed masonry.
3. Cover walls with a waterproof membrane at the end of the workday to prevent mortar joint wash out and entry of water into the completed masonry.
4. Protect site stored stone from mud. Store stone off the ground under protective covering.
5. Careful workmanship should be practiced to prevent excessive mortar droppings. Excess mortar

should be cut off with the trowel as the stone is placed. Joints should be tooled when thumbprint hard. After tooling, excess mortar and dust should be brushed from the surface. Avoid any motion that will result in *rubbing* or *pressing* mortar particles into the stone faces. A medium soft bristle brush is preferable.

Cleaning failures generally fall into one of the three following categories:

1. Failure to thoroughly saturate the masonry surface with water before and after application of chemical or detergent cleaning solutions. Saturation of the surface prior to cleaning reduces the absorption rate, permitting the cleaning solution to stay on the surface rather than be absorbed.
2. Failure to properly use chemical cleaning solutions. Improperly mixed or overly concentrated acid solutions can etch or wash out cementitious materials from the mortar joints.
3. Failure to protect windows, doors, and trim. Many cleaning agents, particularly acid solutions, have a corrosive effect on metal. If permitted to come in contact with metal frames, the solutions can cause pitting of the metal or staining of the masonry surface and trim materials, such as limestone and cast stone.

## *Cleaning stonework*

Clean stonework with a stiff fiber brush and clean water. If stains are difficult to remove, use soapy water and then rinse with clear water. If the stonework is kept clean by sponging during construction, the final cleaning will be easy. Machine cleaning processes should be approved by the supplier before using.

 **Warning!** Acids, wire brushes, and sandblasting are usually not permitted on stonework. Strong acid compounds used for cleaning brick will burn and discolor many types of stone.

# REVIEW QUESTIONS CHAPTER 12

Write all answers on a separate sheet of paper. Do not write in this book.

1. Stone masonry is still used extensively for _____ construction.
2. Name four basic operations that must be mastered before quality work is possible with stone.
3. Describe the proper way to grasp the trowel.
4. Why should one avoid constantly working the mortar?

5. Stone masonry that consists of uncut stones is called _____ masonry.

6. In rubble stone masonry, where should larger stones be placed?

7. A cubic foot of granite weights about _____ lb.

8. What kind of shoes or boots are required when working with stone?

9. An experienced mason uses the trowel to cut stone as well as spread mortar. True or False?

10. How should smooth finished stone panels be stored (orientation)?

11. Stone that has been cut on all sides to specific dimensions is called _____.

12. Mortar used for stone masonry should have the following composition: _____ part(s) nonstaining cement; _____ part(s) hydrated lime; and _____ part(s) clean, washed sand.

13. What is the preferred width of mortar joints for ashlar stone work?

14. Cements with _____ content are called non-staining.

15. The tools used to trim stone are the mason's _____ and _____.

16. Manufactured stone which is a Portland cement product can be cut with a mason's trowel. True or False?

17. What are the three main classes of stone masonry construction (panel construction is not included in this classification).

18. _____ is uncut stone or stone that has not been cut to a rectangular shape.

19. When stone is precut with enough uniformity to allow some regularity in assembly, the wall is generally called _____.

20. What is the maximum length that ashlar stone is usually available in?

21. Stone veneer is generally _____" to _____" thick.

22. Stone veneer may be attached to a masonry or frame backing using _____ ties.

23. When setting a rubble stone veneer wall, mix enough mortar to last for _____ hour(s) of work.

24. What tool may be used for splitting large stones?

25. How can one prevent heavy stones from squeezing mortar out when they are set into place?

26. What simple cleaning procedure is recommended for a completed stone wall?

27. Why should strong acids be avoided when cleaning stone?

28. A solid stone wall should be at least _____ thick.

29. Where was the first organized quarrying effort of record established in 1827?

30. Steel lintels carrying stonework should have ample stiffness to carry the superimposed load with a deflection of less than _____ of the clear span.

31. A minimum clearance of _____ should be maintained between stone panels and all structural members.

32. Why should concrete not be poured against unprotected stonework?

33. Which of the following is not a standard stone panel size?
    A. 2' × 4' × 2"
    B. 4' × 9' × 3"
    C. 5' × 11' × 4"
    D. 5' × 14' × 5"

34. Who generally provides the anchors used in stone panel construction?

35. All anchors in Indiana limestone should be _____, or other noncorrosive metal.

36. Why is pointing cut stone after setting, rather than full bed setting and finishing in one operation, the recommended practice?

37. Why is calcium chloride not used on limestone?

38. Indiana limestone has a low coefficient of expansion, however, a 100' long wall exposed to a temperature variance of 130° F will move 7/16". True or False?

39. The coefficient of expansion of concrete and steel is about the same as that of limestone. True or False?

40. What should you do if stains on stonework are difficult to remove?

Stonework used for garden walls can create a formal or informal atmosphere. Top—This rubble stone wall has a rugged, rustic look as a background for an informal garden. Bottom—The formal atmosphere of this setting is reinforced by a wall that makes use of stone that is uniform in color, cut to size, and laid in even courses.

# Chapter 13
# Construction Details

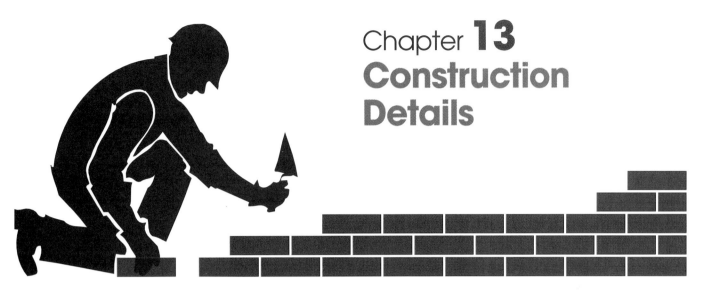

This chapter includes many of the accepted methods of masonry construction for foundation systems, wall systems, fireplaces and chimneys, movement joints, brick masonry soffits, floors and pavements, flashing, steps, sills and lintels, arches, garden walls, caps and copings, corbels, and racking. This material is meant to be a source of help for the mason in selecting appropriate construction methods for many different jobs. Figure 13-1 shows a sample of the subjects covered in the chapter.

## Concrete and Masonry Foundation Systems

One of the first concerns in construction is to prevent the settling of buildings as much as possible. Some settling is to be expected. It cannot be prevented. The worst enemy of buildings is uneven settling. It causes cracks in finished walls and ceilings. It tilts floors, causing doors and windows to bind. It damages woodwork. If severe enough, it can cause total failure of the building.

The foundation is the substructure of a building that resists settling. It supports the weight of the building. Footings transmit (shift) the weight from the foundation walls to the soil. See Figure 13-2.

Footings are usually wider than the foundation wall to provide this extra support that will stop or reduce uneven settling. As a result, the structure remains sound.

The two general types of foundations are spread foundations and pile foundations. Only spread foundations will be covered in this text, because pile foundations do not usually concern the work of the mason.

**Figure 13-1.** A—Foundation, floor, and wall system. B—Forming for slab foundation. C—Concrete pavers. D—Precast stairs. E—Concrete sill. F—Stone cap.

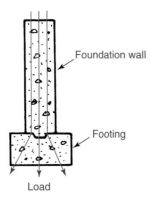

**Figure 13-2.** Footings spread the load of the structure over a broader area.

## Spread foundations

**Spread foundations** are foundations that consist of elements such as walls, pilasters, columns, or piers that rest on a wider base called a footing. Spread foundations distribute the building loads over a wider area of soil. Raft and matt foundations are also classified as spread foundations. They are used over soils with low ability to carry weight.

## Footings

All foundation walls built of masonry units bonded with mortar should rest on footings. These footings are usually made from concrete. They are classified as plain, reinforced, continuous, stepped, and isolated.

Plain footings carry light loads and are not reinforced with steel. They are either continuous, stepped, or isolated.

Steel is embedded in reinforced footings to make them stronger. See Figure 13-3. Reinforcing should be used over weak spots in the soil such as where there have been excavations for sewer, gas, or other connections.

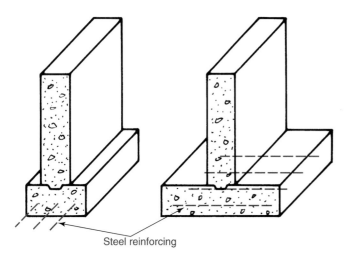

**Figure 13-3.** Two methods of reinforcing footings to increase strength.

Continuous footings are used to support a foundation wall or several columns in a row. See Figure 13-4. A **combined footing** is a single footing supporting more than one column.

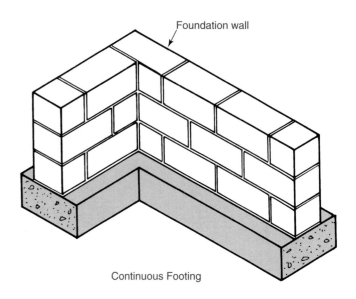

Continuous Footing

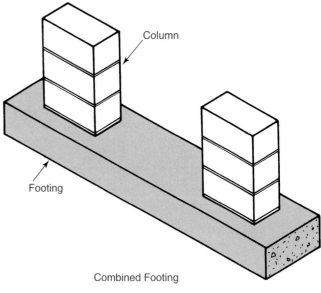

Combined Footing

**Figure 13-4.** Typical applications of a continuous footing.

There are two types of stepped footings. One widens the base to provide added support. The other changes levels to accommodate a sloping grade. See Figure 13-5. Today, reinforced footings have largely replaced the widened footing.

Isolated footings are not part of the foundation. They receive the loads of free standing columns or piers. See Figure 13-6.

Footing size should be related to soil conditions and the weight of the structure. However, the rule-of-thumb size for residential footings placed on average soil is:

1. The width of the footing should be twice the thickness of the wall.
2. The depth of the footing should be equal to the thickness of the wall or a minimum of 6". Figure 13-7 shows the proportions for a standard footing.

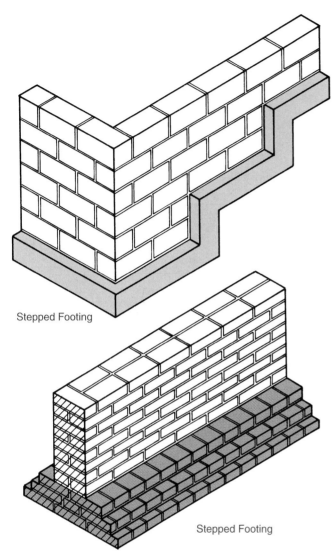

Stepped Footing

**Figure 13-5.** Stepped footings can be used to change levels on a sloping grade or to widen the base for greater support.

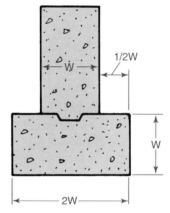

**Figure 13-7.** Typical rule-of-thumb footing proportions.

## Raft and matt foundations

Raft and matt foundations are made of concrete reinforced with steel in such a way that the entire foundation will act as a single unit. Many times they are referred to as *floating foundations*.

A *matt foundation* is a thickened slab that transmits building loads over the entire slab and soil area. See Figure 13-8.

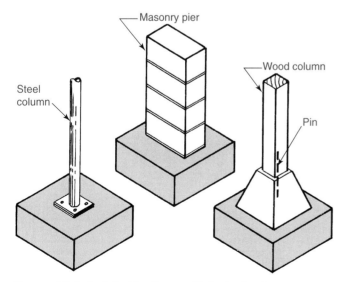

**Figure 13-6.** Isolated footings with typical columns and pier.

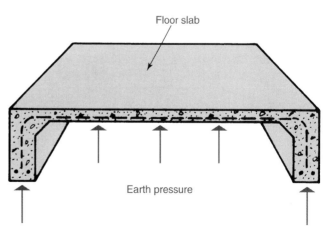

**Figure 13-8.** A matt foundation ties entire floor and footings into one unit.

A *raft foundation* is made with reinforced walls and floor cast into a single unit. See Figure 13-9. Soil is excavated that weighs about the same as the foundation plus the superstructure. This minimizes settling of the structure because the building load on the soil is equal to the weight of the excavated soil.

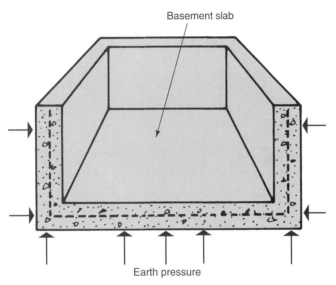

**Figure 13-9.** A raft foundation ties walls and floor into one unit.

## Foundation walls

Like footings, foundation walls are load transferring elements. They must be able to:
1. Support the weight of the building above
2. Resist pressures from the ground
3. Provide anchorage for the superstructure of the building.
4. Be durable and resist moisture penetration

Foundation walls of hollow concrete block should be capped with a course of solid masonry to help distribute the loads from floor beams and serve as a termite barrier. See Figure 13-10.

**Figure 13-10.** Block masonry foundation wall is capped with a course of solid top block. (Portland Cement Association)

The most common type of foundation wall is the T foundation. The name comes from the shape of the foundation and the footings which look like an upside-down T. The footing and foundation wall, in this case, are generally two separate parts, but may be cast as a single unit. Figure 13-11 shows several styles of foundations.

Solid blocks 4" thick are available in some areas of the United States. See Figure 13-12. If stretcher blocks are used, a strip of metal lath is placed under the cores of the top course. The cores are then filled with concrete or mortar and troweled smooth. See Figure 13-13.

## Dampproofing basement walls

The outside of masonry basement walls should be parged with a 1/2" thick coating of Portland cement plaster or mortar. It should be applied in two coats, each 1/4" thick. The wall surface should be clean and damp but not soaked before applying the plaster. The first coat should be troweled firmly over the wall. See Figure 13-14.

When the first coat has partially hardened, it should be roughened with a scratcher. See Figure 13-15. This operation provides a good bond for the next coat.

Keep the first coat moist and allow it to harden for a minimum of 24 hours before applying the second coat shown in Figure 13-16. Cover the wall from the footing to 6" above the finished grade line. Form a cove over the footing to prevent water from collecting at this point. See Figure 13-17. The second coat should be moist cured for at least 48 hours.

Some areas of the United States with certain soil conditions require additional dampproofing. A heavy coat of tar, two coats of a cement-based paint, or a covering of thin plastic film may be used in addition to the parge coat.

Drain tiles of concrete, clay, or plastic are generally placed around footings to carry away ground water. They are connected to a dry well or storm sewer.

The concrete and clay tiles are usually 4" in diameter and are laid with open joints. The new plastic continuous tile have perforated holes that allow the water to enter. All three types are laid in a bed of gravel a foot or more wide and deep. Refer to Figure 13-11, which shows a drain tile in place along the footing.

## Columns, piers, and pilasters

Columns, piers, and pilasters transmit loads to the footing. Columns and piers are freestanding but pilasters are built into the foundation wall.

*Columns* are isolated elements with horizontal dimension, measured at a right angle from the thickness dimension, that does not exceed three times the thickness dimension and has a height that is at least three times its thickness. Brick masonry columns are used to support large axial loads. Axial loads are typically due to the permanent weight of the structure and the transient floor or roof load that is transferred to the column. According to the MSJC Code, columns must be reinforced with a minimum of four reinforcing bars, and the area of reinforcement must be at least 0.0025, but not more than 0.04 times, the

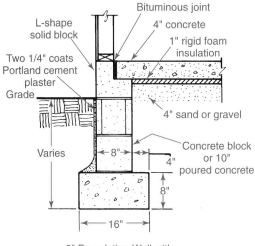

8" Foundation Wall with
Insulated Slab Floor

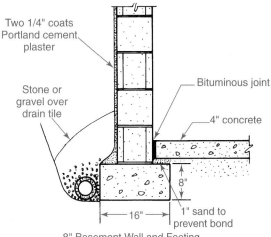

8" Basement Wall and Footing

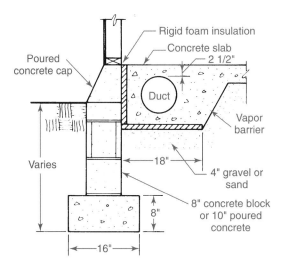

Insulated Slab for Perimeter Heat
with Concrete Block Foundation

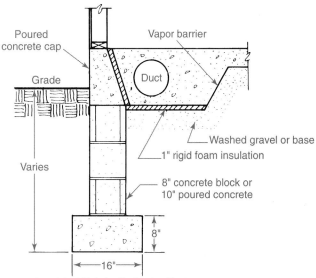

Insulated Slab for Perimeter Heat
with Concrete Block Foundation

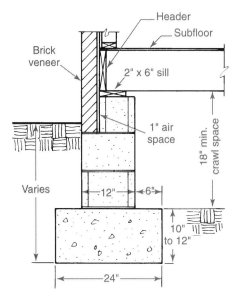

12" Concrete Block Foundation for
Brick Veneer on Frame

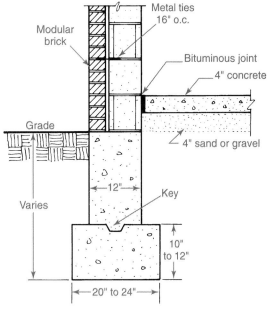

Poured Concrete Foundation for
Composite Brick and Block Wall

**Figure 13-11.** Typical foundation details.

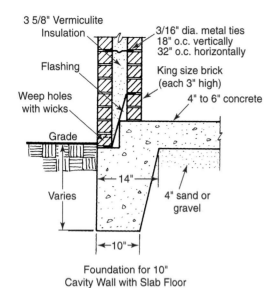

Foundation for 10"
Cavity Wall with Slab Floor

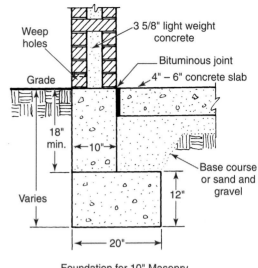

Foundation for 10" Masonry
Bonded Wall with Slab Floor

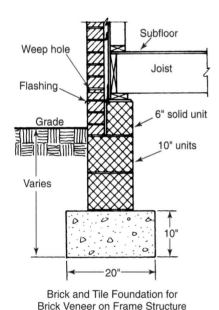

Brick and Tile Foundation for
Brick Veneer on Frame Structure

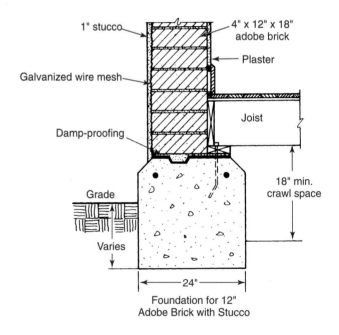

Foundation for 12"
Adobe Brick with Stucco

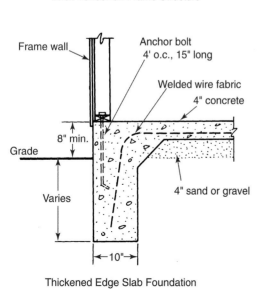

Thickened Edge Slab Foundation
for Frame Wall Construction

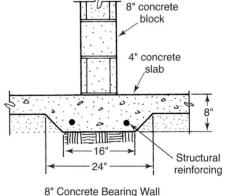

8" Concrete Bearing Wall
Partition on a Slab Floor

**Figure 13-11 (Continued).**

**Figure 13-12.** Foundation wall can be capped with solid block 4" thick.

**Figure 13-15.** A wire scratcher roughens the first plaster coat when it has partially hardened.

**Figure 13-13.** When stretcher blocks are used in the final course of a foundation wall, the cores should be filled with concrete or mortar. (Portland Cement Association)

**Figure 13-16.** The second and final coat of Portland cement plaster is troweled over the roughened surface of the first coat.

**Figure 13-14.** The first coat of parging is being troweled over the surface.

**Figure 13-17.** A cove of mortar is formed over the footing to prevent water from entering. (Portland Cement Association)

column's net cross-sectional area. The minimum nominal dimension of a column is 8" and the ratio of height to least lateral dimension must not exceed 25.

Brick or block columns are vertical supports that are frequently used to hold up wood, steel, or masonry beams. See Figure 13-18. The bonding for a 12" × 12" column is shown in Figure 13-19. Columns built with concrete block can be made using one to three blocks in each course. See Figure 13-20.

**Figure 13-18.** These brick columns are used to support a large wooden beam on this traditional structure.

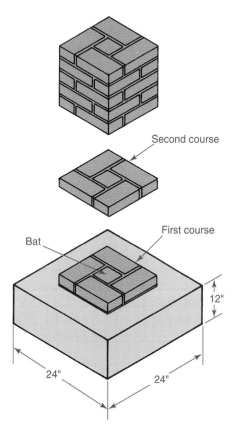

**Figure 13-19.** Bonding for a 12" × 12" brick column. The bat is a piece of brick cut to fill.

A **pier** is an isolated section or column of masonry that may be part of a bearing wall if it is not bonded to masonry at the sides and its length does not exceed four times its thickness. It can support openings in a wall.

Figure 13-21 shows several sections of solid masonry piers. The size of piers, unless governed by a local code, is usually a minimum of 8" × 12" for solid units and 8" × 16" for hollow units. The height should not exceed 10 times their smallest cross-sectional dimension. Footings for piers should be at least 8" thick.

A **pilaster** is a masonry column bonded to a wall used to increase the effective thickness of a wall at a specific location. It has a uniform cross section and serves as a column and/or vertical beam. See Figure 13-22. To work together, the wall and the thickened section must be integrally constructed. The MSJC Code Section 5.10 permits three methods of bonding a pilaster to create integral construction:

1. Interlocking 50% of the masonry units
2. Toothing at 8" maximum offset and attachment with metal ties
3. Providing reinforced bond beams at a maximum spacing of 4' o.c. vertically

The length of the wall or flange that is considered to act integrally with the pilaster from each edge of the pilaster or web is less than six times the thickness of the wall or the actual length of the wall. When pilasters are built with concrete block, the hollow cells are sometimes filled with mortar or grout to increase their strength. Figure 13-23 shows a pilaster being built with concrete blocks.

## Foundation anchorage

The building superstructure should be anchored to the foundation to resist high winds. Sill plates of wood joist floor systems are generally anchored to masonry walls with 1/2" bolts extending at least 15" into the filled cores of masonry units. Anchor bolts should be spaced not more than 8' apart, with one bolt not more than 12" from each end of the sill plate.

Sill plates should be anchored to poured concrete walls with 1/2" anchor bolts embedded 6" with the same maximum spacing as used for masonry walls. Hardened steel studs, driven by power tools, can be used if permitted by local codes. At most, they should be spaced 4' apart.

## Masonry Wall Systems

Masonry walls can be constructed of a single wythe (one unit in thickness) or multiple wythes, and can be reinforced or unreinforced. Masonry walls are classified as solid walls, four-inch RBM (reinforced brick masonry) curtain and panel walls, hollow walls, anchored veneered walls, composite walls, and reinforced walls.

Masonry walls are very popular. They provide excellent structural performance, are easily maintained, and are attractive. They provide an envelope that is uniquely

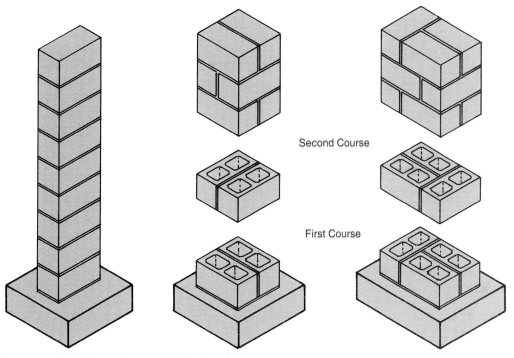

**Figure 13-20.** Three popular styles of concrete block columns.

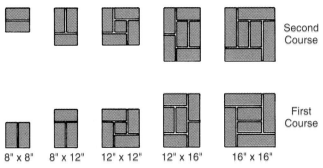

energy efficient due to their high thermal mass. **_Thermal mass_** is the characteristic of heat capacity and surface area capable of affecting building thermal loads by storing heat and releasing it at a later time. Materials with high thermal mass react more slowly to temperature fluctuations and reduce peak energy loads.

The benefits of thermal mass have been known for a long time. However, their inclusion in energy codes and standards is a relatively recent development. All buildings designed today must comply with energy code requirements. Energy performance requirements are found in such documents as the 1992 *Model Energy Code (MEC),*

8" x 8"    8" x 12"    12" x 12"    12" x 16"    16" x 16"

**Figure 13-21.** Bonding for a variety of sizes of solid brick masonry piers.

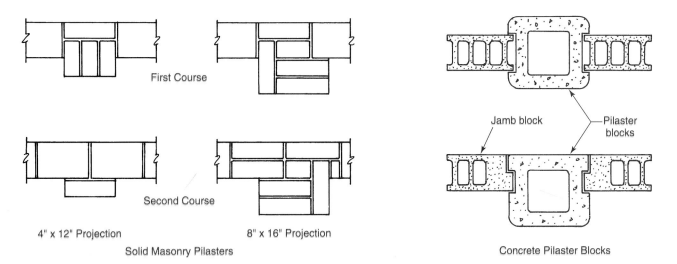

**Figure 13-22.** Typical pilasters made from solid brick masonry and concrete pilaster block. Their purpose is to provide support for a wall. It is usually found at intervals in long wall sections.

**Figure 13-23.** A pilaster being constructed from 6" and 8" concrete block. The 6" and 8" blocks are alternated each course. Left—Pilaster with first course and part of second course in place. Right—Partially completed. You can see how, by alternating positions of 6" and 8" block, the pilaster is tied into the wall.

the ASHRAE/IES Standard 90.1-1989: *Energy Efficient Design of New Buildings Except New Low-Rise Residential Buildings,* and the 1990 edition of the BOCA *National Building Code.* Most energy codes and standards specify a maximum permissible value for the coefficient of thermal transmittance (U-value) for the building envelope as a function of wall weight.

## Solid masonry walls

A **solid masonry wall** is a wall that is built up of masonry units laid close together with all joints between them filled with mortar. Figure 13-24 shows several types. Solid or hollow masonry units or a combination of these materials can be used. Such walls may be either load-bearing or nonloadbearing. The structural bond of these walls is provided by metal ties, masonry headers, or joint reinforcement.

Solid masonry walls can also be classified by the types of units used in their construction. Examples of this classification include:

1. Solid units of brick or concrete brick and block.
2. Hollow units of concrete block or structural clay tile.
3. Composite (faced) walls composed of facing and backup units of different materials bonded in such a way that both facing and backup are load-bearing. Figure 13-24 shows how this wall is constructed. See bottom-center illustration. Header blocks allow masonry headers to overlap block.

Nationally recognized building codes permit the use of exterior loadbearing 6" masonry walls for one-story, single-family dwellings where the wall height does not exceed 9' to the eaves and 15' to the gable peak. Figure 13-25 shows

typical sections of 6" walls. The masonry units used for this type of construction are most commonly the *SCR brick.* It has a nominal thickness of 2 2/3" that produces 16" in six courses. It is a nominal 6" wide and 12" long. This makes it easier to lay in 1/2" bond. This brick is also produced in thicknesses of 3" (four courses in 12"), 3 1/5" (five courses in 16"), 4" (four courses in 16"), and 5 1/3" (three courses in 16").

Clay tiles are also produced in 6" sizes that can be used with the SCR brick.

## Four-inch RBM curtain and panel walls

A **curtain wall** is an exterior nonloadbearing wall not wholly supported at each story. These walls can be anchored to columns, spandrel beams, floors, or bearing walls, but not necessarily built between structural elements. A **panel wall** is an exterior nonloadbearing wall supported at each story. Curtain walls must be capable of supporting their own weight for the height of the wall. Panel walls are required to be self-supporting between stories. Both walls must be able to resist lateral forces such as wind pressures and transfer these forces to adjacent structural members.

Major factors to be considered in the design of 4" curtain or panel walls include:

1. Structural
2. Moisture control
3. Differential movement
4. Special considerations

Since 4" walls do not meet the requirements specified in the SCPI Standard for reinforced walls, they are designed as *partially reinforced.* The horizontal reinforcement in the wall resists the tensile stresses resulting from

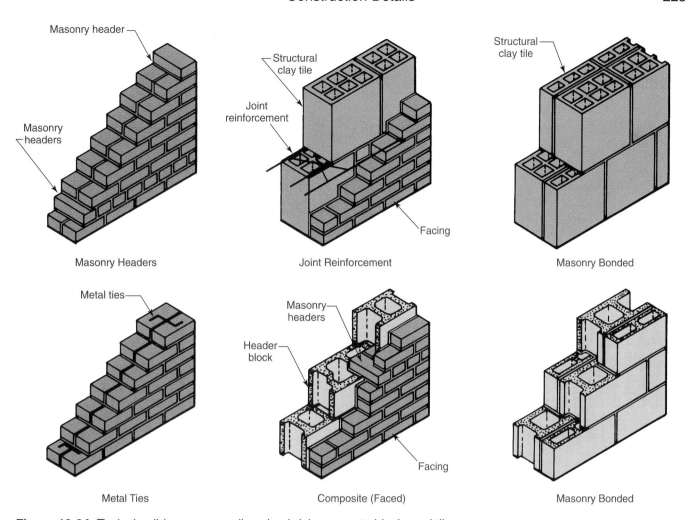

Masonry header

Masonry headers

Structural clay tile

Joint reinforcement

Facing

Structural clay tile

Masonry Headers          Joint Reinforcement          Masonry Bonded

Metal ties

Masonry headers

Header block

Facing

Metal Ties          Composite (Faced)          Masonry Bonded

**Figure 13-24.** Typical solid masonry walls using brick, concrete block, and tile.

lateral pressures as the wall spans horizontally between structural elements such as columns, pilasters, or cross walls. See Figure 13-26.

Since the lateral load due to wind can be either pressure or suction, the wall must be designed to resist the assumed lateral load acting in either direction. It is recommended that the maximum area of reinforcement be provided throughout the length of the wall. Steel reinforcement should be lapped a minimum of 16" to ensure the development of the tensile stresses without exceeding the allowable bond stresses. Since the tables for reinforcement are based on solid walls, walls with openings must be investigated by the designer.

Ladder or truss-type joint reinforcement provides two longitudinal bars connected by cross wires, thus aiding proper positioning and reducing the time and effort involved in placement. See Figure 13-27. Care should be taken to ensure that the reinforcement has mortar coverage both top and bottom as well as proper horizontal placement within the wall. The reinforcement should not be laid directly on top of the brick. Proper vertical placement can be attained by supporting reinforcement wires on small mortar pads placed prior to full bed-joint mortar placement.

The control of moisture through the 4" wall should be considered by the designer. Heavy rains driven by high winds may result in water penetrating the wall. One method of handling the problem is to provide drainage space on the inside of the wall to permit the water to flow to the flashing at the base and to be conducted back to the outside through weep holes. A second method is to provide a water barrier (parging) to the inside of the wall.

When masonry walls are used to enclose a structural frame building, consideration must be given to the method of anchoring the walls to the framing elements. Anchoring should be done in a manner that will permit each to move relative to the other.

Generally, expansion joints may be located at or near corners, offsets, junctures, and openings. Since 4" walls are considered to span horizontally, expansion joints where required must be located at vertical supports. The maximum recommended spacing for expansion joints for these 4" walls is 100' for a straight wall without openings. However, conditions may require expansion joints as close together as 40'.

There are special design possibilities that can be incorporated into 4" reinforced walls. One such

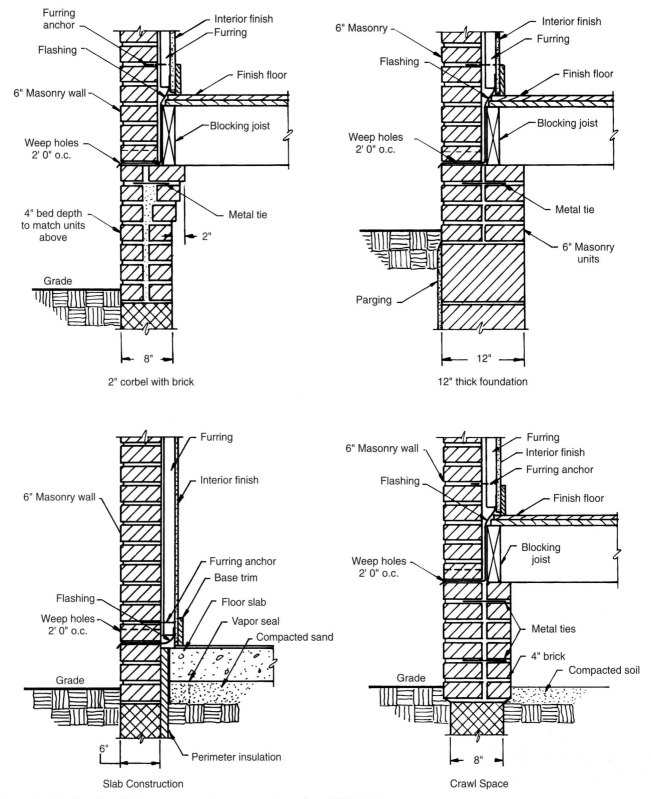

**Figure 13-25.** Details of 6" masonry wall construction using SCR brick.

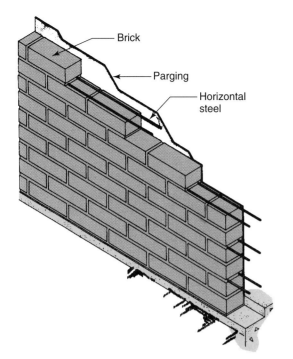

**Figure 13-26.** Notice that horizontal joint reinforcement is used between every other course in this single wythe wall. (Brick Institute of America)

consideration is to design the wall as a plate supported on three or four sides. Another consideration is to utilize the loadbearing capabilities of the 4" wall. The SCPI Standard can be used to design the wall to carry vertical loads in addition to lateral loads.

## Hollow masonry walls

A *hollow masonry wall* is a wall built using solid or hollow masonry units, which are separated to form an inner and an outer wall. They may be either a cavity or masonry bonded type wall.

## Cavity walls

Masonry cavity walls are used extensively throughout the United States in all types of low-rise and high-rise buildings. The primary reasons for their popularity are they exhibit superior rain penetration resistance, excellent thermal capabilities, good sound transmission resistance, and high fire resistance.

A cavity wall is built of masonry units arranged to provide a continuous airspace of 2" to 3" wide. See Figure 13-28. The facing and backing wythes (a vertical section of masonry one unit thick) or tiers are connected with rigid metal ties. The exterior wythe is usually a nominal 4" thick. The interior wythe may be 4", 6", or 8", depending on the load to be supported and the height and length of wall. The exterior masonry wythe is generally solid or hollow brick, and the interior masonry wythe can be brick, structural clay tile, or hollow or solid concrete masonry units depending on the properties required. When a cavity of nominal 2" is maintained, the overall thickness will be 10", 12", or 14".

The cavity provides two advantages:

1. Airspace has insulation value or it can be filled with insulation material for added reduction of heat transfer.
2. It acts as a barrier to moisture.

However, to be effective, the cavity must be kept free of mortar droppings during construction. A board may be used to collect the droppings as shown in Figure 13-29. When weep holes are required at the bottom of a cavity wall, flashing should be used. See Figure 13-30. It keeps any moisture that might collect in the cavity away from the inner wall. See Figure 13-31.

One of the major functions of an exterior wall is to resist rain penetration. A masonry cavity wall, properly designed and built, is totally resistant to rain penetration through the wall assembly. The outside wythe may not be totally resistant to moisture penetration, but the overall design of the cavity wall assembly takes this into account. In a cavity wall, any moisture that may pass through the exterior wythe will run down the cavity face of that wythe. At the bottom of the cavity, it is diverted to the outside by continuous flashing and weep holes. Due to the decreased moisture problem, the chance of efflorescence may be less than with solid walls.

Heat losses and heat gains through masonry walls are minimized by the use of cavity wall construction. The complete isolation of the exterior and interior wythes by the airspace allows a large amount of heat to be absorbed and dissipated in the outer wythe and cavity before reaching the inner wythe and the building interior. This ability is further increased by the use of insulation in the cavity. Thus, considerable energy savings can be realized by proper design, detailing, construction, and use of cavity walls.

Resistance to transmission of sound is accomplished by the use of heavy massive walls, or the use of discontinuous construction. The cavity wall employs both of these, i.e., the weight of the two masonry wythes plus the partial discontinuity of the cavity.

In cavity wall construction, the air cavity provides a partial isolation between the two wythes. Sound on one side of the wythe strikes it and causes it to vibrate, but because of the separation and cushioning effect of the cavity, plus the massiveness of the wythes, the vibration of the other wythe is greatly reduced.

The results of ASTM E 119 fire resistance tests clearly show that masonry cavity walls have excellent fire resistance. Fire resistance ratings of cavity walls range from 2 to 4 h, depending upon the wall thickness and other factors. Due to their high fire resistance properties, brick walls make excellent fire walls for compartmentation in buildings. The spread of fire can be halted by using compartmentation.

Properly designed, detailed, and constructed cavity walls can be used in any building requiring loadbearing or nonloadbearing walls in the same manner as other masonry walls. The increased flexibility afforded by the separation of the wythes and the use of metal ties permits more freedom of differential movement between the

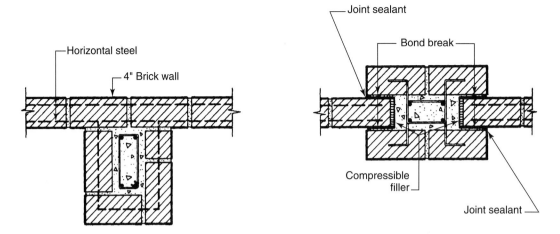

Reinforced Brick Masonry Columns and Pilasters

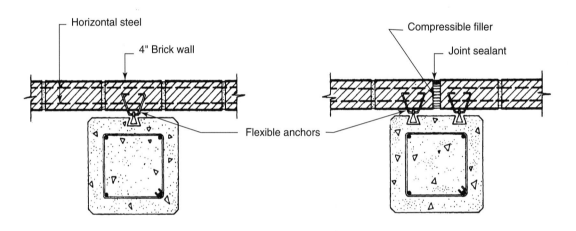

Reinforced Concrete Columns

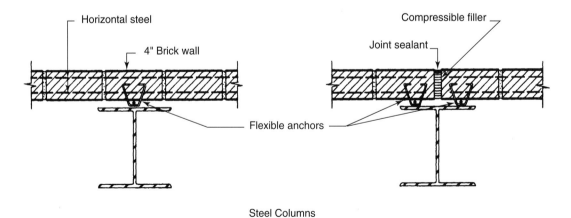

Steel Columns

**Figure 13-27.** Typical 4" reinforced masonry wall details. (Brick Institute of America)

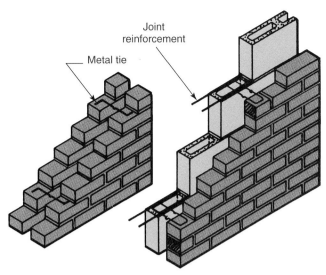

**Figure 13-28.** Method of cavity wall construction which uses metal ties and joint reinforcement.

**Figure 13-30.** Weep holes provide an outlet for moisture that has collected in the space between wythes of masonry.

**Figure 13-29.** A recommended method of keeping the cavity free of mortar droppings is to use a board resting on the joint reinforcement (wire ties). The board is moved up when the next reinforcement level is reached. (Portland Cement Association)

**Figure 13-31.** Flashing should be used to prevent moisture from entering the inner wall from the cavity. (Portland Cement Association)

wythes. This is extremely important in today's construction that makes use of increasingly more combinations of dissimilar materials.

Suitable types of insulation materials for cavity walls are granular fills and rigid boards. Each of these, if properly used, will produce a more thermally efficient wall. See Figure 13-32. Two types of granular fill insulation are recommended—water repellent vermiculite masonry fill and silicone-treated perlite loose fill insulation.

Granular fill insulation is usually poured directly into the cavity from the bag or from a hopper placed on top of the wall. Pours can be made at any convenient interval, but

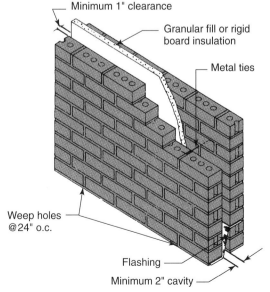

**Figure 13-32.** Rigid board insulation used to insulate a cavity wall. (Brick Institute of America)

the height of any pour should be kept below 20'. Rodding or tamping is not necessary, and could possibly reduce the thermal resistance of the material. The insulation in the wall should be protected from weather during construction.

Expanded or molded polystyrene, expanded polyurethane, rigid urethane, cellular glass, preformed fiberglass, and perlite board have been used as rigid board insulation materials in cavity walls. Rigid board insulation is installed horizontally within the airspace against the cavity face of the backup wythe. A minimum of 1" should be left between the cavity face of the external wythe and the insulation board. The 1" space between the outer wythe and the insulation board facilitates the wall construction and allows for drainage of the cavity. Care should be taken during installation to ensure that all boards are abutted and installed between ties, and fit flush against the inner wythe. An adhesive should be used to hold the insulation in place.

*Bonding.* The facing and backing (adjacent wythes) of cavity walls should be tied together with corrosion resistant 3/16" diameter steel ties, or metal tie wire of equivalent stiffness, embedded in the horizontal mortar joints. There should be at least one metal tie for each 4 1/2 sq ft. of wall area. Individual ties in alternate courses should be staggered; the maximum vertical distance between ties should not exceed 24" and the maximum horizontal distance should not exceed 36". The ends of the ties should be bent to 90° angles to provide hooks not less than 2" long, or ties bent to a rectangular shape should be used. Additional ties should be provided at all openings spaced not more than 3' apart around the perimeter and within 12" of the opening. Where the cavity width exceeds 3 1/2" but is less than 4 1/2", there should be at least one metal tie for each 3 sq. ft. of wall area.

*Flashing and weep holes*. Good flashing details are an absolute necessity in cavity wall construction. In order to divert moisture out of the cavity through the weep holes, continuous flashing should be installed at the bottom of the cavity, and wherever the cavity is interrupted by elements such as shelf angles or lintels.

Flashing should be placed over all wall openings not protected by projecting hoods and eaves. It should also be properly placed at all window sills, spandrels, and parapet walls. Since the purpose of the flashing is to collect moisture so it can be diverted to the outside, weep holes must be provided wherever flashing is used.

Weep holes are located in the joints of the outer wythe immediately above the flashing. Spacing of weep holes should be approximately 2' o.c. maximum, except for those using a wick material that should be 16" o.c. maximum.

*Expansion joints.* The movement of the outer brick wythe due to thermal and moisture expansion can be greater than the movement in solid or composite walls exposed to the same environment. This is due to the greater differences between the mean maximum and mean minimum temperatures of the outer wythe of the cavity wall, and the absence of restraint usually provided by dead and live loads, masonry bonders, or filled collar joints in solid walls. For this reason, it is recommended that expansion joints be provided through the outer wythe of the cavity wall on each side of an external corner where the walls are 50' or more in length. Where possible, this joint can be placed in line with the jambs of the windows nearest the corner.

## Hollow masonry bonded walls

Hollow masonry bonded walls are walls used for foundation and exterior loadbearing walls. They are popular in some parts of the United States. Though economical they are not resistant to high moisture. See Figure 13-33 for two examples of hollow masonry bonded walls.

### Anchored veneered walls

Brick and stone masonry units are widely used as a facing veneer. See Figure 13-34. In this application, loadbearing properties of the materials are not used. The veneer is attached to the backing, but does not act structurally with the rest of the wall.

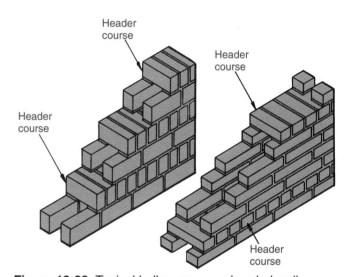

**Figure 13-33.** Typical hollow masonry bonded walls.

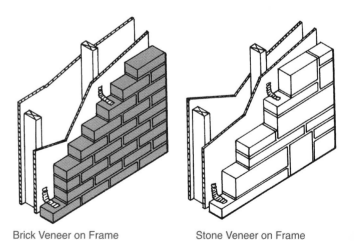

Brick Veneer on Frame          Stone Veneer on Frame

**Figure 13-34.** Brick and stone veneer attached to frame construction using metal ties. These veneers must rest on an extension of the foundation.

Anchored brick veneer construction consists of a nominal 3" or 4" thick exterior brick wythe anchored to a backing system with metal ties in such a way that a clear airspace is provided between the veneer and the backing system. The backing system is either wood frame, steel frame, concrete, or masonry. A **veneer wall** is a wall having a facing of masonry units, or other weather-resisting, noncombustible materials, securely attached to the backing, but not so bonded as to intentionally exert common action under load. The brick veneer is designed to carry loads due to its own weight; no other loads are to be resisted by the veneer. Bricks should conform to ASTM C 62, C 216, or C 652 for Building Brick, Facing Brick, and Hollow Brick, respectively.

 **Trade Tip.** The use of salvaged brick is not recommended for use as brick veneer.

For many years brick veneer construction was limited principally to wood frame houses. It is now being used on low-rise commercial and institutional construction and is used frequently for high-rise buildings, especially with concrete masonry or steel stud backing systems.

A paperbacked welded wire mesh, which is attached directly to the studding and grouted, is also sometimes used to attach veneer to frame construction. This technique eliminates the need for sheathing.

Veneer wall construction looks much like cavity wall construction in its final form. However, construction and function are considerably different. For example, the supporting structure is completed before attaching the veneer that is not loadbearing. Both wythes in a cavity wall are loadbearing and are built at the same time.

## Moisture resistance

Brick veneer wall assemblies are drainage-type walls. Walls of this type provide good resistance to rain penetration. It is essential to maintain the clear airspace between the brick veneer and the backing to ensure proper drainage. Flashing and weep holes work with the airspace to provide moisture penetration resistance. Brick veneer with wood or metal frame backing is typically built with a 1" minimum airspace.

## Foundations for brick veneer

Brick veneer on a frame backing must transfer the weight of the veneer to the foundation. Typical foundation details for brick veneer are shown in Figure 13-35. It is recommended that the foundation brick ledge supporting the brick veneer be at least equal to the total thickness of the brick veneer wall assembly.

Foundations must extend beneath the frost line as required by the local building code. Design of the foundation should consider differential settlement and the effect of concentrated loads such as those from columns or fire-

places. Brick walls which enclose crawl spaces must have openings to provide adequate ventilation. Openings should be located to achieve cross ventilation.

## Ties

Ties typically used with wood framing are shown in Figure 13-36. There should be one tie for every 2 2/3 sq. ft. of wall area with a maximum spacing of 24" o.c. in either direction. The nail attaching a corrugated tie must be located within 5/8" of the bend in the tie. The best location of the nail is at the bend in the corrugated tie, and the bend should be 90°.

Wire ties must be embedded at least 5/8" into the bed joint from the airspace and must have at least 5/8" cover of mortar to the exposed face. Corrugated ties must penetrate to at least half the veneer thickness and have at least 5/8" cover. Ties should be placed so that the portion within the bed joint is completely surrounded by the mortar.

## Flashing and weep holes

Flashing and weep holes should be located above the grade, but as near as possible to the bottom of the wall, above all openings, and beneath sills. Weep holes must be located in the head joints immediately above all flashing. Clear, open weep holes should be spaced no more than 24" o.c. Weep holes formed with wick materials or with tubes should be spaced at a maximum of 16" o.c. If the veneer continues below the flashing at the base of the wall, the space between the veneer and the backing should be grouted to the height of the flashing. Flashing should be securely fastened to the backing system and extend through the face of the brick veneer. The flashing should be turned up at least 8".

## Lintels, sills, and jambs

Brick veneer backed by wood or metal frame must always be supported by lintels over openings unless the masonry is self-supporting. Loose steel, stone, or precast lintels should bear at least 4" at each jamb. All lintels should have space at the end of the lintel to allow for expansion. The clear span for 1/4" thick steel angles varies between 5' and a maximum of 8', depending on the size of the angle selected. Steel lintels with spans greater than 8' may require lateral bracing for stability. Concrete, cast stone, and stone lintels must be appropriately sized to carry the weight of the veneer. Reinforced brick lintels are also a viable option.

## Expansion joints

To allow for horizontal movement, expansion joints may be required. These joints are needed in brick veneer when there are long walls, walls with returns, or large openings.

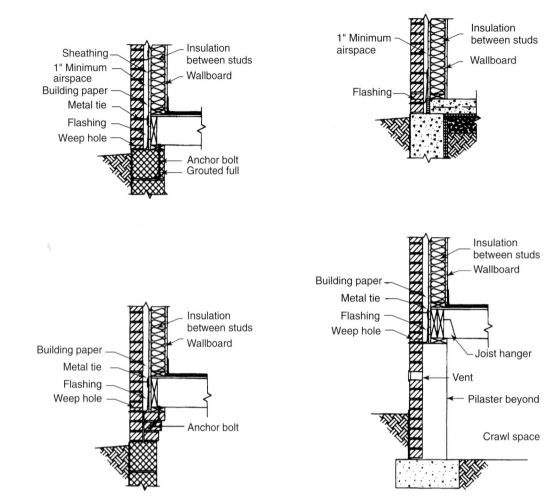

**Figure 13-35.** Brick veneer details. (Brick Institute of America)

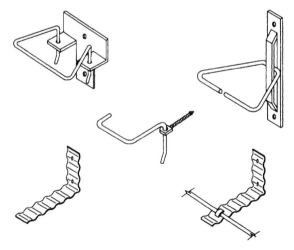

**Figure 13-36.** Typical ties used with brick veneer on frame construction. (Brick Institute of America)

## Composite walls

A composite wall is two wythes bonded together with masonry or wire ties. See Figure 13-37. The two wythes are joined together in a continuous mass using a vertical collar joint. See Figure 13-38. This joint prevents the passage of water through the wall. A **collar joint** is the narrow space between the facing units and the backup units in a wall.

When building an 8" composite wall, the first course of facing may be either headers, as shown in Figure 13-39, or stretchers. It is important that all facing courses be laid in a full mortar bed with the head joints filled completely. Mortar that is extruded (squeezed out) on the backside of the facing units should be cut flush with the trowel before it hardens. See Figure 13-40. Parging the backside of the facing is another method of bonding the wythes across the collar joint. See Figure 13-41. Facing headers are laid every seventh course in an 8" composite wall.

A 12" composite wall is constructed in a manner similar to the 8" composite wall. Figure 13-42 shows the facing header course being laid overlapping the header block. The header block can be laid with the recessed notch up or down, depending on construction requirements. Figure 13-43 shows the seventh course backup wythe being constructed using stretcher block. Header blocks were used for the sixth course bonding.

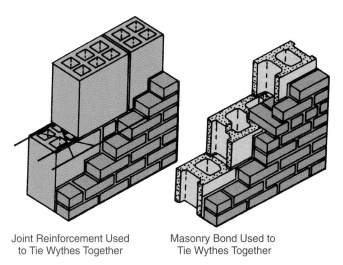

Joint Reinforcement Used
to Tie Wythes Together

Masonry Bond Used to
Tie Wythes Together

**Figure 13-37.** Composite walls can be bonded together with masonry or joint reinforcement.

**Figure 13-39.** Building a composite wall with six courses of brick between each header course. (Portland Cement Association)

**Figure 13-38.** The outer and inner wythes of masonry are bonded together by the mortar in the collar joint between them. (Portland Cement Association)

**Figure 13-40.** Well placed facing units will have extruded mortar on the backside in a composite wall. This must be cut flush with the trowel.

## Resisting moisture condensation

Moisture, formed by the condensation of water vapor, can cause many problems in masonry walls and should be minimized where possible. When a vapor pressure differential exists, water vapor will move independently of air. The vapor movement through common building materials is at a relatively high rate for common pressure differentials. When vapor passes through pores of homogenous walls, which are warm on one side and cold on the other, it may reach its dew point and condense into water within the wall. But if the flow of vapor is impeded by a vapor-resistant material in the wall, the vapor may not reach that point in the wall at which the temperature is low enough to cause condensation.

Condensation problems are most frequent during the heating season when buildings of tight, highly insulated construction have occupancies and/or heating systems which produce humidity. This gain in moisture content of the interior air increases the interior vapor pressure sub-

**Figure 13-41.** Bonding a wall across a collar joint by parging the backside of the facing. (Portland Cement Association)

stantially above that existing in the outdoor atmosphere. This tends to drive vapor outward from the building through any vapor-porous materials that comprise the wall

**Figure 13-42.** Header course being laid in composite wall.

**Figure 13-43.** Stretcher block being placed as the backup wythe. (Portland Cement Association)

assembly. This can be controlled either by the use of a properly placed vapor barrier or by decreasing the vapor pressure differential across the wall section by the use of ventilation.

Many building materials are affected by water. For example, wood expands with increasing moisture content. If conditions of varying humidity occur in different parts of the cross section of a single wood framing member, there will be a tendency to warp. High humidity can also cause the decay of wood. Water promotes the corrosion of metal, and many insulating materials show permanent change over the course of time when in contact with water. The insulating value of most materials is greatly reduced by the presence of free water. Volumetric changes in fired clay

masonry units and concrete masonry units due to gains in moisture content are to be expected and should be given consideration in the design process. Alternate freezing and thawing of masonry units when saturated may lead to eventual deterioration, such as cracking and spalling. If soluble salts are present in or in contact with brick masonry, moisture caused by condensation may contribute to efflorescence.

## Reinforced masonry walls

**Reinforced masonry walls** are walls built with steel reinforcement embedded with the masonry units. See Figure 13-44. The walls are structurally bonded by grout which is poured into the cavity (collar joint) between the wythes of masonry. The grout core seals the space between the wythes of masonry and bonds the reinforcing steel. Full bed joints are used for reinforced walls.

Grouting techniques vary in different areas of the country, but the *Uniform Building Code* specifies that:
"All longitudinal vertical joints shall be grouted and shall be not less than three-fourths inch (3/4") in thickness. In members (walls or columns) of three or more tiers in thickness, interior bricks shall be not less than three-fourths inch (3/4") in thickness. In members of three or more tiers in thickness, interior bricks shall be embedded into the grout so that at least three-fourths inch (3/4") of grout surrounds the sides and end of each unit. One exterior tier may be carried up twelve inches (12") before grouting, but the other exterior tier shall be grouted in lifts not to exceed four inches (4") or one unit, whichever is greater. If the work is stopped for one hour or longer, the horizontal construction joints shall be formed by stopping all tiers at the same elevation and with the grout one inch (1") below the top."

Reinforced masonry walls should be reinforced with an area of steel not less than 0.002 times the cross-sectional area of the wall. Not more than two-thirds of this area

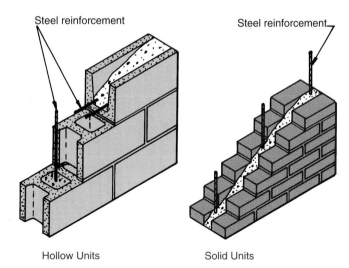

Steel reinforcement                        Steel reinforcement

Hollow Units                        Solid Units

**Figure 13-44.** Reinforced masonry walls of brick and concrete block.

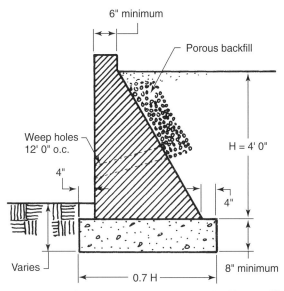

may be used in either direction. Maximum spacing of principal reinforcement should not exceed 48". Horizontal reinforcement should be placed in the top of footings, the bottom and top of wall openings, roof and floor levels, and at the top of parapet walls.

The primary use of steel reinforcement is in vertical members (such as columns and walls), lintels, and bond beams. Typical methods of reinforcing concrete masonry are shown in Figure 13-45.

## Retaining walls

Reinforced masonry is ideal for the construction of retaining walls that must stand lateral earth pressures. By adding reinforcement, the mass can be greatly reduced and the strength maintained.

Two common types of retaining walls are the gravity and cantilever designs. A gravity-type retaining wall depends primarily on its own weight to hold back the earth pressure. See Figure 13-46.

A cantilever-type retaining wall uses the weight of the soil together with its own strength to get the same results. See Figure 13-47. For walls over 3' in height, the cantilever- type retaining wall provides the best solution. The same principles of good construction should be followed in building a retaining wall as in any other reinforced masonry wall.

**Figure 13-46.** Gravity-type retaining wall with specifications. The width of the base must be at least 1/2 to 3/4 of the wall height. This is expressed in the drawings as "0.7H" or "0.7 of the height."

Figure 13-48 provides specifications for building reinforced concrete masonry retaining walls up to 9'-4" in height.

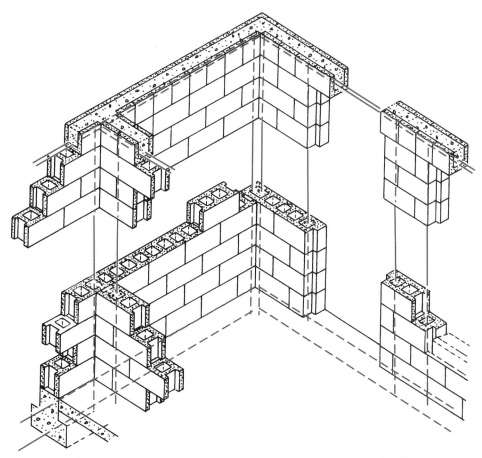

**Figure 13-45.** Typical locations of reinforcement for concrete block masonry construction.

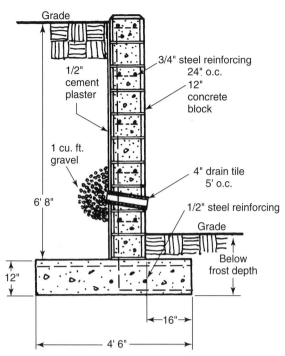

**Figure 13-47.** Cantilever-type retaining wall constructed of reinforced concrete block.

## Grouted masonry walls

*Grouted masonry walls* are very similar to reinforced masonry walls except that they do not contain reinforcements. Grout is added to the cores in loadbearing masonry walls to provide added strength. Other uses include filling bond beams and the collar joint in a two-wythe wall.

Single-wythe walls are constructed using hollow masonry units laid with face-shell mortar bedding. The vertical cores are aligned to form continuous vertical spaces that can receive grout. Two-core block are generally

preferred over three-core block because of the ease of placing grout.

Multi-wythe walls generally have two wythes. The wythes are spaced 1" to 6" apart and can be constructed of hollow or solid units depending on the building requirements. The space between the wythes can be grouted solid. Grouting is usually done as the wall is built.

## Thin brick veneer

*Thin brick veneers* are thin brick units formed from kiln-fired shale and/or clay, and are much like facing bricks, except they are approximately 1/2" to 1" thick. The face sizes are normally the same as conventional bricks and therefore, when in place, give the appearance of a conventional brick masonry wall.

Thin bricks are available in various sizes, colors, and textures. The most commonly found face size is standard modular with nominal dimensions of 2-2/3" × 8". The actual face dimensions vary slightly among manufacturers, but are typically 3/8" to 1/2" less than the nominal dimensions. The economy modular face size, 4" × 12", is popular for use in large buildings because productivity is increased, and the unit's size decreases the number of visible mortar joints. Other sizes, such as Norwegian, 3-inch, non-modular, oversize, etc., may be available. It is advisable to check with individual manufacturers or distributors regarding sizes available in a particular area. Figure 13-49 illustrates the various types of thin brick units.

Thin brick veneer is classified as an adhered veneer. Adhered veneer relies on a bonding agent between the thin brick units and the substrate. Adhered veneer construction is classified as either thin bed set or thick bed set.

### Thin bed set

The thin bed set procedure typically utilizes an epoxy or organic adhesive, and is normally used on interior

| 8" Walls | | | | | |
|---|---|---|---|---|---|
| Height of Wall | Width of Footing | Thickness of Footing | Dist. to Face of Wall | Size & Spacing of Vertical Rods in Wall | Size & Spacing of Horizontal Rods in Footing |
| 3' - 4" | 2' - 4" | 9" | 8" | 3/8" @ 32" | 3/8" @ 27" |
| 4' - 0" | 2' - 9" | 9" | 10" | 1/2" @ 32" | 3/8" @ 27" |
| 4' - 8" | 3' - 3" | 10" | 12" | 5/8" @ 32" | 3/8" @ 27" |
| 5' - 4" | 3' - 8" | 10" | 14" | 1/2" @ 16" | 1/2" @ 30" |
| 6' - 0" | 4' - 2" | 12" | 15" | 3/4" @ 24" | 1/2" @ 25" |
| 12" Walls | | | | | |
| 6' - 8" | 4' - 6" | 12" | 16" | 3/4" @ 24" | 1/2" @ 22" |
| 7' - 4" | 4' - 10" | 12" | 18" | 7/8" @ 32" | 5/8" @ 26" |
| 8' - 0" | 5' - 4" | 12" | 20" | 7/8" @ 24" | 5/8" @ 21" |
| 8' - 8" | 5' - 10" | 14" | 22" | 7/8" @ 16" | 3/4" @ 26" |
| 9' - 4" | 6' - 4" | 14" | 24" | 1" @ 8" | 3/4" @ 21" |

**Figure 13-48.** Specifications for reinforced concrete masonry retaining walls up to 9'-4" high.

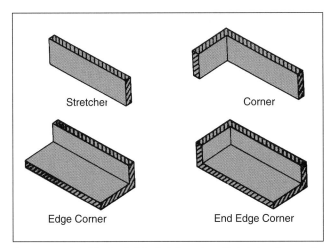

**Figure 13-49.** Typical thin brick units. (Brick Institute of America)

surfaces only. For areas subject to dampness, only clear and dry masonry surfaces or concrete surfaces should be used for backup. For dry locations, the backing material (substrate) may be wood, wallboard, masonry, etc. A cross section depicting a wood frame wall upon which thin brick veneer (thin set procedure) is installed is shown in Figure 13-50.

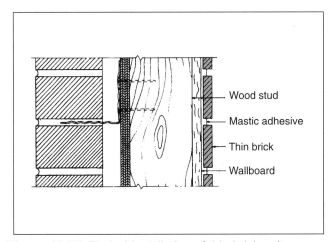

**Figure 13-50.** Typical installation of thin brick units on wallboard. (Brick Institute of America)

## Thick bed set

The thick bed set procedure is used on interior and exterior surfaces. The backing material is either masonry, concrete, steel, or wood stud framing. For applications over steel studs, procedures are similar to those used for concrete or masonry backup. However, wallboard and building felt must be installed over the studs before the lathe and mortar bed are placed. See Figure 13-51.

### Wall openings

Wall openings can extend through the wall as required for a door or window, or they may take the form of

a chase or recess. **Chases** and **recesses** are horizontal or vertical spaces left in a wall for the purpose of containing plumbing, heating ducts, electrical wiring, or other equipment. Figure 13-52 shows several chases and recesses in masonry walls.

Chases and recesses are formed by the mason as the wall is built. Chases are generally located on the inside of the wall and vary in size from 4" to 12" in width. Recesses reduce the wall thickness and strength and are usually limited to one-third the thickness of the wall.

## Lintels

A **lintel** is a structural member placed over an opening in a wall used to support the loads above that opening. Masonry above a wall opening should be supported by a lintel made from a structural steel shape, reinforced concrete or masonry, precast concrete, or by a brick masonry arch.

Steel lintels should be supported on either side of the opening for a distance of at least 4". Reinforced concrete lintels should have a minimum bearing of at least 8" at each end. Longer lintels carrying heavier loads should have greater bearing surfaces at the ends. The lintel should be stiff enough to resist bending in excess of 1/360th of the span. Figure 13-53 shows several types of lintels used in masonry construction. Specifications for one-piece reinforced concrete lintels are given in Figure 13-54. Design data for reinforced concrete split lintels are shown in Figure 13-55. Figure 13-56 gives the size requirements for concrete reinforced lintels with stirrups. These lintels are made for openings supporting wall and floor loads. Lintels made of angle steel should meet the following minimum specifications for 4" masonry veneer:

**Steel Angles to Support 4" Masonry Walls**

| Span | Size of Angle |
|------|---------------|
| 0' – 5' | 3" × 3" × 1/4" |
| 5' – 9' | 3-1/2" × 3-1/2" × 5/16" |
| 9' – 10' | 4" × 4" × 5/16" |
| 10' – 11' | 4" × 4" × 3/8" |
| 11' – 15' | 6" × 4" × 3/8" |
| 15' – 16' | 6" × 4" × 1/2" |

Reinforced masonry lintels (namely brick and tile) are becoming more popular because the steel is completely protected from the elements and the initial cost is less because less steel is required. Figure 13-57 shows several sizes of reinforced brick masonry lintels. Figure 13-58 shows a detail of a brick masonry lintel integrated into the wall.

Slight movement often occurs at the location of lintels. For this reason, control joints are often located at the ends of lintels. A noncorroding metal plate is placed under the ends of lintels where control joints occur. See Figure 13-59. The metal plate permits the lintel to slip and prevent uncontrolled cracking.

A full bed of mortar should be used over the plate to distribute the lintel load uniformly. When the mortar has hardened sufficiently, it should be raked out to a depth of 3/4" and filled with caulking.

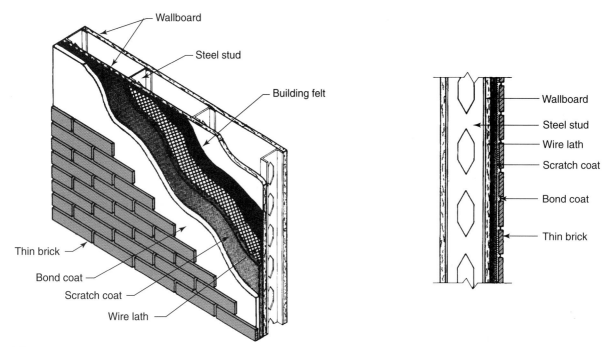

**Figure 13-51.** Typical installation of thin brick units over steel stud framing using the thick set method. (Brick Institute of America)

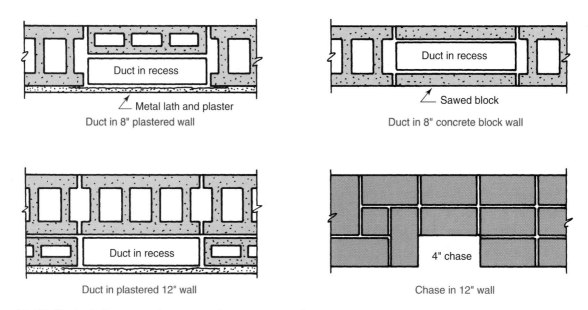

**Figure 13-52.** Typical chases and recesses in masonry walls.

## Arches

Arches have been used for centuries to span openings. Some have been built which span distances of more than 130'. Several standard types are easily recognized in modern construction. Figure 13-60 shows most of the popular types of arches used in contemporary construction. The following descriptive statements are presented to further define the various types of arches:

- **Blind arch.** An arch with the opening filled with masonry.

- **Bull's–eye arch.** An arch with an intrados (the curve that bounds the lower edge of the arch) that is a full circle. Also known as a *circular arch*.
- **Elliptical arch.** An arch with two centers and continually changing radii.
- **Fixed arch.** An arch with a skewback that is fixed in position and inclination. Masonry arches are fixed arches by nature of their construction.
- **Gauged arch.** An arch formed with tapered voussoirs (wedge-shaped pieces) and thin mortar joints.

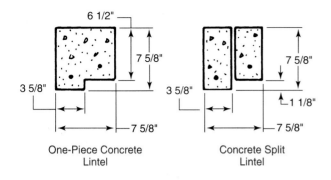

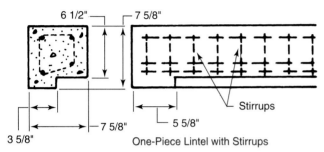

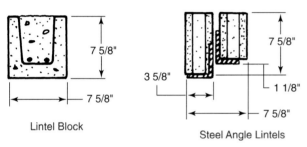

**Figure 13-53.** Various types of lintels used in masonry construction.

| Concrete Reinforced One-Piece Lintels with Wall Load Only | | | | |
|---|---|---|---|---|
| Size of Lintel | | | Bottom Reinforcement | |
| Height in. | Width in. | Clear Span of Lintel ft. | Size Designation of Bars | Size of Bars |
| 5 3/4 | 7 5/8 | Up to 7 | No. 2 | 3/8" round deformed |
| 5 3/4 | 7 5/8 | 7 to 8 | No. 2 | 5/8" round deformed |
| 7 5/8 | 7 5/8 | Up to 8 | No. 2 | 3/8" round deformed |
| 7 5/8 | 7 5/8 | 8 to 9 | No. 2 | 1/2" round deformed |
| 7 5/8 | 7 5/8 | 9 to 10 | No. 2 | 5/8" round deformed |

**Figure 13-54.** Specifications for one-piece reinforced concrete lintels. (Portland Cement Association)

| Concrete Reinforced Split Lintels with Wall Load Only | | | | |
|---|---|---|---|---|
| Size of Lintel | | | Bottom Reinforcement | |
| Height in. | Width in. | Clear Span of Lintel ft. | Size Designation of Bars | Size of Bars |
| 5 3/4 | 3 5/8 | Up to 7 | No. 1 | 3/8" round deformed |
| 5 3/4 | 3 5/8 | 7 to 8 | No. 1 | 5/8" round deformed |
| 7 5/8 | 3 5/8 | Up to 8 | No. 1 | 3/8" round deformed |
| 7 5/8 | 3 5/8 | 8 to 9 | No. 1 | 1/2" round deformed |
| 7 5/8 | 3 5/8 | 9 to 10 | No. 1 | 5/8" round deformed |

**Figure 13-55.** Specifications for reinforced concrete split lintels. (Portland Cement Association)

- *Gothic arch.* An arch with relatively large rise-to-span ratio, with sides that consist of arcs of circles, which have centers that are at the level of the spring line. Also referred to as a *Drop, Equilateral,* or *Lancet arch,* depending upon whether the spacings of the centers are respectively less than, equal to, or more than the clear span.
- *Horseshoe arch.* An arch with an intrados that is greater than a semicircle and less than a full circle. Also known as an *Arabic* or *Moorish arch.*
- *Jack arch.* A flat arch with zero or little rise.
- *Multicentered arch.* An arch with a curve that consists of several arcs of circles that are normally tangent at their intersections.
- *Relieving arch.* An arch built over a lintel, jack arch, or smaller arch to divert loads, thus relieving the lower arch or lintel from excessive loading. Also known as a *Discharging* or *Safety arch.*
- *Segmental arch.* An arch with an intrados that is circular but less than a semicircle. See Figure 13-61
- *Semicircular arch.* An arch with an intrados that is a semicircle (half circle).
- *Slanted arch.* A flat arch that is constructed with a keystone with sides that are sloped at the same angle as the skewback and uniform width brick and mortar joints.
- *Triangular arch.* An arch formed by two straight, inclined sides.
- *Tudor arch.* A pointed, four-centered arch of medium rise-to-span ratio with four centers that are all beneath the extrados (the curve that bounds the upper edge of the arch) of the arch.
- *Venetian arch.* An arch formed by a combination of jack arch at the ends and semicircular arch at the middle. Also known as a *Queen Anne arch.*

| Concrete Reinforced Lintels with Stirrups for Wall and Floor Loads | | | | | |
|---|---|---|---|---|---|
| Size of Lintel | | | Reinforcement | | Web Reinforcement No. 6 Gauge Wire Stirrups. Spacings from End of Lintel–Both Ends the Same |
| Height in. | Width in. | Clear Span of Lintel ft. | Top | Bottom | |
| 7 5/8 | 7 5/8 | 3 | None | 2 — 1/2" round | No stirrups required |
| 7 5/8 | 7 5/8 | 4 | None | 2 — 3/4" round | 3 stirrups, SP. :2, 3, 3 in. |
| 7 5/8 | 7 5/8 | 5 | 2 — 3/8" round | 2 — 7/8" round | 5 stirrups, SP. :2, 3, 3, 3, 3 in. |
| 7 5/8 | 7 5/8 | 6 | 2 — 1/2" round | 2 — 7/8" round | 6 stirrups, SP. :2, 3, 3, 3, 3, 3 in. |
| 7 5/8 | 7 5/8 | 7 | 2 — 1" round | 2 — 1" round | 9 stirrups, SP. :2, 3, 3, 3, 3, 3, 3, 3, 3 in. |

**Figure 13-56.** Specifications for reinforced concrete lintels with stirrups.

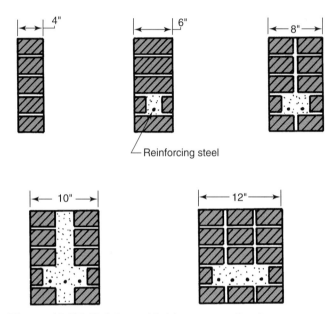

**Figure 13-57.** Reinforced brick masonry lintels.

**Figure 13-59.** A noncorroding metal plate is placed under the ends of lintels where control joints occur. A full bed of mortar will be placed over the plate. (Portland Cement Association)

An arch is normally classified by the curve of its intrados and by its function, shape, or architectural style. Arches are also classified as major and minor arches. Minor arches have spans that do not exceed 6' with maximum rise-to-span ratio of 0.15. Jack, segmental, semicircular, and multicentered arches are the most common types used for building arches. See Figure 13-62. For very long spans and for bridges, semicircular arches are often used because of their structural efficiency.

Mainly due to their variety of components and elements, arches have developed their own set of terminology. The following terms are used in the discussion of arches:

- **Abutment.** The masonry or combination of masonry and other structural members that support one end of the arch at the skewback.
- **Arch.** A form of construction in which masonry units span an opening by transferring vertical loads

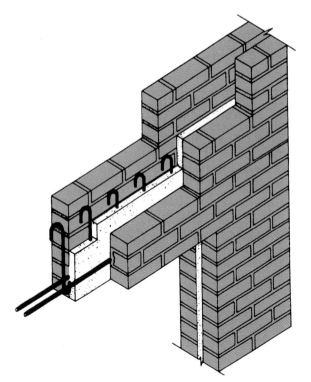

**Figure 13-58.** Detail of a brick masonry lintel with stirrups. (Brick Institute of America)

A. Jack  B. Segmental  C. Semicircular

D. Bull's–eye  E. Horseshoe  F. Multicentered

G. Venetian  H. Tudor

I. Triangular  J. Gothic

**Figure 13-60.** Arches used in contemporary construction. (Brick Institute of America)

**Figure 13-61.** A typical segmental arch in modern construction.

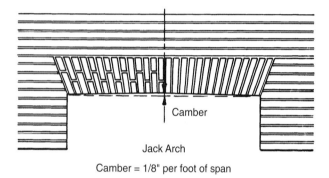

Camber

Jack Arch

Camber = 1/8" per foot of span

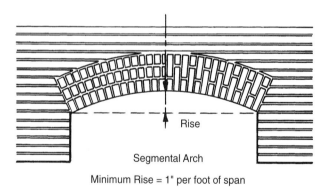

Rise

Segmental Arch

Minimum Rise = 1" per foot of span

**Figure 13-62.** The jack and segmental arches are frequently used for short spans.

laterally to adjacent voussoirs and, thus, to the abutments.

- **Camber.** The relatively small rise of a jack arch.
- **Centering.** Temporary shoring used to support an arch until the arch becomes self-supporting.
- **Crown.** The apex of the arch's extrados. In symmetrical arches, the crown is at the midspan.
- **Depth.** The dimension of the arch at the skewback which is perpendicular to the arch axis, except that the depth of a jack arch is taken to be the vertical dimension of the arch at the springing.

- **Extrados.** The curve that bounds the upper edge of the arch.
- **Intrados.** The curve that bounds the lower edge of the arch. The distinction between soffit and intrados is that the intrados is a line, while the soffit is a surface.
- **Keystone.** The voussoir located at the crown of the arch. Also called the *key*.
- **Label course.** A ring of projecting brickwork that forms the extrados of the arch.
- **Rise.** The maximum height of the arch soffit above the level of its spring line.
- **Skewback.** The surface on which the arch joins the supporting abutment.
- **Skewback angle.** The angle from horizontal made by the skewback.
- **Soffit.** The lower surface of an arch or vault.
- **Span.** The horizontal clear dimension between abutments.
- **Spandrel.** The masonry contained between a horizontal line drawn through the crown and a vertical line drawn through the uppermost point of the skewback.
- **Springing.** The point where the skewback intersects the intrados.
- **Springer.** The first voussoir from a skewback.
- **Spring line.** A horizontal line that intersects the springing.
- **Voussoir.** One masonry unit of an arch.

## Function of arches

Structural efficiency is attributed to the curvature of the arch, which transfers vertical loads laterally along the arch to the abutments at each end. The curvature of the arch and the restraint of the arch by the abutments cause a combination of flexural stress and axial compression. The arch depth, rise, and configuration can be manipulated to keep stresses primarily compressive. Brick masonry is very strong in compression, so brick masonry arches can support considerable loads.

Historically, arches have been constructed with unreinforced masonry. Most brick masonry arches continue to be built with unreinforced masonry. Very long span arches and arches with a small rise may require steel reinforcement to resist tensile stresses. Also, reduction in abutment size and arch thickness for economy may require incorporation of reinforcement for adequate load resistance.

If an unreinforced or reinforced brick masonry arch is not structurally adequate, the arch will require support. Typically, this support is provided by a steel angle. The steel angle is bent to the curvature of the intrados of the arch. Curved sections of steel angle are welded to horizontal steel angles to form a continuous support. See Figure 13-63.

Arches are constructed with the aid of temporary shoring or centering. The centering carries the load of the arch until it has developed sufficient strength to support itself and the imposed load. The centering should not be

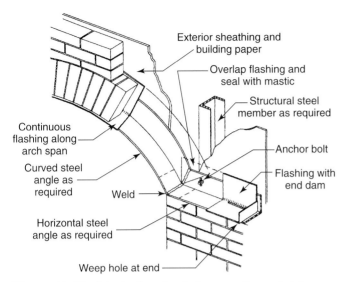

**Figure 13-63.** A typical method of supporting a brick masonry arch. (Brick Institute of America)

removed for at least seven days after the arch has been completed.

## Arch design

Arches can be built using a variety of arch depths, brick sizes and shapes, and bonding patterns. Usually, an odd number of units are used for aesthetic purposes. Arch voussoirs are typically laid in radial orientation and are most often of similar size and color to the surrounding brickwork. However, the arch can be formed with brick which are thinner or wider than the surrounding brickwork and of a different color for variation. Another variation is to project or recess rings of multiple-ring arches to provide shadow lines or a label course.

Selection of tapered or rectangular brick can be determined by the arch type, arch dimensions, and the appearance desired. Some arch types require more unique shapes and sizes of brick if uniform mortar joint thickness is desired. For example, the brick in a traditional jack arch or elliptical arch are different sizes and shapes from the abutment to the keystone. Conversely, the voussoirs of a semicircular arch are all the same size and shape.

The arch span should also be considered when selecting the arch brick. For short arch spans, use of tapered brick is recommended to avoid excessively wide mortar joints at the extrados. Larger span arches require less taper of the voussoirs and, consequently, can be formed with rectangular brick and tapered mortar joints. The thickness of mortar joints between arch brick should be a maximum of 3/4" and a minimum of 1/8".

The arch depth will depend upon the size and orientation of the brick used to form the arch. Typically, the arch depth is a multiple of the brick's width. For structural arches, a minimum arch depth is determined from the structural requirements. If the arch is supported by a lintel, any arch depth can be used.

The depth of the arch should also be based on the scale of the arch in relation to the scale of the building and surrounding brickwork. To provide proper visual balance and scale, the arch depth should increase with increasing arch span.

The following rules-of-thumb will help provide an arch with proper scale. For segmental and semicircular arches, the arch depth should equal or exceed 1" for every foot of arch span or 4", whichever is greater. For jack arches, the arch depth should equal or exceed 4" plus 1" for every foot of arch span or 8", whichever is greater. For example, the minimum arch depth for an 8' span should be 8" for segmental arches and 12" for jack arches. Typically, the depth of a jack arch will equal the height of 3, 4, or 5 courses of the surrounding brickwork, depending upon the course height. The keystone is either a single brick, multiple brick, stone, precast concrete, or terra cotta.

## Arch construction

Two general methods are used in the construction of brick masonry arches. One involves the use of special shapes that provide for uniform joint thickness. The second method uses regular units of uniform thickness while the joint thickness is varied to obtain the desired curvature. The arch dimensions and appearance desired will determine the method to be used.

It is important that all mortar joints be completely filled because the arch is a structural member. The crown of the arch is generally laid in a soldier bond or rowlock header bond and it is frequently difficult to completely fill the joints. See Figure 13-64.

**Jack arch construction.** The jack arch is relatively weak and should be supported by steel if the opening is over 2' wide. The steel must be bent to the camber which is generally 1/8" per foot of span. For best appearance, each joint should be the same width as the entire length of the joint. Special bricks are required or regular bricks must be cut to have a perfect job. The bricks should be shaped so that the end joints are horizontal rather than

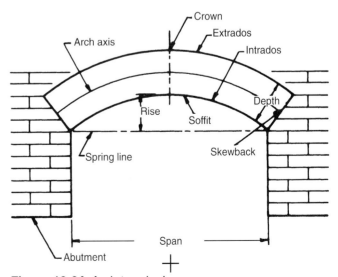

**Figure 13-64.** Arch terminology.

perpendicular to the radius of the arch. See Figure 13-65. When constructing any arch, the bricks should be laid out to determine the number required, inclination, and spacing.

**Segmental arch construction.** The rise and span are the most important features to be considered when constructing a segmental arch. The length of the arch must be found in order to determine the number of courses and size of bricks needed. The number of courses of bricks in an arch is determined by finding the length of the extrados. Remember, the extrados is the upper portion of the arch that is between the crown and the skewback. The size of the brick may be determined by finding the length of the intrados. Remember, the intrados is the lower surface of the arch. The rise of a segmental arch should be 1/6, 1/8, 1/10, or 1/12 of the span. Figure 13-66 shows the key dimensions of one segmental arch example.

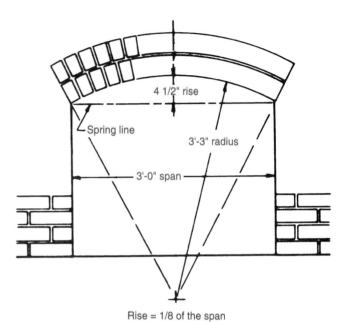

**Figure 13-66.** Construction layout for segmental arch.

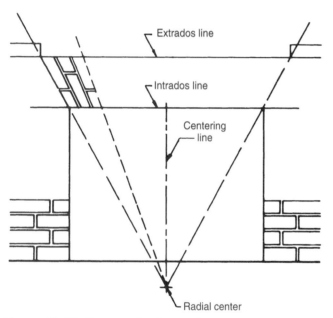

**Figure 13-65.** Construction layout for a jack arch.

## Window and door details

There are many types of windows and doors which could be used in a structure. Details of the windows and doors being used in a given situation should be provided so the mason can plan ahead. Details shown in Figure 13-67 and Figure 13-68 are typical and illustrate the relationship of the head, jambs, and sill to the masonry wall. The **head** is the top of a window or door. The **jambs** are the vertical sides of a door or window. The **sill** is the bottom piece connecting the jambs.

## Sills

The primary function of a sill is to channel water away from the building. The sill either consists of a single unit or multiple units. It can be built in place or prefabricated. It can be built of various materials. See Figure 13-69. Single unit sills are either slip sills or lug sills. See Figure 13-70. Slip sills are the same length as the window opening, but lug sills extend into the masonry on either side of the opening. The slip sill can be left out when the opening is being laid and set at a later time. The lug sill, however, must be set when the masonry is up to the bottom of the window or door opening. Sills are usually made of cut stone, brick, concrete, or metal. See Figure 13-71.

## Movement joints

Because all materials in a building experience changes in volume, a system of movement joints is necessary to allow these movements to occur. Failure to permit these movements can result in cracks in masonry construction. The various types of movement joints in buildings include expansion joints, control joints, building expansion joints, and construction joints. Each type of movement joint is designed to perform a specific task, and they should not be used interchangeably.

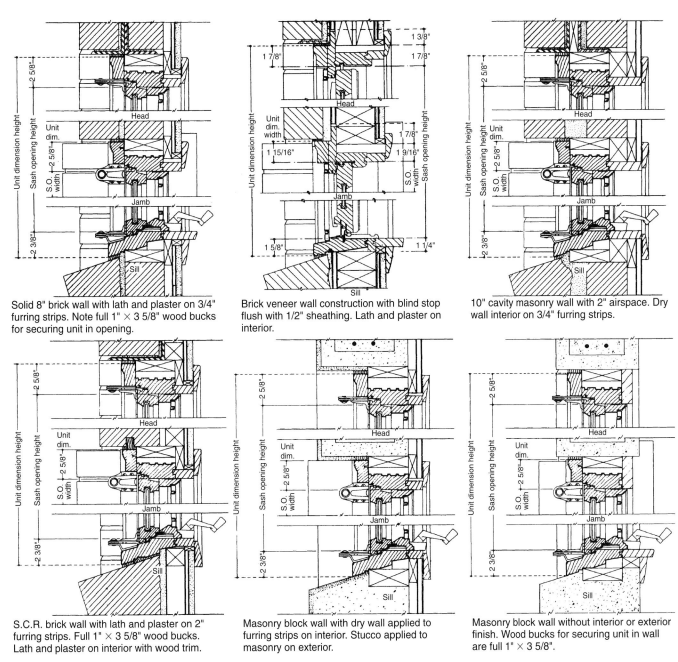

Solid 8" brick wall with lath and plaster on 3/4" furring strips. Note full 1" × 3 5/8" wood bucks for securing unit in opening.

Brick veneer wall construction with blind stop flush with 1/2" sheathing. Lath and plaster on interior.

10" cavity masonry wall with 2" airspace. Dry wall interior on 3/4" furring strips.

S.C.R. brick wall with lath and plaster on 2" furring strips. Full 1" × 3 5/8" wood bucks. Lath and plaster on interior with wood trim.

Masonry block wall with dry wall applied to furring strips on interior. Stucco applied to masonry on exterior.

Masonry block wall without interior or exterior finish. Wood bucks for securing unit in wall are full 1" × 3 5/8".

**Figure 13-67.** Window details for masonry construction. (Anderson Corporation)

## Expansion joints

An *expansion joint* is used to separate brick masonry into segments to prevent cracking due to changes in temperature, moisture expansion, elastic deformation due to loads, and creep. Expansion joints are either horizontal or vertical. The joints are formed of highly *elastic* materials placed in a continuous, unobstructed opening, through each wythe. This allows the joints to close as a result of an increase in size of the assemblage. Expansion joints must be located so that the structural integrity of the masonry is not compromised.

Although the primary purpose of expansion joints is to accommodate movement, the joint must also resist water penetration and air infiltration. The various ways of forming vertical expansion joints is shown in Figure 13-72. Expansion joints should be formed as the wall is built.

## Control joint

A *control joint* is used in concrete or concrete masonry to create a plane of weakness that is used in conjunction with reinforcement or joint reinforcement. It controls the location of cracks due to volume changes resulting from shrinkage and creep. A control joint is usually a vertical opening through the concrete masonry wythe and may be formed of *inelastic* materials. A control joint will open rather than close. Control joints must be located so that the structural integrity of the concrete masonry is not affected.

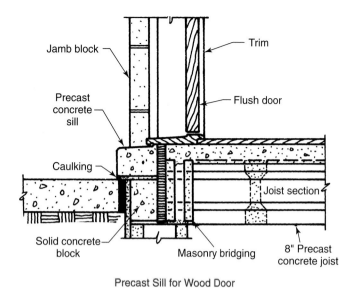

Precast Sill for Wood Door

Sill as Part of Basement Floor

**Figure 13-68.** Typical door sill construction in masonry walls.

**Figure 13-69.** Stone slip sill.

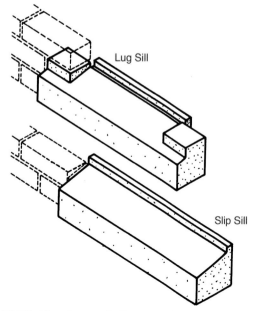

**Figure 13-70.** How lug and slip sills of stone or cast concrete fit into a brick wall.

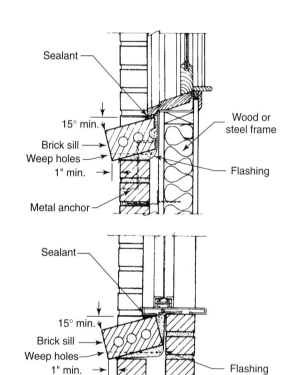

**Figure 13-71.** Top —Brick sill in veneer construction. Bottom—Brick sill in cavity wall construction. (Brick Institute of America)

## Building expansion joint

A *building expansion (isolation) joint* is used to separate a building into discrete sections so that stresses developed in one section will not affect the integrity of the entire structure. The isolation joint is a through-the-building joint.

## Construction joint

A *construction joint (cold joint)* is used primarily in concrete construction where construction work is interrupted. Construction joints are located where they will least impair the strength of the structure.

## Bond breaks

Concrete and concrete masonry have moisture and thermal movements which are considerably different than those of masonry. Also, floor slabs and foundations are usually under different states of stress due to loading than are walls. Therefore, it is important to separate these elements by bond breaks such as building paper or flashing. Bond breaks are used between foundations and walls, between floor slabs and walls, and between concrete and clay masonry. As a result, each element will be able to move somewhat independently while still providing the necessary support. Typical methods of breaking the bond between walls and slabs, and walls and foundations are shown in Figure 13-73.

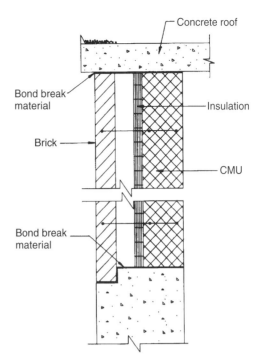

**Figure 13-73.** Typical method of forming bond breaks in a loadbearing cavity wall. (Brick Institute of America)

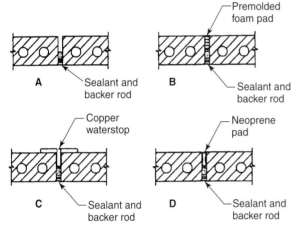

**Figure 13-72.** Methods of forming vertical expansion joints in a brick wall. (Brick Institute of America)

## *Brick masonry soffits*

The principal function of a brick masonry soffit is to enclose the building and provide a pleasing appearance. Large soffits that are constructed in place often require expensive forming and shoring. But, if the project involves only a small soffit area, using a brick masonry soffit may be the most efficient method.

Brick masonry soffits are usually reinforced and grouted in some fashion whether they are built in place or prefabricated. Each soffit involves unique problems with respect to configuration, support available from the structure, space restrictions, and construction sequencing. When a soffit is constructed in place, it frequently requires a complicated system of centering and falsework that must be left in place for a number of days. The responsibility for developing the appropriate details that allow the elements of a soffit to perform their primary functions as well as possible, rests with the designer.

# Floors, Pavements, and Steps

Because of color, pattern, and high resistance to abrasion, brick, tile, and stone are very desirable surfaces for roadways, walks, patios, and interior floors. These materials can be used over the ground, over a floor, or they can be incorporated within a structural floor system.

Several factors affect the selection of bricks and jointing materials. They include traffic, exposure to moisture and freezing, and appearance desired. Hard-burned bricks are very resistant to wear and a Grade SW will prove satisfactory for residential and nonindustrial floors. For exterior pavements, brick's resistance to freezing and thawing damage is just as important as its resistance to abrasion. Bricks having an average saturation coefficient of 0.78 or less are recommended for severe weathering conditions. Refer to ASTM Specifications C-216 and C-62, Grade SW.

Finished appearance of the surface depends largely on the color, size of units, and pattern in which they are laid. When bricks are laid flat with the largest plane surface horizontal, noncored bricks are generally used. Cored bricks can be used if they are placed on edge. Figure 13-74 and Figure 13-75 show several pattern bonds that can be used for floors and pavements.

Interior masonry floors are usually laid over concrete slabs. See Figure 13-76. The primary elements of an unreinforced masonry floor are:

1. The base, that provides the support
2. The cushion, an intermediate layer that aids in leveling

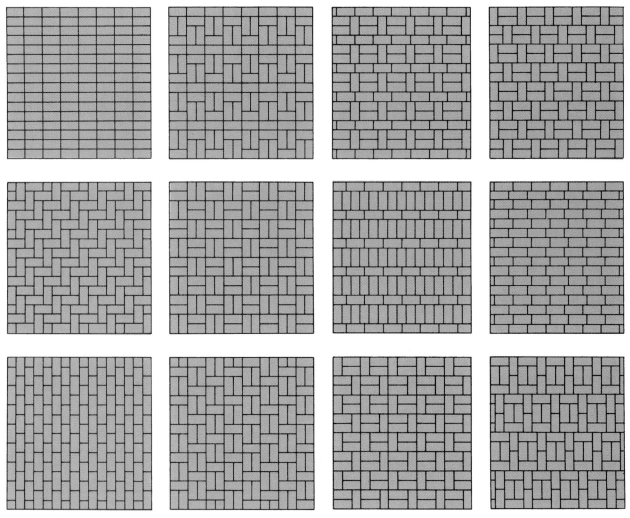

**Figure 13-74.** Brick pattern bonds commonly used for floors and pavements.

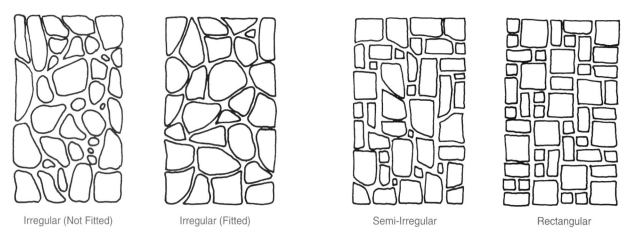

Irregular (Not Fitted)       Irregular (Fitted)       Semi-Irregular       Rectangular

**Figure 13-75.** Stone pattern bonds frequently used for floors and pavements.

3. The wearing surface of brick, tile, or stone
4. The joints

Floors or pavements can be laid without mortar joints, but most interior applications have mortar joints. The procedure for laying a masonry floor on a slab is simple. First, decide on a pattern and arrange the materials for easy access. Form a cushion of mortar for each unit as it is laid and leveled. Strike the joints and brush away excess mortar with burlap rags as it begins to harden. The joints can be tooled for maximum hardness when the mortar has sufficiently hardened.

Exterior pavings for walks, patios, and drives are similar to interior floors. However, more care must be taken in selecting materials which will withstand the weather. See Figure 13-77. Also, drainage must be considered. Walks should be sloped to one or both sides to shed surface water. See Figure 13-78. Slope patios away from buildings, retaining walls, or other obstructions that will hold water on the surface. The slope is generally 1/8" to 1/4" per foot for walks and patios.

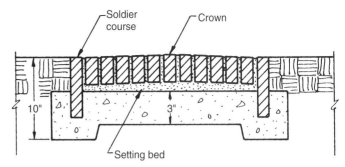

**Figure 13-78.** Brick sidewalk on a concrete foundation, sloped to both sides.

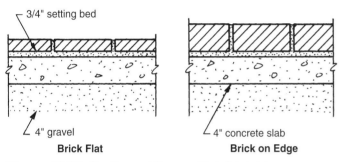

**Figure 13-76.** Sections of floors with brick laid over a concrete slab.

Below-surface drainage can also be required for large paved areas or locations with a high water table. A layer of gravel under the slab, or masonry units if no slab is used, will usually provide the necessary drainage and stop the upward capillary flow of moisture. An expansion joint should be installed between the slab and building walls to provide for expansion.

Pavings tend to spread or shift at the edges under normal traffic and weather conditions. Therefore, an edging is desirable. Figure 13-79 shows a soldier course of bricks used as an edging. The edging is generally placed first and can be used as a guide for elevation and slope.

Several companies produce *paver bricks* that are 1-1/4" × 3 5/8" × 7 5/8". They are approximately 1" thinner than regular brick and are designed especially for paving.

## Steps

Steps made of brick, block, or stone offer a practical solution to a sloping terrain. They also give a particular appearance in an interior setting. A good approach to building steps or stairs with masonry units is to pour a concrete foundation for the structure and cap the treads and risers with the material selected. Figure 13-80 shows a typical set of steps capped with brick.

In designing the steps, use care to ensure that each tread is the same size and each riser is identical. Any change in height of riser or depth of tread will invite an

**Figure 13-77.** Left—Cobblestone paving on a rigid base will perform well as a driveway. (Stone Products Corporation) Right —Clay pavers in a variety of colors. (Glen-Gery Corporation)

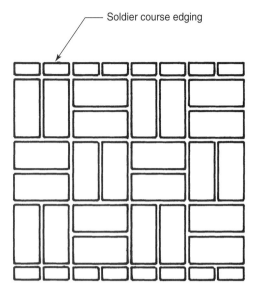

Soldier course edging

**Figure 13-79.** This sidewalk uses a soldier course of brick as an edging.

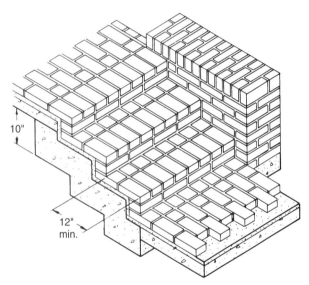

10"

12" min.

**Figure 13-80.** Typical brick steps laid over a concrete foundation.

accident. Treads for all outside steps are usually 12". A riser height equal to two regular brick plus mortar joints is an appropriate height. Treads should be sloped about 1/4" to the front edge to aid in water drainage.

# Fireplace and Chimney Construction

The design and construction of fireplaces and chimneys has a long history. Much data has been collected regarding the most efficient designs for these structures. It is wise to follow recommended practices and procedures if the finished product is to perform satisfactorily.

## *Design and construction*

There are several basic designs of fireplaces which are currently in use. They include the single face, the two face adjacent, the two face opposite, and the three face. See Figure 13-81.

Other fireplaces with exotic shapes and designs are also possible, but they require special engineering to function properly. Because of this, these fireplaces are not covered in this text. All fireplaces, regardless of their design, have basically the same component parts. Figure 13-82 illustrates these components.

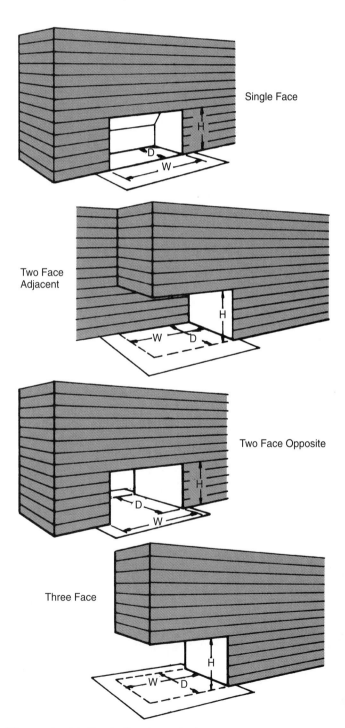

Single Face

Two Face Adjacent

Two Face Opposite

Three Face

**Figure 13-81.** Basic fireplace designs.

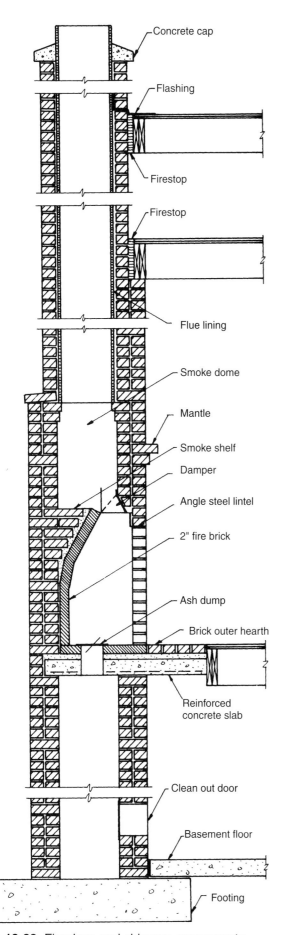

**Figure 13-82.** Fireplace and chimney components.

Concrete cap
Flashing
Firestop
Firestop
Flue lining
Smoke dome
Mantle
Smoke shelf
Damper
Angle steel lintel
2" fire brick
Ash dump
Brick outer hearth
Reinforced concrete slab
Clean out door
Basement floor
Footing

## Sizes of fireplaces

The size of a fireplace is important from an aesthetic (how it looks) as well as functional (how it works) point of view. If it is too small, it may work well, but not provide enough heat. If too large, it would be too hot for the room. Also, it would require a large chimney and an abnormal amount of air to supply its needs for combustion (burning). Research has shown that a room with 300 sq. ft. of floor area can be adequately served by a fireplace with an opening 30" to 36" wide.

Fireplace opening sizes, hearth dimensions, and flue sizes are shown in Figure 13-83. Fireplace openings are seldom greater than 36". The dimensions shown in Figure 13-83 can be changed slightly to meet brick courses and joints if necessary, but should not be changed significantly.

| Fireplace Dimensions | | | |
|---|---|---|---|
| Fireplace Type | Opening Height H in in. | Hearth Size W by D, in. | Modular Flue Size, in. |
| Single face | 29 | 30 by 16 | 12 by 12 |
| | 29 | 36 by 16 | 12 by 12 |
| | 29 | 40 by 16 | 12 by 16 |
| | 32 | 48 by 18 | 16 by 16 |
| Two face-adjacent | 26 | 32 by 16 | 12 by 16 |
| | 29 | 40 by 16 | 16 by 16 |
| | 29 | 48 by 20 | 16 by 16 |
| Two face-opposite | 29 | 32 by 28 | 16 by 16 |
| | 29 | 36 by 28 | 16 by 20 |
| | 29 | 40 by 28 | 16 by 20 |
| Three face | 27 | 36 by 32 | 20 by 20 |
| | 27 | 36 by 36 | 20 by 20 |
| | 27 | 44 by 40 | 20 by 20 |

**Figure 13-83.** Specifications for fireplace opening height, hearth size, and flue size.

## Combustion chamber

The shape and depth of the combustion chamber affect both the draft and the amount of heat radiated into the room. See Figure 13-83. The slope of the back forces the flame forward and forces the gases with increasing velocity into the throat. The combustion chamber should be lined with fire bricks that are laid with thin joints of fire clay mortar. The back and end walls of an average size fireplace are generally 8" thick. Large sizes will have thicker walls to support the load above.

## Throat

The throat is a critical spot in the fireplace and should not be less than 6" and preferably 8" above the highest point of the fireplace opening. The sloping back is the same height and supports the back of the damper.

The metal damper is placed in the throat and extends the full width of the fireplace opening. The best designs open upward toward the back and form a barrier to stop downdrafts. Typical damper sizes are shown in Figure 13-84.

| Steel Dampers | | | | | |
|---|---|---|---|---|---|
| Width of Fireplace in Inches | Damper Dimensions in Inches | | | | |
| | A | B | C | D | E |
| 24 to 26 | 28 1/4 | 26 3/4 | 13 | 24 | 9 1/2 |
| 27 to 30 | 32 1/4 | 30 3/4 | 13 | 28 | 9 1/2 |
| 31 to 34 | 36 1/4 | 34 3/4 | 13 | 32 | 9 1/2 |
| 35 to 38 | 40 1/4 | 38 3/4 | 13 | 36 | 9 1/2 |
| 39 to 42 | 44 1/4 | 42 3/4 | 13 | 40 | 9 1/2 |
| 43 to 46 | 48 1/4 | 46 3/4 | 13 | 44 | 9 1/2 |
| 47 to 50 | 52 1/4 | 50 3/4 | 13 | 48 | 9 1/2 |
| 51 to 54 | 56 1/4 | 54 3/4 | 13 | 52 | 9 1/2 |
| 57 to 60 | 62 1/2 | 60 3/4 | 13 | 58 | 9 1/2 |

| Cast Iron Dampers | | | | | |
|---|---|---|---|---|---|
| Width of Fireplace in Inches | Damper Dimensions in Inches | | | | |
| | A | B | C | D | E |
| 24 to 26 | 28 | 21 | 13 1/2 | 24 | 10 |
| 27 to 30 | 34 | 26 3/4 | 13 1/2 | 30 | 10 |
| 31 to 34 | 37 | 29 3/4 | 13 1/2 | 33 | 10 |
| 35 to 38 | 40 | 32 3/4 | 13 1/2 | 36 | 10 |
| 39 to 42 | 46 | 38 3/4 | 13 1/2 | 48 | 10 |
| 43 to 46 | 52 | 44 3/4 | 13 1/2 | 48 | 10 |
| 47 to 50 | 57 1/2 | 50 1/2 | 13 1/2 | 54 | 10 |
| 51 to 54 | 64 | 56 1/2 | 14 1/2 | 60 | 11 1/2 |
| 57 to 60 | 76 | 58 | 14 1/2 | 72 | 11 1/2 |

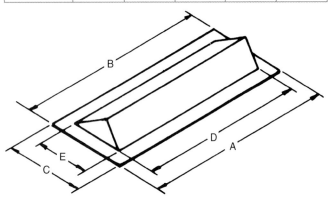

**Figure 13-84.** Steel and cast iron damper sizes for fireplace construction. (Donley Brothers Company)

## Smoke shelf and chamber

The smoke shelf is located in the throat and should be directly under the bottom of the flue. It extends the full width of the throat and is constructed horizontally. The smoke chamber is the space above the smoke shelf. The back and front walls of the chamber are vertical with the side walls sloping to the bottom of the flue lining.

Metal lining plates can be purchased for the smoke chamber that give it the proper form with smooth surfaces. This simplifies the bricklaying.

## Fireplace flue

Each fireplace should have its own clay flue that is free of openings and connections. The first section of lining must start at the center line of the fireplace opening. If the lining is offset to one side, the draft will not be uniform over the full width of the fireplace. The lining should be sup-

ported on at least three sides by a ledge of projecting bricks that is flush with the inside of the lining. Flue lining sizes are given in Figure 13-83.

## Hearth

The modern hearth is usually constructed by building a reinforced brick masonry cantilevered slab shown in Figure 13-82, or a reinforced concrete slab. The simplest method of constructing this slab is to cover the entire area of the chimney base the width of the hearth. Openings for the ash chute and basement flue (if required) are formed in the hearth.

## Chimney construction

Masonry chimneys should be supported on foundations of masonry or reinforced concrete. The foundation wall should be a minimum of 8" thick. The footing for a chimney is generally 12" thick and extends 12" beyond the outside dimensions of the foundation. Reinforcing in the footing is recommended.

The chimney should not change in size or shape as it passes through the roof. Changes must not be made within 6" above or below the roof joists or rafters.

Chimney walls cannot be less than 4" thick for residential structures and lined with approved fire clay fire liners not less than 5/8" thick. The flue liners should be installed ahead of construction of the chimney as it is laid up. The joints must be close fitting and smooth on the inside. In masonry chimneys with walls less than 8" thick, liners are separate from the chimney wall. The space between the liner and masonry is not filled. Liners should extend from the throat of the fireplace to the top of the chimney as nearly straight up as possible.

When a chimney has two flues not separated by masonry, the joints of adjacent flue linings should be staggered at least 7". If more than two flues are located in the same chimney, masonry wythes at least 4" wide should be built between them so that no more than two are grouped together. See Figure 13-85.

The tops of chimneys above the roof must meet certain specifications. The flue lining should project 4" above the chimney cap. The cap should be finished with a straight or concave slope to direct air current upward and drain water from the top of the chimney. The top of the flue liner should extend a minimum of 3' above the highest point where the chimney passes through the roof and at least 2' above any part of the building within 10' of the chimney. See Figure 13-86.

At the intersection of the chimney and roof, the connection should be weathertight. Flashing and counter flashing should be used. See Figure 13-87. Copper or other rust-resisting metal can be used.

## *Prefabricated steel heat circulating fireplaces*

Prefabricated steel heat circulating fireplaces are becoming very popular in residential construction. They

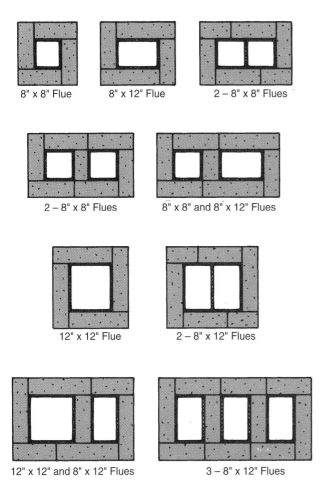

8" x 8" Flue     8" x 12" Flue     2 – 8" x 8" Flues

2 – 8" x 8" Flues     8" x 8" and 8" x 12" Flues

12" x 12" Flue     2 – 8" x 12" Flues

12" x 12" and 8" x 12" Flues     3 – 8" x 12" Flues

**Figure 13-85.** Masonry chimney section details.

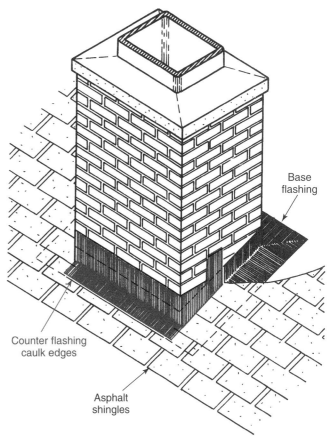

**Figure 13-87.** Recommended flashing for chimneys at the roof intersection.

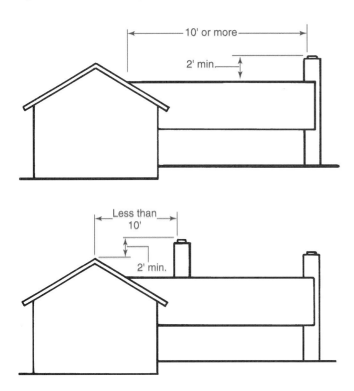

**Figure 13-86.** Minimum recommended chimney heights above the roof.

include not only the firebox and heating chamber, but also the throat, damper, smoke shelf, and smoke chamber. See Figure 13-88. These units are very efficient because the sides and back are double-walled where air is heated and returned through vents. Figure 13-89 shows another model. Design specifications are given in Figure 13-90.

Fireplace inserts are popular because some of them can burn corn, rye, wheat, and wood pellets. See Figure 13-91. Venting can be through a typical chimney or through the wall.

## Stone Quoins

Quoins are large squared stones used at corners and around openings of buildings for ornamental purposes. See Figure 13-92. The height of each stone should be 3, 5, 7, or any odd number of brick courses. This permits the brick above and below the stone to have a full lap in bond. The length of each stone should be equal to one or more of the masonry units used in the wall.

## Garden Walls

There are many types of garden walls (free standing structures) and new variations are constantly being

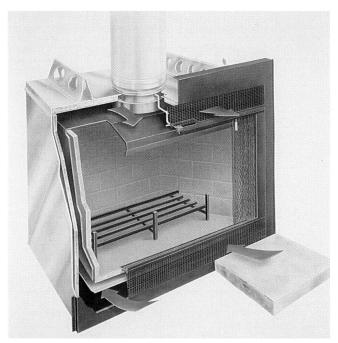

**Figure 13-88.** Prefabricated steel heat circulating fireplace with complete firebox, damper, and smoke shelf. (Heatilator, Inc.)

**Figure 13-89.** A prefabricated steel heat circulating fireplace. (Vega Inc.)

**Prefabricated Steel Heat Circulating Fireplace Sizes (in inches)**

| Overall Dimensions | | | Finished Opening | | Flue size |
| Width | Height | Depth | Width | Height | (Modular) |
|---|---|---|---|---|---|
| 34 3/4 | 44 | 18 | 28 | 22 | 8 x 12 |
| 38 1/2 | 48 | 19 | 32 | 24 | 8 x 12 |
| 42 1/2 | 51 | 20 | 36 | 25 | 12 x 16 |
| 47 3/4 | 55 | 21 | 40 | 27 | 12 x 16 |
| 55 | 60 | 22 | 46 | 29 | 12 x 16 |
| 63 | 65 | 23 | 54 | 31 | 16 x 16 |

**Figure 13-90.** Design specifications for prefabricated fireplace. (The Majectic Co.,Inc.)

**Figure 13-91.** Example of a clean burning fireplace insert. (Eneco Corporation)

designed. Perforated walls can be used with any of the typical types. Perforated walls are sometimes used rather than solid walls because they admit light, breezes, and do not block all vision. Figure 13-93 shows a typical perforated brick wall.

## Straight walls

This form of garden wall depends on the texture and color of the masonry for its character. The thickness must be sufficient to provide lateral stability against wind and impact loads.

It is recommended that for 10 pounds per square foot (psf) wind pressure, the height above grade not exceed 3/4 of the wall thickness squared. For example, an 8" thick wall

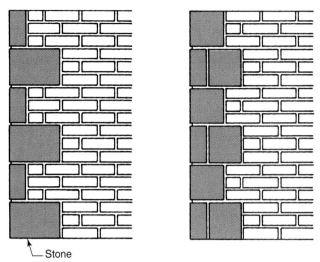

**Figure 13-92.** Quoins used at corners for ornamental purposes.

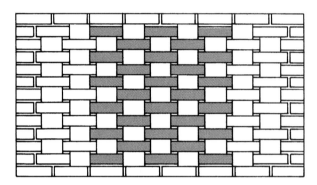

**Figure 13-93.** Perforated brick wall with each unit bonding one quarter its length on either end. Patterns are limited only to one's imagination.

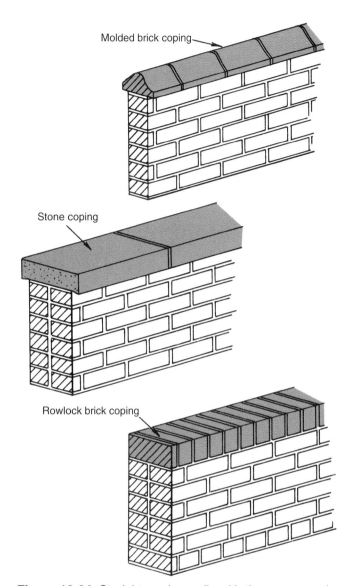

**Figure 13-94.** Straight garden walls with three commonly used copings.

could be a maximum height of 48" (3/4 × 8² = 48). This formula does not depend on a bond between the foundation and the wall. Therefore, reinforcing will greatly increase the height that can be attained for a given wall thickness.

Figure 13-94 shows three typical straight walls. Each has a different coping.

## Pier and panel walls

A **pier and panel wall** is a wall that is composed of a series of relatively thin panels 4" thick, which are braced by masonry piers. See Figure 13-95. This type of wall is relatively easy to build and is economical because of the reduced panel thickness. It is also ideal for uneven terrain. Foundations are only required for the piers. These should extend below the frost depth.

The panels may be constructed with 2 × 4 forms under the first course. The forms are removed when the masonry has cured a minimum of seven days. Tables in Figure 13-96, Figure 13-97, and Figure 13-98 give the specifications for building pier and panel walls for varying heights and lengths.

## Serpentine walls

Serpentine walls have been used successfully for hundreds of years. The serpentine shape provides lateral strength so that it can normally be built 4" thick. Since the shape of the wall provides the strength, it is important that the degree of curvature be sufficient. The radius of curvature of a 4" wall should be no more than twice the height of the wall above the grade. The depth of curvature should be no less than 1/2 of the height. Figure 13-99 shows the details of a typical serpentine wall.

# Caps and Copings

The primary function of caps and copings is to channel water away from the building. *Cap* refers to a covering within the height of the wall, normally where there is a

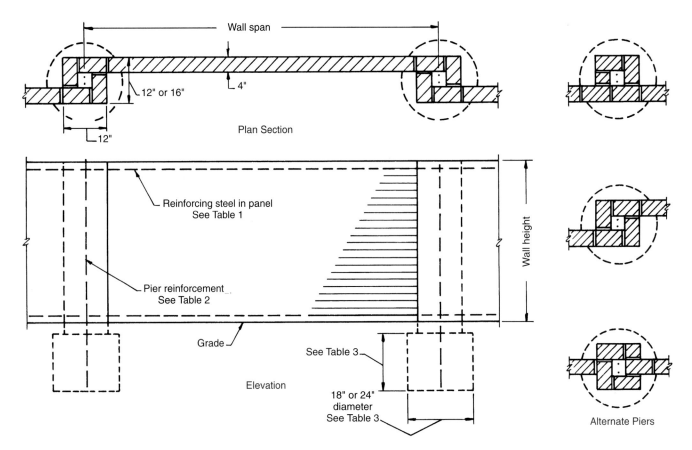

**Figure 13-95.** Details of a pier and panel garden wall. See Figure 13-96, 13-97, and 13-99 for specifications.

Table 1
Panel Wall Reinforcing Steel

| Wall Span ft. | Vertical Spacing, in. | | | | | | | | |
|---|---|---|---|---|---|---|---|---|---|
| | Wind Load, 10 psf | | | Wind Load, 15 psf | | | Wind Load, 20 psf | | |
| | A | B | C | A | B | C | A | B | C |
| 8 | 45 | 30 | 19 | 30 | 20 | 12 | 23 | 15 | 9.5 |
| 10 | 29 | 19 | 12 | 19 | 13 | 8.0 | 14 | 10 | 6.0 |
| 12 | 20 | 13 | 8.5 | 13 | 9.0 | 5.5 | 10 | 7.0 | 4.0 |
| 14 | 15 | 10 | 6.5 | 10 | 6.5 | 4.0 | 7.5 | 5.0 | 3.0 |
| 16 | 11 | 7.5 | 5.0 | 7.5 | 5.0 | 3.0 | 6.0 | 4.0 | 2.5 |

Note:    A = 2–No. 2 bars
            B = 2–3/16" diam. wires
            C = 2–9 gauge wires

**Figure 13-96.** Specifications for panel wall reinforcing steel. (Brick Institute of America)

Table 2
Pier Reinforcing Steel (1)

| Wall Span ft. | Wind Load, 10 psf Wall Height, ft. | | | Wind Load, 15 psf Wall Height, ft. | | | Wind Load, 20 psf Wall Height, ft. | | |
|---|---|---|---|---|---|---|---|---|---|
| | 4 | 6 | 8 | 4 | 6 | 8 | 4 | 6 | 8 |
| 8 | 2#3 | 2#4 | 2#5 | 2#3 | 2#5 | 2#6 | 2#4 | 2#5 | 2#5 |
| 10 | 2#3 | 2#4 | 2#5 | 2#4 | 2#5 | 2#7 | 2#4 | 2#6 | 2#6 |
| 12 | 2#3 | 2#5 | 2#6 | 2#4 | 2#6 | 2#6 | 2#4 | 2#6 | 2#7 |
| 14 | 2#3 | 2#5 | 2#6 | 2#4 | 2#6 | 2#6 | 2#5 | 2#5 | 2#7 |
| 16 | 2#4 | 2#5 | 2#7 | 2#4 | 2#6 | 2#7 | 2#5 | 2#6 | 2#7 |

(1) Within heavy lines 12 × 16" pier required. All other values obtained with 12 × 12" pier.

**Figure 13-97.** Specifications for pier reinforcing steel.

Table 3
Required Embedment for Pier Foundation (1)

| Wall Span ft. | Wind Load, 10 psf Wall Height, ft. | | | Wind Load, 15 psf Wall Height, ft. | | | Wind Load, 20 psf Wall Height, ft. | | |
|---|---|---|---|---|---|---|---|---|---|
| | 4 | 6 | 8 | 4 | 6 | 8 | 4 | 6 | 8 |
| 8 | 2'-0" | 2'-3" | 2'-9" | 2'-3" | 2'-6" | 3'-0" | 2'-3" | 2'-9" | 3'-0" |
| 10 | 2'-0" | 2'-6" | 2'-9" | 2'-3" | 2'-9" | 3'-3" | 2'-6" | 3'-0" | 3'-3" |
| 12 | 2'-3" | 2'-6" | 3'-0" | 2'-3" | 3'-0" | 3'-3" | 2'-6" | 3'-3" | 3'-6" |
| 14 | 2'-3" | 2'-9" | 3'-0" | 2'-6" | 3'-0" | 3'-3" | 2'-9" | 3'-3" | 3'-9" |
| 16 | 2'-3" | 2'-9" | 3'-0" | 2'-6" | 3'-3" | 3'-6" | 2'-9" | 3'-3" | 4'-0" |

(1) Within heavy lines 24" diam. foundation required. All other values obtained with 18" diam. foundation.

**Figure 13-98.** Specifications for embedment for pier foundations.

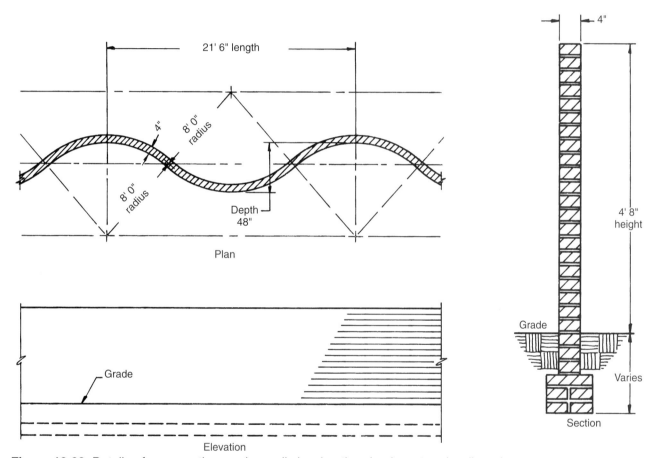

**Figure 13-99.** Details of a serpentine garden wall showing the plan layout and wall sections.

change in wall thickness. **Coping** applies to the covering at the top of a wall. The cap or coping are either a single unit or multiple units. The tops can slope in one direction or both directions. Additionally, where caps are discontinuous, a minimum slope from the ends of 1/8" in 12" should be provided. Caps and copings normally do not serve any structural function, and do not present any major problems in their construction.

Caps and copings can be constructed of several materials: brick, pre-cast or cast-in-place concrete, stone, terra cotta, or metal. See Figures 13-100 and 13-101.

Because of their location in the structure, caps and copings are exposed to climatic extremes and, therefore, brick masonry may not be the best choice of materials. This is because brick requires more joints than other materials.

Concrete, stone, and terra cotta all have thermal expansion properties similar to those of brick masonry. They normally present no extreme problems with differential movement when applied as caps and copings.

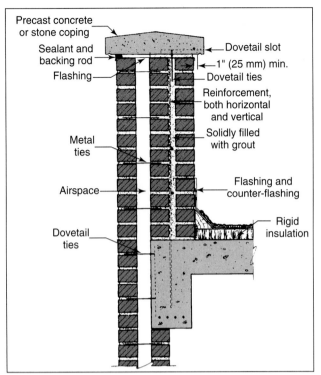

**Figure 13-100.** Stone or precast concrete coping for parapet wall. (Brick Institute of America)

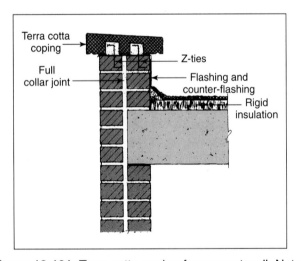

**Figure 13-101.** Terra cotta coping for parapet wall. Note how the coping slopes toward the roof to deflect water from the front of the building. (Brick Institute of America)

# Corbels and Racking

A **corbel** is a shelf or ledge formed by projecting successive courses of masonry *out* from the face of the wall. Corbelling of brick masonry is done to achieve the desired aesthetics, or to provide structural support. **Racking** is masonry in which successive courses are stepped *back* from the face of the wall.

Generally, the total horizontal projection should not exceed 1/2 the thickness of a solid wall, or 1/2 the thickness of the veneer of a veneered wall. It is also required that the projection of a single course not exceed 1/2 of the

unit height or 1/3 of the unit bed depth, whichever is less. From these limitations, the minimum slope of the corbelling can be established (angle measured from the horizontal to the face of the corbelled surface) as 63°26'. See Figure 13-102.

**Trade Tip.** When racking back to achieve the desired dimensions, care must be exercised. Since there is no limitation on the distance each unit can be racked, be sure the cores of the units are not exposed.

# Mortarless Retaining Walls

A recent development in materials designed for retaining walls, tree rings, planters, and edgings, is a concrete masonry unit that incorporates a lip on the lower back edge that provides proper alignment and prevents forward wall movement. See Figure 13-103. Typical products include straight and beveled front units with uniform or tapered widths. See Figure 13-104. The units are placed so that the lip locks behind the lower course to provide proper alignment and prevent forward movement exerted by earth pressure.

Typical sizes of these units include: 6" × 16" × 12", 6" × 17 1/4" × 12", and 3" × 17" × 12". Sizes vary from one manufacturer to another.

## *Typical guidelines for installation*

Before beginning construction, prepare a layout plan that considers the topography, drainage patterns, soil conditions, and local code restrictions. The following procedures are typical:

1. Excavate a trench. For walls up to 4' high, one course is generally buried below grade. The trench should be at least 14" wide and deep enough to accommodate the amount of wall to be buried and require compacted base material.
2. Prepare the subbase. Compact the soil in the bottom of the trench and place 6" of compactable base aggregate in the bottom of the trench and compact it.
3. Install the base course. Install the first course of units (without lip) on the prepared base. Level each unit as it is placed from side to side and front to back. Backfill with base aggregate to top edge of the course.
4. Install the next course. Stagger the vertical joints. Pull each unit forward until the lip aligns with the back edge of the course below.
5. Install the drain tile and drainage aggregate. Place the drain tile at grade level behind the wall

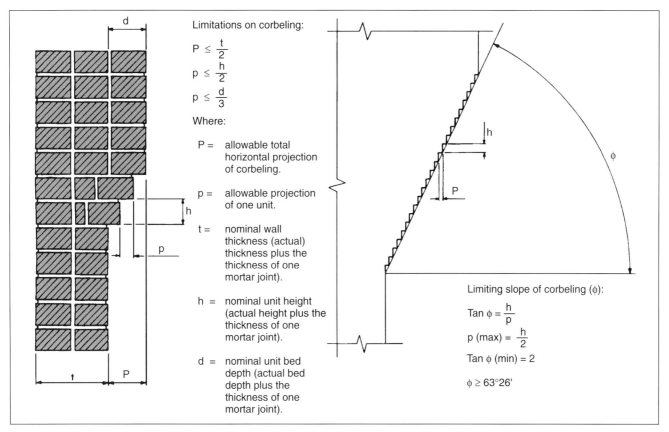

Limitations on corbeling:

$$P \leq \frac{t}{2}$$

$$p \leq \frac{h}{2}$$

$$p \leq \frac{d}{3}$$

Where:

P = allowable total horizontal projection of corbeling.

p = allowable projection of one unit.

t = nominal wall thickness (actual) thickness plus the thickness of one mortar joint).

h = nominal unit height (actual height plus the thickness of one mortar joint).

d = nominal unit bed depth (actual bed depth plus the thickness of one mortar joint).

Limiting slope of corbeling (φ):

$$Tan\ \phi = \frac{h}{p}$$

$$p\ (max) = \frac{h}{2}$$

$$Tan\ \phi\ (min) = 2$$

$$\phi \geq 63°26'$$

**Figure 13-102.** Corbelling limitations. (Brick Institute of America)

**Figure 13-103.** This mortarless masonry unit provides a significant amount of flexibility in laying up straight or curved retaining walls. (Anchor Wall Systems, Inc.)

**Figure 13-104.** The tapered sides of the units shown in this illustration make possible straight or curved walls for a variety of designs. (Anchor Wall Systems, Inc.]

to remove moisture behind the wall. Use 3/4" free-drainage aggregate 12" directly behind the wall in 6" layers. Fill the void between units with aggregate or sand.

6. Install the backfill. Fill any remaining voids between the drainage aggregate and the embankment with backfill material (native soil). Repeat steps 5, 6, and 7 until the wall reaches the desired height.

7. Compact the backfill. Firmly compact the backfill material behind the drainage aggregate.

 **Trade Tip.** To create partial units, use a masonry saw, splitter, or hammer and chisel to cut each unit to desired shape and size.

# REVIEW QUESTIONS
# CHAPTER 13

Write all answers on a separate sheet of paper. Do not write in this book.

1. The footing is usually wider than the foundation wall to provide _____ and _____.
2. Any foundation not placed on bedrock will settle. True or False?
3. There are two general types of foundations. They are _____ foundations and _____ foundations.
4. _____ footings carry light loads and are not reinforced with steel.
5. Footings that are independent and receive the loads of freestanding columns or piers are called _____ footings.
6. The usual width of a footing for an 8" concrete block foundation wall is _____.
7. The primary reason for parging a masonry basement wall is to _____ the wall.
8. Columns, piers, and pilasters are classified as similar. The _____ is built into the foundation wall.
9. The minimum length of an anchor bolt used in a concrete block masonry foundation wall is _____".
10. Solid masonry walls can be built of solid or hollow masonry units or a combination of the two. True or False?
11. The *SCR brick* is designed to build a solid masonry wall _____" thick.
12. A(n) _____ wall is an exterior nonloadbearing wall not wholly supported at each story.
13. A hollow masonry wall can be built using solid or hollow masonry units. True or False?
14. Two types of hollow masonry walls are _____ and _____.
15. A cavity wall is usually bonded together with _____.
16. Suitable types of insulation materials for cavity walls are granular _____ and rigid _____.
17. In order to divert moisture out of the cavity in cavity wall construction through the weep holes, _____ should be installed at the bottom of the cavity.
18. In cavity wall construction, weep holes are located a maximum of _____' apart.
19. A(n) _____ wall is one that does not use the facing material as a loadbearing agent.
20. Brick veneer wall assemblies are drainage type walls. True or False?

21. Brick veneer with wood or metal frame backing is typically built with a(n) _____" minimum airspace.
22. Brick veneer on a frame backing must transfer the weight of the veneer to the _____.
23. Metal ties that are typically used with wood framing and brick veneer require 1 tie for every _____ square feet of wall area.
24. Brick veneer backed by wood or metal frame must always be supported by _____ over openings unless the masonry is self-supporting.
25. Reinforced masonry walls should be reinforced with an area of steel not less than _____ times the cross-sectional area of the wall.
    A. 0.002
    B. 0.02
    C. 0.2
    D. 2.0
26. Name two types of retaining walls.
27. Thin brick veneer is classified as a(n) _____ veneer.
28. Masonry above a wall opening can be supported by a(n) _____ or _____.
29. Identify three types of lintels.
30. Arches that do not exceed 6' are classified as _____ arches.
31. Arches are constructed with the aid of temporary shoring or _____.
32. A(n) _____ arch is an arch with an intrados that is circular but less than a semicircle.
33. An arch is normally classified by the curve of its _____ (the curve that bounds the lower edge of the arch) and by its function, shape, or architectural style.
34. The _____ arch is relatively weak and should be supported by steel if the opening is over 2' wide.
35. The _____ is the horizontal support placed over the top of a window or door.
36. _____ sills are generally the same length as the window opening.
37. A(n) _____ joint is used to separate brick masonry into segments to prevent cracking due to changes in temperature, moisture expansion, elastic deformation due to loads, and creep.
38. The _____ joint is a through-the-building joint.
39. A(n) _____ joint is used primarily in concrete construction where construction work is interrupted.
40. The proper slope for walks and patios is _____" to _____" per foot.
41. Treads for all outside steps are generally _____" wide.
42. Name four basic types of fireplaces.
43. What is the purpose of the smoke shelf?

44. The minimum wall thickness of a residential chimney is _____ ".

45. Approved fire clay flue liners cannot be less than _____ " thick.

46. The flue lining should project _____ " above the chimney cap.

47. Large squared stones used at corners and around openings of buildings for ornamental purposes are called _____.

48. Which type of garden wall provides lateral support through its shape?

49. The primary function of _____ and _____ is to channel water away from the building.

50. A _____ is defined as a shelf or ledge formed by projecting successive courses of masonry out from the face of the wall.

51. What device on mortarless retaining wall masonry units prevents forward movement of the wall?

Top—Metal tie and joint reinforcement used to bond and strengthen this brick and block masonry wall. Bottom—Truss joint reinforcement used to bond brick and concrete wythes together. (Wire Reinforcement Institute)

# SECTION IV
# CONCRETE

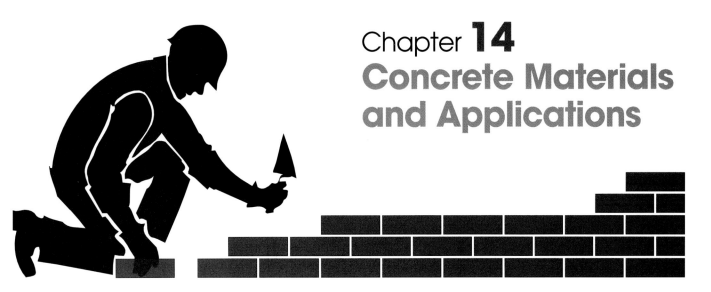

# Chapter 14
# Concrete Materials and Applications

Concrete is one of our most important building materials. It is used in almost every type and size of architectural structure. See Figure 14-1. In footings and foundations, in exterior and interior walls, floors and roof systems, walks, and driveways, concrete is a major material. The plastic quality of concrete lends itself to almost unlimited design possibilities in form, pattern, and texture. Hardened concrete is durable, needs little maintenance, and can be used in many ways.

## Basic Materials of Concrete

Concrete is a mixture of four basic materials in proper amounts (proportions). The basic materials are Portland cement, fine aggregate (sand or finely crushed stone), coarse aggregate (gravel or crushed stone), and water.

Lightweight, manufactured materials are also used to produce lightweight concrete. Sometimes other ingredients

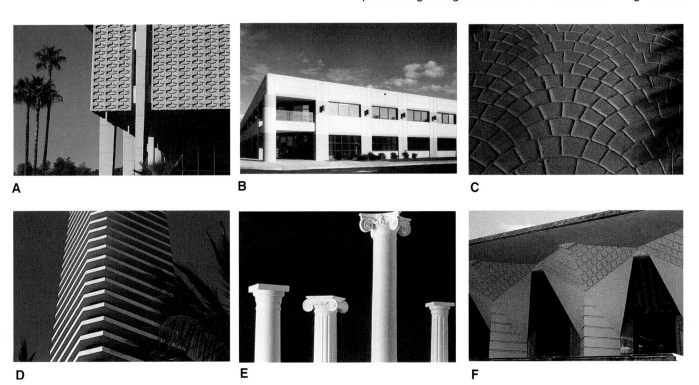

**Figure 14-1.** A—The primary building material for this contemporary Frank Lloyd Wright designed structure is cast concrete. B—Tilt-up wall construction (concrete panels) is the system used to construct this building. (Tilt-up Concrete Association) C—Scored concrete paving. D—This elegant building was made with cast concrete. E—Columns such as these are made using white Portland cement. F—Most any shape or texture can be produced using cast concrete.

are added to the basic concrete mix. Their purpose is to produce certain properties in the finished product. These ingredients are called admixtures.

## Portland cement

Portland cement is hydraulic because it sets and hardens by reacting with water. This chemical reaction is called **hydration**. The reaction forms the cement into a stone-like mass.

Portland cement must contain varying amounts of lime, silica, alumina, and iron components. The raw materials are pulverized and combined into a mixture that has the desired chemical makeup.

Manufacturers of Portland cement use their own brand names. However, all Portland cements are manufactured to meet the **American Society for Testing and Materials** Standard Specifications, ASTM Designation:

- C 150 *Portland Cement, Types I through V*
- C 175 *Air-Entraining Portland Cement*
- C 205 *Portland Blast Furnace Slag Cement*
- C 340 *Portland-Pozzolan Cement*

### Types of Portland cement

Portland cement is produced to meet various physical and chemical requirements for specific purposes. There are several types.

- **Type I—Normal Portland Cement.** This general purpose cement is suitable for all uses when the special properties of the other types are not required. It is used where it will not be subject to attack by sulfate from the soil or water or to an objectionable temperature rise from the heat generated by hydration. It may be used for pavements, sidewalks, reinforced concrete structures, bridges, tanks, reservoirs, culverts, water pipe, and masonry units.
- **Type II—Modified Portland Cement.** This type of cement produces less heat due to hydration and can resist sulfate attack better. This cement can be used in structures such as piers, heavy abutments, and heavy retaining walls.
- **Type III —High Early Strength Portland Cement.** A cement producing high strength quickly, usually a week or less. It is used when forms are to be removed as soon as possible or when the structure must be used quickly. Its controlled curing period may also be reduced in cold weather.
- **Type IV—Low-Heat Portland Cement.** Use when the heat generated by hydration must be kept low and when a slower developing strength is not objectionable. It is designed for use in massive concrete structures such as dams.
- **Type V—Sulfate-Resistant Portland Cement.** Used only in construction exposed to severe sulfate action such as in some western states with water and soil of high alkali content. It gains strength more slowly than normal cement.

- **Air-Entraining Portland Cement.** Designated as IA, IIA, and IIIA. They correspond, in composition, to Types I, II, and III except that small quantities of air-entraining materials are added to the cement during production. These cements improve resistance to freeze-thaw action and scaling caused by chemicals used in snow and ice removal. Concrete made with these cements contains billions of tiny, well-distributed and completely separated air bubbles.
- **White Portland Cement.** Manufactured to conform to the specifications of ASTM Designation C 150 and C 175. The manufacturing process is controlled so that the finished product will be white instead of gray. It is used mainly for architectural purposes such as precast wall and facing panels, terrazzo, stucco, cement paint, tile grout, and decorative concrete. It is recommended wherever white or colored concrete or mortar is desired.
- **Portland Blast-Furnace Slag Cements.** In these cements, granulated (like sand) blast-furnace slag is ground and mixed with Portland cement. They can be used in general concrete construction when the specific properties of other types are not required.
- **Portland-Pozzolan Cements.** In these cements, pozzolan is blended with ground Portland cement. They are used principally for large hydraulic structures such as bridge piers and dams.

The strength of concretes made with different cements is shown in Figure 14-2. Type I or normal Portland cement concrete is used as the basis for comparison.

# AGGREGATES

Sand, gravel, crushed stone, and air-cooled blast furnace slag are the most commonly used aggregates in concrete. See Figure 14-3. The aggregates usually make up 60% to 80% of the volume of concrete and can be considered to be filler material.

Aggregates must meet certain requirements. They must be clean, hard, strong, durable particles free of chemicals and coatings of clay which could affect the bond. Since aggregates are such a large proportion of the volume of concrete, they greatly affect the quality of the finished product. Requirements for concrete aggregates are specified in ASTM C 33. These specifications place limits on allowable amounts of damaging substances. They also cover requirements as to grading, strength, and soundness of particles.

## Fine aggregate

Fine aggregate contains sand or other acceptable fine materials. A desirable concrete sand should contain particles varying uniformly in size from very fine up to 1/4" in diameter. Fine particles are necessary for good workability and smooth surfaces. These fine particles help fill the

| Type of Portland Cement | | Compressive Strength, Percent of Strength of Type I or Normal Portland Cement Concrete | | | |
|---|---|---|---|---|---|
| ASTM Designation | Canadian Standards Assoc. | 1 Day | 7 Days | 28 Days | 3 Months |
| I | Normal | 100 | 100 | 100 | 100 |
| II | | 75 | 85 | 90 | 100 |
| III | High early strength | 190 | 120 | 110 | 100 |
| IV | | 55 | 55 | 75 | 100 |
| V | Sulfate resisting | 65 | 75 | 85 | 100 |

**Figure 14-2.** Chart shows relative strengths of concretes made with different cements. These concretes were moist-cured until tested. (Portland Cement Association)

| Concrete Weight with Various Aggregates | | |
|---|---|---|
| Aggregate Used in Concrete | | Unit Weight, pcf |
| Normal weight concrete | Sand and gravel concrete | 130 - 145 |
| | Crushed stone and sand concrete | 120 - 140 |
| | Air-cooled slag concrete | 100 - 125 |
| Lightweight concrete | Coal cinder concrete | 80 - 105 |
| | Expanded slag concrete | 80 - 105 |
| | Pelletized-fly ash concrete | 75 - 125 |
| | Scoria concrete | 75 - 100 |
| | Expanded clay, shale, slate, and sintered fly ash concretes | 75 - 90 |
| | Pumice concrete | 60 - 85 |
| | Cellular concrete | 25 - 44 |

**Figure 14-3.** The weight of a cubic foot of concrete varies from 25 pounds per cubic foot to 145 pounds per cubic foot depending on the aggregate used in the mix. (Portland Cement Association)

spaces between the larger particles. But an excess of fine particles requires more cement-water paste to attain the right strength and will increase the cost.

## Coarse aggregate

Coarse aggregate consists of gravel, crushed stone, or other acceptable materials larger than 1/4" in diameter. Coarse aggregate suitable for use in concrete should be sound, hard, and durable. It should not be soft, flaky, or wear away rapidly. It should also be clean and free of loam, clay, or vegetable matter. These foreign materials prevent the cement paste from properly binding the aggregate particles together and will reduce the strength of the concrete.

The maximum size of coarse aggregate is usually 1 1/2" in diameter. This large size aggregate could be used in a thick foundation wall or footing. As a rule of thumb, the largest piece should never be greater than one-fifth the thickness of a finished wall section, one-third the thickness of a slab, or three-fourths the width of the narrowest space through which the concrete will be required to pass during placing.

Well-graded, coarse aggregate will range uniformly from 1/4" up to the largest size that is satisfactory for the kind of work to be performed. The use of the maximum allowable particle size usually results in less drying shrinkage and lower cost.

## Mixing water for concrete

The primary purpose for using water with cement is to cause hydration (a chemical reaction) of the cement. Water usually constitutes (makes up) from 14% to 21% of the total volume of concrete. The amount of water used definitely affects the quality of the concrete.

Concrete mixing water should be clean and free of oil, alkali, and acid. Generally, water that is good for drinking is suitable for mixing concrete. However, water with too many sulfates should not be used even though it may be potable (good for drinking).

# Admixtures

**Admixtures** are any materials added to the concrete batch before or during mixing other than Portland cement, water, and aggregates. Admixtures are classified as air-entraining, water-reducing, retarding, accelerating, pozzolans, workability agents, and miscellaneous.

The desirable qualities of concrete can often be attained conveniently and economically through the proper design of the mix, and the selection of suitable materials without using admixtures (except air-entraining admixtures).

## Air-entraining admixtures

**Air-entraining admixtures** are materials used to improve the durability of concrete exposed to moisture during cycles of freezing and thawing. Concrete resists surface scaling caused by ice-removal agents far better with air-entraining admixtures.

The workability of fresh concrete is also improved significantly. Segregation (separating of materials) and bleeding are reduced or stopped. Entrained air can be produced in concrete through the use of an air-entraining cement or through an air-entraining admixture.

## Water-reducing admixtures

**Water-reducing admixtures** are materials that reduce the quantity of mixing water required to produce concrete of a given consistency. These materials increase the slump of concrete with less water content. **Slump** is the relative consistency or stiffness of the plastic concrete.

Some water-reducing admixtures also retard the setting time of concrete. Others are modified to give varying degrees of retardation and some do not affect the setting time much. Still others can entrain air in the concrete.

An increase in strength is usually gained with water-reducing admixtures if the water content is reduced for a given mix, and if the cement content and slump are kept the same. Even though water content is reduced with these admixtures, increases in drying shrinkage are probable.

## Retarding admixtures

A **retarding admixture** is a material used to retard (slow down) the setting time of concrete. High temperatures of plastic concrete often cause it to harden too fast. This increased hardening rate makes placing and finishing of concrete difficult. One of the most practical ways to combat this difficulty is to reduce the temperature of the concrete by cooling the mixing water and/or aggregates. The retarders will not lower the initial temperature of the concrete.

Retarders can be used when it is desirable to offset rapid setting up (hardening) of concrete due to hot weather. It is also used to delay the initial set of concrete when difficult conditions of placement are encountered.

Retarders usually reduce the strength of concrete for the first one to three days. Effect on other properties such as shrinkage may not be predictable. It is advisable to test the retarders under actual job conditions to be sure of results.

## Accelerating admixtures

An **accelerating admixture** is a material used to speed up the setting and the strength development of concrete. However, most of the commonly used accelerators cause an increase in the drying shrinkage of concrete. The development of concrete strength can also be increased by:
1. Using Type III Portland cement
2. Lowering the water-cement ratio
3. Increasing the cement content or curing at higher temperatures

Calcium chloride is the most commonly used accelerating admixture. The amount added should not be more than 2% of the cement weight.

 **Warning!** Calcium chloride should not be used in prestressed concrete, in concrete containing embedded aluminum, where galvanized steel will remain in contact with the concrete, or in concrete subjected to alkali or sulfates.

## Pozzolans

A **pozzolan** is, by ASTM C 219, a siliceous or siliceous and aluminous material, which in itself, possesses little or no cementitious value but will, in finely divided form and in the presence of moisture, chemically react with calcium hydroxide at ordinary temperatures to form compounds possessing cementitious properties. These materials are sometimes used in concrete to help control high internal temperatures. These temperature buildups can occur from slow loss of heat generated by hydration in massive structures. However, these high temperatures can often be controlled by using Type II cement and/or by lowering the temperature of the mixing water and aggregates.

The use of pozzolans can substantially reduce the early strength of concrete during the first 28 days. The effect on concrete varies considerably. Tests should be performed before using pozzolans in construction.

## Workability agents

Plastic concrete is often harsh due to aggregate characteristics such as particle shape and poor grading. This condition can be improved by increasing the amount of fine aggregate or entraining air. Entrained air is considered the best and acts like a lubricant. Excessive amounts of these fine materials will most likely reduce the strength and increase shrinkage of the concrete.

# Concrete Mixtures

The proportioning of concrete mixtures will determine the qualities of the plastic and hardened concrete. Desirable qualities of plastic concrete are consistency, workability, and finishability. Figure 14-4 shows concrete with the proper consistency. Desirable qualities of hardened concrete are strength, durability, water tightness, wear resistance, and economy.

In properly proportioned concrete, each particle of aggregate (no matter how large or small) will be completely surrounded by the cement-water paste. The spaces between aggregate particles will also be filled completely with paste. To lower costs, the proportioning should minimize the amount of cement required without reducing concrete quality.

**Figure 14-4.** This concrete has the proper consistency. (Portland Cement Association)

## Concrete mix characteristics

Before a concrete mixture can be proportioned you must know the size and shape of concrete members, the concrete strength required, and the conditions of exposure.

Most of the desired properties of hardened concrete depend upon the quality of the cement paste. The first step then in determining the mix proportions, is to select the right water-cement ratio.

## Water-cement ratio

The **water-cement ratio** is the proportion of water to cement. This ratio is stated in gallons of water per bag (94 lb.) of cement. The binding properties of the cement paste are due to chemical reactions between the cement and water.

The reactions require time and favorable conditions of temperature and moisture. They proceed very rapidly at first and then much more slowly for a long period of time if conditions are satisfactory. Only a small amount of water (about 3 1/2 gallons per bag of cement) is necessary to complete the chemical reactions. However, more water (4 gallons or 9 gallons per bag of cement) is used to increase workability and aid placement. Also, more water increases the amount of aggregate which can be used. This makes the mix even less expensive.

Good cement paste requires a proper proportion of water to cement. If too much water is used, the paste will be thin and weak when it hardens. Too little water in the mix will make placing and finishing difficult.

Cement paste made with the proper amount of water will be strong, watertight, and durable. The concrete will be strong and durable if the aggregates and cement paste are strong and durable. The quality of the concrete is therefore largely determined by the water-cement ratio that is so important a factor in the cement paste.

Strength of concrete is directly related to the water-cement ratio. Figure 14-5 shows the results of many tests

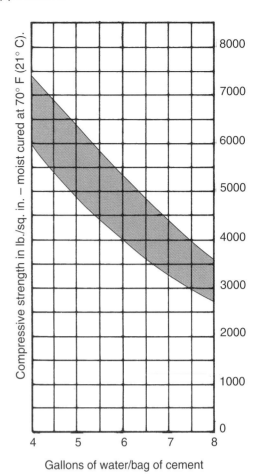

Gallons of water/bag of cement

**Figure 14-5.** The area within the curves represents the age-compressive strength relationship for Type I Portland cement after 28 days moist curing at 70°F (21°C).

performed on cured concrete. The results were similar for other types of cements. More water resulted in less strength and less water resulted in greater strength.

After concrete is in place, the strength continues to increase as long as moisture and satisfactory temperatures are present. Highest curing temperature is 73°F (23°C). When concrete is permitted to dry, the chemical reactions cease and lower strength results. Concrete should therefore be kept constantly moist as long as possible after placing.

There is also a direct relationship between the durability of concrete and the relative quantities of water and cement in the mixture. Freezing and thawing are the most destructive natural forces of weathering on wet or moist concrete. During this process, water expands as it changes to ice. If the water-cement ratio was high during mixing, the concrete is more porous and is more subject to damage by freezing and thawing.

Watertightness, likewise, depends on the water-cement ratio in the mix. The more water used in the mix, the less likely the concrete is to be watertight.

## Selecting concrete mixtures

The right concrete mixture for any job is the one that is most economical and workable while giving the desired properties. Figure 14-6 shows concrete mixtures for various construction applications. Column 1 specifies the construction application. Column 2 recommends the approximate maximum water-cement ratio. Column 3 shows the appropriate consistency as measured by inches of slump. Column 4 indicates the maximum size of aggregate. Column 5 specifies the cement content. Column 6 shows the probable approximate compressive strength after 28 days.

Rule-of-thumb approximation of volumes such as 1:2:3 of cement to sand to gravel are not recommended for quality concrete mixtures. Most fine aggregates contain some water. Therefore, allowance should be made for this factor in determining the amount of water to be added to the mix. A simple test will determine whether sand is *damp*, *wet*, or *very wet*. Press some sand in your hand. If it falls apart after you open your hand, it is *damp*; if it forms a ball that holds its shape, it is *wet*; if it sparkles and wets your hand, it is *very wet*.

Figure 14-7 shows adjustments to the water-cement ratio that should be made when the fine aggregate is damp, wet, or very wet. The amount of water specified in the chart plus the water in the fine aggregate will equal the amount of water required for the mix. If the sand is dry, then use the full amount indicated in Figure 14-6.

## Mix consistency

**Slump** is the mix consistency or degree of stiffness of plastic concrete. Slump is measured in inches.

| If the mix calls for a water-cement ratio of: | Use these amounts of water (gals.) when the sand is: | | |
|---|---|---|---|
| | Damp | Wet | Very wet |
| 5 Gal/Bag | 4 3/4 | 4 1/2 | 4 1/4 |
| 6 Gal/Bag | 5 1/2 | 5 | 4 1/2 |
| 7 Gal/Bag | 6 1/4 | 5 1/2 | 4 3/4 |

**Figure 14-7.** Recommended adjustment to the water-cement ratio when the fine aggregate is damp, wet, or very wet.

Very fluid (wet) mixes are called **high-slump concrete**. Stiff (dry) mixes are called **low-slump concrete**. Slump is related primarily to water-cement ratio. A high-water content causes a high slump. Generally, a low-slump concrete will produce a better concrete product.

A slump test can be performed in the field by using a sheet metal slump cone. See Figure 14-8. The cone is 8" in diameter at the bottom, 4" in diameter at the top, and 12" high. It is placed on a clean dry surface and filled one third full with the concrete being tested. The concrete is rodded 25 times with a metal rod that is 24" long and 5/8" in diameter. Concrete is then added until the cone is 2/3 full and is again rodded 25 times with the rod penetrating the lower layer. Finally, the cone is filled, the top raked off level, and the mixture rodded 25 more times. The cone is then removed by lifting it straight up. The slump is measured immediately. The rod is placed across the top of the cone extending over the concrete sample. The distance from the bottom edge of the rod to the concrete is measured with a rule. This distance is the slump rating for the batch.

The slump test can be used as a rough measure of the consistency of concrete, but should not be considered

| Construction Application | Water/cement ratio gal. per bag | Consistency (amount of slump) | Maximum size of aggregate | Appropriate cement content bags per yd. | Probable 28th day strength (psi) |
|---|---|---|---|---|---|
| Footings | 7 | 4" to 6" | 1 1/2" | 5.0 | 2800 |
| 8" basement wall moderate ground water | 7 | 4" to 6" | 1 1/2" | 5.0 | 2800 |
| 8" basement wall severe ground water | 6 | 3" to 5" | 1 1/2" | 5.8 | 3500 |
| 10" basement wall moderate ground water | 7 | 4" to 6" | 2" | 4.7 | 2800 |
| 10" basement wall severe ground water | 6 | 3" to 5" | 2" | 5.5 | 3500 |
| Basement floor 4" thickness | 6 | 2" to 4" | 1" | 6.2 | 3500 |
| Floor slab on grade | 6 | 2" to 4" | 1" | 6.2 | 3500 |
| Stairs and steps | 6 | 1" to 4" | 1" | 6.2 | 3500 |
| Topping over concrete floor | 5 | 1" to 2" | 3/8" | 8.0 | 4350 |
| Sidewalks, patios, driveways, porches | 6 | 2" to 4" | 1" | 6.2 | 3500 |

**Figure 14-6.** Recommended concrete mixtures for various construction applications.

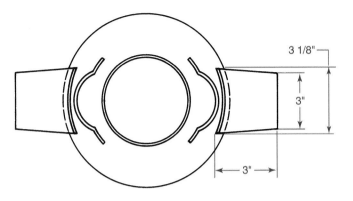

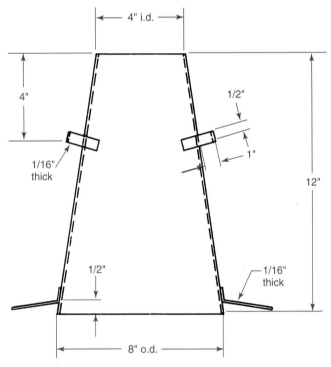

**Figure 14-8.** A sheet metal slump cone showing dimensions.

a measure of workability. It should not be used to compare mixes of entirely different proportions or mixes containing different aggregates. The slump test does, however, help to signal changes in grading or proportions of the aggregate or water content in the mix.

# Measuring Materials

The ingredients of each batch of concrete must be measured accurately if uniform batches of proper proportions and consistency are to be produced. The problem of varying amounts of moisture in the aggregate has already been discussed. This must be taken into account if accurate control is to be obtained.

### Measuring cement

If bagged cement is used, the batches should be of such a size that only full bags are used. If this is not possible, then the proper amount should be weighed out each time. Volume is not an accurate method of measuring cement. Bulk cement (not bagged) should always be weighed for each batch.

### Measuring water

The effects of the water-cement ratio on the qualities of concrete make it just as necessary to measure accurately the water used as any other materials. Water is measured in gallons and any method that will ensure accuracy is acceptable.

### Measuring aggregates

Measurement of aggregates by weight is the recommended practice and should be required on all jobs that require a high degree of consistency in each batch. Measurement of fine aggregate by volume is not accurate because a small amount of moisture is nearly always present. This moisture causes the fine aggregate to bulk or fluff up. The degree of bulking depends on the amount of moisture present and the fineness of the sand. A fine sand with a 5% moisture content will increase in volume about 40% above its dry volume.

Another problem with aggregates, especially coarse aggregates, is separation of the various sizes. This problem can be reduced by weighing the coarse aggregate in two or more sizes.

# Mixing

All concrete should be mixed thoroughly until it is uniform in appearance and all ingredients are uniformly distributed. Mixing time will depend on the following factors:
1. The speed of the machine
2. The size of the batch
3. The condition of the mixer

Generally, the mixing time should be at least one minute for mixtures up to 1 cu. yd. with an increase of 15 seconds for each 1/2 cu. yd. or fraction thereof. Mixing time should be measured from the time all materials are in the mixer.

Mixers should not be loaded above their capacity. They should be run at the speed for which they were designed. Mixers coated with hardened concrete or having badly worn blades will not perform as efficiently as they should. These conditions should be corrected.

Generally, about 10% of the mixing water is placed in the mixer before the aggregate and cement are added. Water should then be added uniformly along with the dry materials. The last 10% of the water is added after all the dry materials are in the mixer.

## Ready-mixed concrete

Ready-mixed concrete is purchased directly from a central plant. It is convenient and usually of high quality. In some ready-mixed operations, the materials are dry-batched at the central plant and then mixed enroute to the site in truck mixers.

Another method is to mix the concrete in a stationary mixer at the central plant just enough to intermingle the ingredients. The mixing is then completed in a truck mixer enroute to the job site. Most truck mixers have a capacity of 1 to 5 cu. yd. and carry their own water supply. ASTM C 94 requires that the concrete must be delivered and discharged from the truck mixer within 1 1/2 hours after the water has been added to the mix.

## Remixing concrete

Initial set of concrete usually takes place two or three hours after the cement is mixed with water. During this time the concrete may dry out and stiffen if left standing. It can be remixed if it becomes sufficiently plastic and workable. No water can be added to make it more workable (retempering). This will reduce the quality of the concrete just as a larger amount of water in the original mix.

# Reinforced Concrete ▬▬▬▬

Steel reinforcing is frequently used in footings, foundations, slabs, and other concrete work to add strength and control stresses. *Reinforced concrete* is a composite material that utilizes the concrete in resisting compression forces, and some other material, usually steel bars or wires, to resist the tension forces.

Concrete has great compressive strength. *Compressive strength* is the ability to support great loads placed directly upon it. However, it has very little strength to resist stresses or forces that tend to bend or pull it apart. The compressive strength of concrete is about 10 times its tensile strength. The use of steel reinforcing may be used to greatly increase its tensile strength.

## Types of reinforcement

Steel is regarded as the best reinforcing material for concrete. It has almost the same contraction and expansion rate due to temperature changes as concrete. Steel can also be purchased in many sizes and forms. The two main types of steel reinforcing used in concrete are reinforcing bars and welded wire reinforcement. See Figure 14-9. Bars are either smooth or deformed. Smooth bars are generally smaller in diameter. Deformed bars have ridges along the sides. These ridges increase the bond between the concrete and steel.

Bars are produced in standard sizes. They are designated by a number. Figure 14-10 shows bars from 1/4" to 1" in diameter with standard number, area, and approxi-

**Figure 14-9.** Reinforcing bars and welded wire reinforcement are being used to strengthen the slab joist in this skip pan joist system. The floor also has welded wire reinforcement. (Wire Reinforcement Institute)

| Bar Number | Bar Diameter in Inches | Bar Area in Sq. In. | Approximate Weight of 100 Ft. |
|---|---|---|---|
| 2 | 1/4 | .05 | 17 |
| 3 | 3/8 | .11 | 38 |
| 4 | 1/2 | .20 | 67 |
| 5 | 5/8 | .31 | 104 |
| 6 | 3/4 | .44 | 150 |
| 7 | 7/8 | .60 | 204 |
| 8 | 1 | .79 | 267 |

**Figure 14-10.** Size, area, and weight of reinforcing bars from 1/4" to 1" in diameter.

mate weight per 100' of bar. Larger bars are made for extremely heavy construction. Bars usually come from the mill in 60' lengths, but most local suppliers stock 20' and 40' lengths.

The size of bar required for the job will depend on the amount of tensile strength needed. This is usually calculated by engineers. In commercial construction, professional steelworkers normally make the installation.

Welded wire reinforcement is manufactured in many types and sizes. It is a prefabricated reinforcement consisting of parallel series of high-strength, cold-drawn wires welded together in square or rectangular grids. Smooth wires, deformed wires, or a combination of both can be used in welded wire reinforcement.

Welded wire reinforcement provides more uniform stress distribution and more effective crack control in slabs and walls than larger diameter bars more widely spaced. The range of wire sizes and spacings available is broad. Thus, it is possible to furnish almost exactly the cross-sectional steel area required for a specific job.

Cross-sectional area is the basic element used in specifying wire size. Smooth wire sizes are specified by the letter *W*, and deformed wires are specified by the letter *D*. This letter is followed by a number indicating the cross-sectional area of a wire in hundredths of a square inch. For example, W10 is a smooth wire with a cross-sectional area of 0.10 sq. in. This is a new method of designating wire sizes. It is a complete departure from the steel wire gauge system previously used. See Figure 14-11.

Spacings and sizes of wires in welded wire reinforcement are identified by *style*. A typical style designation is: 6 × 12—W16 × W8. This specifies a welded wire reinforcement which has a 6" spacing for horizontal wires, a 12" spacing for transverse wires, longitudinal wires (W16) of 0.16 sq. in., and transverse wires (W8) of 0.08 sq. in.

A welded deformed wire reinforcement style would be specified the same way except that D-number wire sizes would be substituted for the W-number wire sizes. The terms *longitudinal* and *transverse* are related to the manufacturing process. They bear no relationship to the position of the wires in a concrete structure. Some of the stock sizes of welded wire reinforcement used in light construction are shown in Figure 14-12.

Spacing of wires in welded wire reinforcement provides a wide variety of combinations. Some of the many spacing combinations possible are indicated in the following chart:

| | | | |
|---|---|---|---|
| 2" × 4" | 3" × 3" | 4" × 4" | 6" × 6" |
| 2" × 6" | 3" × 12" | 4" × 8" | 6" × 12" |
| 2" × 12" | 3" × 16" | 4" × 12" | 12" × 12" |
| 2" × 16" | 4" × 16" | 16" × 16" | |

Recommended styles of welded wire reinforcement for some specific applications are shown in Figure 14-13.

### Placing steel reinforcement

Placing steel reinforcement is a very important operation and is performed at the job site. See Figure 14-14. Reinforcement should be placed so that it is protected by an adequate coverage of concrete. See Figure 14-15.

Reinforcing steel carries the tensile load in concrete. It must, therefore, be lapped at a splice. The overlap distance for deformed bars should be at least 24 bar diameters and never less than 12". Smooth bars should be lapped a greater distance than deformed bars. See Figure 14-16. Welded wire reinforcement should be lapped at least one full stay spacing plus 2". An 8" × 8" material would require a minimum of 10" overlap.

## Lightweight Concrete

The weight of concrete is dependent on the type of aggregate used. Aggregates used in normal-weight concrete include sand and gravel, crushed stone, and air-cooled blast-furnace slag. Lightweight concrete includes aggregates of expanded shale, clay, and slate; expanded blast furnace slag; sintered fly ash; coal cinders; and natural lightweight materials such as pumice and scoria. See Figure 14-13. Generally, local availability determines the use of a given aggregate.

A cubic foot of concrete is considered to be *lightweight* if it has a density or weight between 105 and 125 pcf (pounds per cubic foot). Normal-weight concrete weighs more than 125 pcf. The weight of concrete is based on the density or oven-dry weight per cubic foot of concrete. The use of lightweight aggregates can reduce the weight of concrete about 20% to 45% compared with the weight of concrete made with normal-weight aggregates, without compromising the structural properties. Study comparisons have also shown that worker efficiency increases with lightweight masonry units when compared with normal-weight units.

## Metric Wire Area, Diameter & Mass With Equivalent U.S. Customary Units

| Metric Units* | | | | U.S. Customary Units** | | | | Gage Guide |
|---|---|---|---|---|---|---|---|---|
| Size ★ (mw = Plain) (mm²) | Area (mm²) | Diameter (mm) | Mass (kg/m) | Size ★ (w = Plain) (in²x100) | Area (in²) | Diameter (in) | Weight (lb/ft) | |
| MW200 | 200 | 16.0 | 1.57 | W31 | .310 | .628 | 1.054 | |
| MW130 | 130 | 12.9 | 1.02 | W20.2 | .202 | .507 | .687 | 7/0 |
| MW120 | 120 | 12.4 | .941 | W18.6 | .186 | .487 | .632 | 6/0 |
| MW100 | 100 | 11.3 | .784 | W15.5 | .155 | .444 | .527 | 5/0 |
| MW90 | 90 | 10.7 | .706 | W14.0 | .140 | .422 | .476 | |
| MW80 | 80 | 10.1 | .627 | W12.4 | .124 | .397 | .422 | 4/0 |
| MW70 | 70 | 9.4 | .549 | W10.9 | .109 | .373 | .371 | 3/0 |
| MW65 | 65 | 9.1 | .510 | W10.1 | .101 | .359 | .343 | |
| MW60 | 60 | 8.7 | .470 | W9.3 | .093 | .344 | .316 | 2/0 |
| MW55 | 55 | 8.4 | .431 | W8.5 | .085 | .329 | .289 | |
| MW50 | 50 | 8.0 | .392 | W7.8 | .078 | .314 | .263 | 1/0 |
| MW45 | 45 | 7.6 | .353 | W7.0 | .070 | .298 | .238 | |
| MW40 | 40 | 7.1 | .314 | W6.2 | .062 | .283 | .214 | 1 |
| MW35 | 35 | 6.7 | .274 | W5.4 | .054 | .262 | .184 | 2 |
| MW30 | 30 | 6.2 | .235 | W4.7 | .047 | .245 | .160 | 3 |
| MW26 | 26 | 5.7 | .204 | W4.0 | .040 | .226 | .136 | 4 |
| MW25 | 25 | 5.6 | .196 | W3.9 | .039 | .223 | .133 | |
| MW20 | 20 | 5.0 | .157 | W3.1 | .031 | .199 | .105 | |
| MW19 | 19 | 4.9 | .149 | W2.9 | .029 | .192 | .098 | 6 |
| MW15 | 15 | 4.4 | .118 | W2.3 | .023 | .171 | .078 | |
| MW13 | 13 | 4.1 | .102 | W2.0 | .020 | .160 | .068 | 8 |
| MW10 | 10 | 3.6 | .078 | W1.6 | .016 | .143 | .054 | |
| MW9 | 9 | 3.4 | .071 | W1.4 | .014 | .135 | .048 | 10 |

*Metric wire sizes can be specified in 1 mm² increments.
**U.S. customary sizes can be specified in .001 in² increments.

Note ⬩ — For other available wire sizes, consult other WRI Publications or discuss with WWF manufacturers.

Note ★ — Wires may be deformed, use prefix MD or D, except where only MW or W is required by building codes (usually less than MW26 or W4).

**Figure 14-11.** This chart shows metric wire area, diameter, and mass with equivalent U. S. Customary units. (Wire Reinforcement Institute)

### Common Styles of Metric Welded Wire Reinforcement (WWR) With Equivalent US Customary Units[3]

| | $A_s$ (mm²/m) | Metric Styles (MW = Plain wire)[2] | Wt. (kg/m²) | Equivalent Inch-Pound Styles (W = Plain Wire)[2] | $A_s$ (in²/ft) | Wt. (lbs/CSF) |
|---|---|---|---|---|---|---|
| A[1 & 4] | 88.9 | 102x102 - MW9xMW9 | 1.51 | 4x4 - W1.4xW1.4 | .042 | 31 |
| | 127.0 | 102x102 - MW13xMW13 | 2.15 | 4x4 - W2.0xW2.0 | .060 | 44 |
| | 184.2 | 102x102 - MW19xMW19 | 3.03 | 4x4 - W2.9xW2.9 | .087 | 62 |
| | 254.0 | 102x102 - MW26xMW26 | 4.30 | 4x4 - W4.0xW4.0 | .120 | 88 |
| | 59.3 | 152x152 - MW9xMW9 | 1.03 | 6x6 - W1.4xW1.4 | .028 | 21 |
| | 84.7 | 152x152 - MW13xMW13 | 1.46 | 6x6 - W2.0xW2.0 | .040 | 30 |
| | 122.8 | 152x152 - MW19xMW19 | 2.05 | 6x6 - W2.9xW2.9 | .058 | 42 |
| | 169.4 | 152x152 - MW26xMW26 | 2.83 | 6x6 - W4.0xW4.0 | .080 | 58 |
| B[1] | 196.9 | 102x102 - MW20xMW20 | 3.17 | 4x4 - W3.1xW3.1 | .093 | 65 |
| | 199.0 | 152x152 - MW30xMW30 | 3.32 | 6x6 - W4.7xW4.7 | .094 | 68 |
| | 199.0 | 305x305 - MW61xMW61 | 3.47 | 12x12 - W9.4xW9.4 | .094 | 71 |
| | 362.0 | 305x305 - MW110xMW110 | 6.25 | 12x12 - W17.1xW17.1 | .171 | 128 |
| C[1] | 342.9 | 152x152 - MW52xMW52 | 5.66 | 6x6 - W8.1xW8.1 | .162 | 116 |
| | 351.4 | 152x152 - MW54xMW54 | 5.81 | 6x6 - W8.3xW8.3 | .166 | 119 |
| | 192.6 | 305x305 - MW59xMW59 | 8.25 | 12x12 - W9.1xW9.1 | .091 | 69 |
| | 351.4 | 305x305 - MW107xMW107 | 9.72 | 12x12 - W16.6xW16.6 | .166 | 125 |
| D[1] | 186.3 | 152x152 - MW28xMW28 | 3.22 | 6x6 - W4.4xW4.4 | .088 | 63 |
| | 338.7 | 152x152 - MW52xMW52 | 5.61 | 6x6 - W8xW8 | .160 | 115 |
| | 186.3 | 305x305 - MW57xMW57 | 3.22 | 12x12 - W8.8xW8.8 | .088 | 66 |
| | 338.7 | 305x305 - MW103xMW103 | 5.61 | 12x12 - W16xW16 | .160 | 120 |
| E[1] | 177.8 | 152x152 - MW27xMW27 | 3.08 | 6x6 - W4.2xW4.2 | .084 | 60 |
| | 317.5 | 152x152 - MW48xMW48 | 5.52 | 6x6 - W7.5xW7.5 | .150 | 108 |
| | 175.7 | 305x305 - MW54xMW54 | 3.08 | 12x12 - W8.3xW8.3 | .083 | 63 |
| | 317.5 | 305x305 - MW97xMW97 | 5.52 | 12x12 - W15xW15 | .150 | 113 |

[1]Group A - Compares areas of WWR at $f_y$ = 60,000 psi with other reinforcing at $f_y$ = 60,000 psi
Group B - Compares areas of WWR at $f_y$ = 70,000 psi with other reinforcing at $f_y$ = 60,000 psi
Group C - Compares areas of WWR at $f_y$ = 72,500 psi with other reinforcing at $f_y$ = 60,000 psi
Group D - Compares areas of WWR at $f_y$ = 75,000 psi with other reinforcing at $f_y$ = 60,000 psi
Group E - Compares areas of WWR at $f_y$ = 80,000 psi with other reinforcing at $f_y$ = 60,000 psi

[2]Wires may also be deformed, use prefix MD or D, except where only MW or W is required by building codes (usually less than a MW26 or W4). Also wire sizes can be specified in 1mm² (metric) or .001 in.² (inch-pound) increments.
[3]For other available styles or wire sizes, consult other WRI publications or discuss with WWR manufacturers.
[4]Styles may be obtained in roll form. Note: It is recommended that rolls be straightened and cut to size before placement.

**Figure 14-12.** Common styles of metric welded wire reinforcement with equivalent U.S. Customary units. (Wire Reinforcement Institute)

| Type of construction | Recommended style | Remarks |
|---|---|---|
| Basement floors | 6 x 6 - W1.4 x W1.4<br>6 x 6 - W2.1 x W2.1<br>6 x 6 - W2.9 x W2.9 | For small areas (15' foot maximum side dimension) use 6 x 6 - W1.4 x W1.4. As a rule of thumb, the larger the area or the poorer the subsoil, the heavier the gauge. |
| Driveways | 6 x 6 - W2.9 | Continuous reinforcement between 25' to 30' contraction joints. |
| Foundation slabs (Residential only) | 6 x 6 - W1.4 x W1.4 | Use heavier gauge over poorly drained subsoil, or when maximum dimension is greater than 15'. |
| Garage floors | 6 x 6 - W2.9 x W2.9 | Position at midpoint of 5" or 6" thick slab. |
| Patios and terraces | 6 x 6 - W1.4 x W1.4 | Use 6 x 6 - W2.1 x W2.1 if subsoil is poorly drained. |
| Porch floor<br>A. 6" thick slab up to 6' span<br>B. 6" thick slab up to 8' span | 6 x 6 - W2.9 x W2.9<br><br>4 x 4 - W4 x W4 | Position 1" from bottom form to resist tensile stresses. |
| Sidewalks | 6 x 6 - W1.4 x W1.4<br>6 x 6 - W2.1 x W2.1 | Use heavier gauge over poorly drained subsoil. Construct 25' to 30' slabs as for driveways. |
| Steps (Free span) | 6 x 6 - W2.9 x W2.9 | Use heavier style if more than five risers. Position reinforcement 1" from bottom of form. |
| Steps (On ground) | 6 x 6 - W2.1 x W2.1 | Use 6 x 6 - W2.9 x W2.9 for unstable subsoil. |

**Figure 14-13.** Recommended styles of welded wire reinforcement for specific concrete applications. (Wire Reinforcement Institute)

**Figure 14-14.** Flat sheets of welded wire reinforcement will strengthen the floor of this large structure. The welded wire reinforcing is 6 × 6—W4 × W4. (Wire Reinforcement Institute)

**Figure 14-16.** Sheets of welded wire reinforcing are overlapped at least one full stay spacing plus 2". (Wire Reinforcement Institute)

Recommended Protection for Reinforcement

| Application | Minimum Concrete Protection |
|---|---|
| Footings | 3" |
| Concrete surface exposed to weather | 1 1/2" for no. 5 bars and smaller |
| Slabs and walls | 3/4" |
| Beams and girders | 1 1/2" |
| Joists | 3/4" |
| Columns | Not less than 1 1/2" or 1 1/2 times the maximum size aggregate |
| Corrosive atmosphere or severe exposures | Protection shall be suitably increased |

From: "Building code requirements for reinforced concrete," ACI 318

**Figure 14-15.** The American Concrete Institute recommends the above minimum concrete protection for reinforcement. It means that rods used for reinforcing must be covered by that much concrete.

**Warning!** Contact with wet (plastic) concrete, cement, mortar, grout, or cement mixtures can cause skin irritation, severe chemical burns, or serious eye damage. Wear waterproof gloves, a long-sleeved shirt, full-length trousers, and proper eye protection when working with these materials. If you must stand in wet concrete, wear high top waterproof boots. Wash wet concrete, mortar, grout, cement, or cement mixtures from your skin immediately. Flush eyes with clear water immediately upon contact. Seek medical attention if you experience a reaction to contact with these materials.

# REVIEW QUESTIONS
# CHAPTER 14

Write all answers on a separate sheet of paper. Do not write in this book.

1. What are the four basic materials that are used to make concrete?
2. Portland cement is _____ because it sets and hardens by reacting with water. This chemical reaction is called _____.
3. A general purpose cement suitable for all uses when special properties are not required is _____.
   A. Type I
   B. Type II
   C. Type III
   D. Type IV
   E. Type V
4. What type of cement produces high strength quickly, usually in a week or less?
5. Air-entrained cements improve resistance to freeze-thaw action and scaling caused by chemicals used in snow and ice removal. True or False? Why?
6. Aggregates usually constitute about _____ to _____ percentage of the volume of concrete.
7. Aggregates do not affect the quality of finished concrete very much. True or False? Why?
8. Fine aggregate or sand should vary uniformly in size from very fine to _____".
9. The maximum size of coarse aggregate is usually _____" in diameter.
10. What is the result of using the maximum allowable particle size?
11. Water usually constitutes from _____% to _____% of the total volume of concrete.
12. _____ are any materials added to the concrete batch before or during mixing other than Portland cement, water, and aggregates.

13. Identify four types of admixtures used in concrete.
14. What is the purpose of adding air-entraining admixtures to concrete?
15. What type of admixture is used to slow down the setting time of concrete?
16. What is the most commonly used type of accelerating admixture?
17. List three desirable qualities of plastic concrete.
18. List five desirable qualities of hardened concrete.
19. The proportion of water to cement, stated in gallons of water per bag (94 lb.) of cement, is called the _____.
20. The ideal curing temperature of concrete is _____ °F.

21. Stronger concrete results if it is kept moist for many days as opposed to only a couple. True or False?
22. The mix consistency or degree of stiffness of plastic concrete is called _____.
23. Generally, how long should concrete be mixed?
24. ASTM C 94 requires that ready-mixed concrete must be delivered and discharged from the truck mixer within _____ hours after the water has been added to the mix.
25. What are the two main types of steel reinforcing used in concrete?
26. Identify five types of aggregates used in lightweight concrete.

A power trowel like this one is used to quickly smooth large concrete slabs, such as the floors of commercial and office buildings.

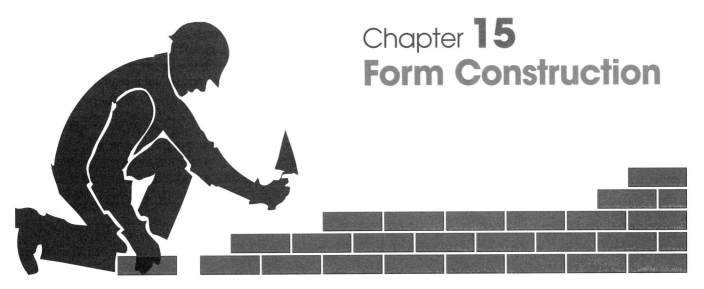

# Chapter 15
# Form Construction

Concrete structures require forms to provide the desired shape and surface texture. See Figure 15-1. Forms are made of wood, steel, fiberglass, hardboard, plastics, or other materials. The forms must be strong enough to resist the forces developed by the plastic (liquid) concrete. Regular concrete weighs about 150 pounds per cubic feet (lb/ft$^3$). Poured in a form 8' high, it would create a pressure of about 1200 pounds per square feet (lb/ft$^2$) along the bottom side of the form. Also, the forms should give the desired surface texture to the concrete structure.

Safe construction procedures must be followed when constructing forms for concrete. Two excellent sources of information are the **American Concrete Institute (ACI)** and the **Occupational Safety and Health Administration (OSHA).**

## Form Materials

Wood is the most popular form material. Both lumber and plywood are generally used in form construction. Construction lumber is used for frames, bracing, and shoring while plywood is used for the form surface. Boards may be used instead of plywood if certain surface patterns are desired. However, plywood is manufactured in more than 40 surface textures, ranging from glass-smooth to board-and-batten panels. Its large, smooth surface, resistance to change in shape when wet, and its ability to resist splitting has made it the most popular material for form construction.

**Figure 15-1.** A— Insulated form for concrete wall. (American Polysteel Forms) B—Soil as a form material for footing. C— Using typical dimension lumber for footing form. D—1 1/8" thick plywood forming system designed especially for residential and low industrial concrete walls. E—Steel panel forms. (Symons Corporation) F—Bronze strips used to form terrazzo sections. (National Terrazzo and Mosaic Association, Inc.)

Virtually any exterior-type plywood can be used for concrete formwork because it is made with waterproof glue. However, the plywood industry markets a product called Plyform, which is specially designed for concrete forming.

*Plyform* is an exterior-type plywood made from special wood species and veneer grades to assure high performance. This product is available in Plyform Class I and Plyform Class II

Class I is stronger and stiffer than Class II. Either can be purchased with a high-density overlaid surface on each side. This provides a very smooth, grainless surface that resists abrasion and moisture penetration.

Structural 1 Plyform is also available in some areas of the United States. It is especially designed for engineered applications. It is stronger and stiffer than either Plyform Class I or II. This plywood is also made with high-density overlay faces. All Plyform panels are made in thicknesses of 19/32", 23/32", and 3/4". Panel size is 4' × 8'.

Plywood is also valued as a form material because of its ability to bend. Curved surfaces can be formed easily if the thickness of the sheet and the direction of face grain is considered. Figure 15-2 shows the minimum bending radii for plywood panels.

The rate of bending can be increased for a given thickness sheet by sawing kerfs across the inner face at right angles to the curve. See Figure 15-3. Curves can also

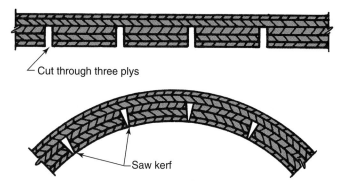

Figure 15-3. Saw kerfs through three plies help bend short radius curves.

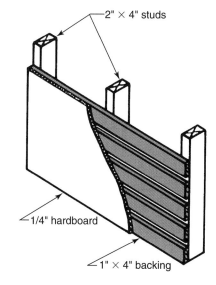

Figure 15-4. Recommended backing for hardboard lined forms.

| Plywood Thickness | Medium Bend Radius | |
| --- | --- | --- |
| | Perpendicular to Face Grain | Parallel to Face Grain |
| 1/4" | 16" | 48" |
| 5/16" | 24" | 60" |
| 3/8" | 36" | 72" |
| 1/2" | 72" | 96" |
| 5/8" | 96" | 120" |
| 3/4" | 120" | 144" |

Figure 15-2. Minimum bending radii for plywood panels used in form construction.

be formed by using two or more thinner sheets bent one at a time and then fastened together.

Hardboard that is tempered, specially treated, and 1/4" thick can be used as a form facing material. This material is coated with plastic to prevent water penetration. It produces a very smooth surface, but is essentially a form liner. It must be supported by a backing of lumber. Figure 15-4 shows the recommended method of backing for hardboard liners.

Steel or aluminum are frequently used for form construction, both frames and form facing. See Figures 15-5 and 15-6. Steel or aluminum angles and other structural shapes are used for the frames of forms. See Figure 15-7. They provide greater strength and support heavier loads than wood frames. Metal frame forms can be faced with either plywood or sheet metal. Metal forms, as you might

expect, have a longer life than wood forms. This somewhat offsets their greater cost.

Fiberglass form facing is increasing in use. This material is usually prefabricated in the size and shape desired. It is being used extensively in precast concrete plants and pan forms for ribbed concrete floors are frequently made of fiberglass as well.

Insulating board and rigid foam can be used as form liners but have little or no strength. Most often they are left attached to the concrete when the form is removed. Clips can be used for this purpose.

# Form Designs

Forms should be designed so that they are practical and economical. They should be in the correct shape, width, and height. They must be strong enough to resist the pressure of the plastic concrete. They must be able to retain their shape during the pouring and curing phases.

**Figure 15-5.** Steel-Ply ™ is the industry's most popular modular forming system. More than 80 panel and filler sizes make it ideal for handset or gang forms. (Symons Corporation)

**Figure 15-6.** Symons Silver ™ is a lightweight aluminum system that makes residential forming operations very efficient and productive. (Symons Corporation)

Wet concrete should not leak from joints and cause fins and ridges.

Forms should be simple as possible to build and use in sizes that can be easily handled and stored. They must be designed so that they can be removed without damaging the concrete. Forms should also be safe for those who work around them.

The primary consideration in form design is usually its strength. It must support its own weight, and the weight of the liquid concrete and any other loads that may be placed upon it such as wind, workers, and equipment. For most general form requirements, concrete that is made with natural sand and gravel aggregates weighs about 150 lb/ft³. Designs presented in this chapter are based on that weight of concrete and average conditions.

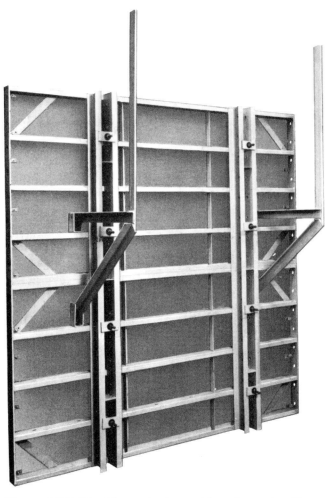

**Figure 15-7.** Steel-framed, plywood-faced panels with standard walkway. (Symons Corporation)

## Form elements

Forms generally have five elements. See Figure 15-8. These elements are sheathing, studs, wales, braces or supports, and ties and/or spreaders.

*Sheathing* gives the surface its shape and texture. It keeps the concrete in place until it hardens.

*Studs* support the sheathing and prevent bowing. Some types of forms do not require studs, but most wooden forms do.

*Wales* are used to align the forms, secure the ties, and support the studs.

*Braces and supports* provide lateral support against wind and other forces. One type of brace that is easy to use is the turnbuckle. See Figures 15-9 and 15-10. It can be used to align the form and is only required on one side of the form. These braces are usually spaced 8' to 10' apart.

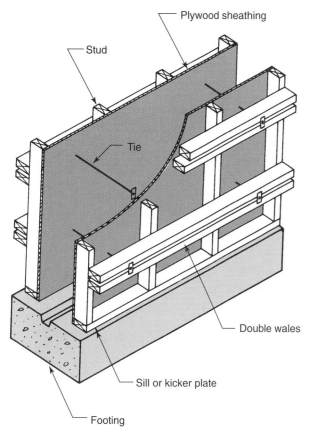

Plywood sheathing

Stud

Tie

Double wales

Sill or kicker plate

Footing

**Figure 15-8.** Parts of the concrete form.

***Ties and spreaders*** prevent the sides of the form from spreading or moving together. See Figure 15-11.

Wooden spreaders are removed from the inside of the form as the concrete is poured to their level. Ties remain in the concrete and thus become a permanent part of the structure.

When the concrete has cured and the forms are removed, the ties will project out from the surface of the concrete. They may be broken off or disconnected internally depending on the type. Figure 15-12 shows two types of ties which may be taken apart to remove most parts from the wall. Figure 15-13 shows another type of tie.

When specifying ties, it is necessary to know the type required, the wall thickness, the actual size of sheathing, studs and wales, and the required break back. The ***break back*** is the depth the tie is broken off in the concrete. Some ties have spreader washers and these take the place of wood spreaders.

A variety of patented ties are available to secure and hold form walls apart. The typical patented tie has a rod that passes through the wall with a holding device at each end. Two broad classes are popular: *continuous single-member* ties and *internal disconnecting* types. They can support loads ranging from 1000 lb. to 50,000 lb.

The most common type of continuous single-member tie is the *snap tie*. It can be used with either single or double waler systems and is made for wall thicknesses from 6"

to 26". See Figure 15-11. Wedges or wedge brackets are used to secure these ties which are nothing more than a metal rod that extends from the outside surfaces of the opposing wales. Small plastic cones or metal washers along the rod act as spreaders to hold the walls at the proper spacing. See Figure 15-11. The protruding sections of the ties are snapped off after the forms are removed.

Another type of continuous single-member tie that is used is the *loop and tie*. It is secured with a tapered steel wedge driven through the loop at the end of the tie and against a plate that is outside the waler. These ties are frequently used with prefabricated metal forms.

Internal disconnecting ties are designed for use with heavier construction work where heavier loads are expected. These ties have external sections that screw into an internal threaded section. Internal disconnecting ties are used for wall thicknesses from 8" to 36".

## Prefabricated forms

Prefabricated forms (panels) are pre-built usually in modular sizes. They can be fastened together on the job to produce the desired wall or column size. Prefab forms can be purchased, rented, or constructed by a contractor.

The most common type of prefab form has a metal frame with a plywood facing. See Figures 15-14 and 15-15. Special connectors are used to attach the modular units together. Some have hinged corners which aid in removing the forms. The most common size of module is 2' by 8'. The modules can be combined to make gang forms.

Gang forming is used on large jobs where repetitive forming is required. These are generally lifted with a crane since they are too heavy to be lifted by hand.

## Slip forms

Slip forms are used for casting very large structures of great height. They are raised slowly as the concrete is poured, using jacks. An average speed is about 15" per hour.

Figure 15-16 shows a section through a typical slip form. The form facing is 3/4" high-density overlaid plywood. The wales are lumber and the yokes are steel. The form is pulled upward by the jacks that move up jackrods embedded in the concrete. Jackrods can be recovered after the pouring is finished by pulling them from a recovery pipe.

Slip forms are also used for casting long, low walls such as median walls between lanes of traffic.

# Form Applications

Forms used in building construction can be grouped into those used for footings, walls, slabs, steps, beams or columns, and masonry support. The major attention in this text will be directed toward forms that can be built on site.

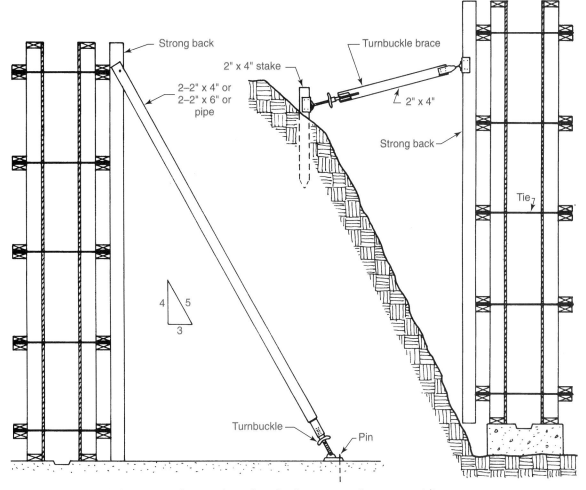

**Figure 15-9.** The turnbuckle brace can be used to align the form as well as support it.

**Figure 15-10.** Here a turnbuckle brace is used to support one side of the form. (Jack Klasey)

## Footing forms

Footings are not usually visible and therefore appearance is of little concern. However, they must be located accurately and built to the specified dimensions. If the soil is firm and not too porous, it may serve as the form for foot-

ings. See Figure 15-17. However, inferior results are possible if the soil absorbs too much water from the concrete or soil falls into the concrete. Figure 15-18 shows two types of earth footing forms.

Concrete footing forms are usually constructed from 2" construction lumber the same width as the footing thickness. The boards are placed on edge and held in place by stakes and cross-spreaders. See Figure 15-19. Soil may be used to fill cracks around the forms and provide some support.

**Trade Tip.** Remember that footings must be poured on undisturbed or well-compacted earth. Do not fill in low spots under the footing with loose earth.

Occasionally, a site will require stepped footings. Figure 15-20 illustrates one method of forming stepped footings. The height and length of the steps should coincide with the length and course height of one or more masonry units being used. Always study the plans carefully.

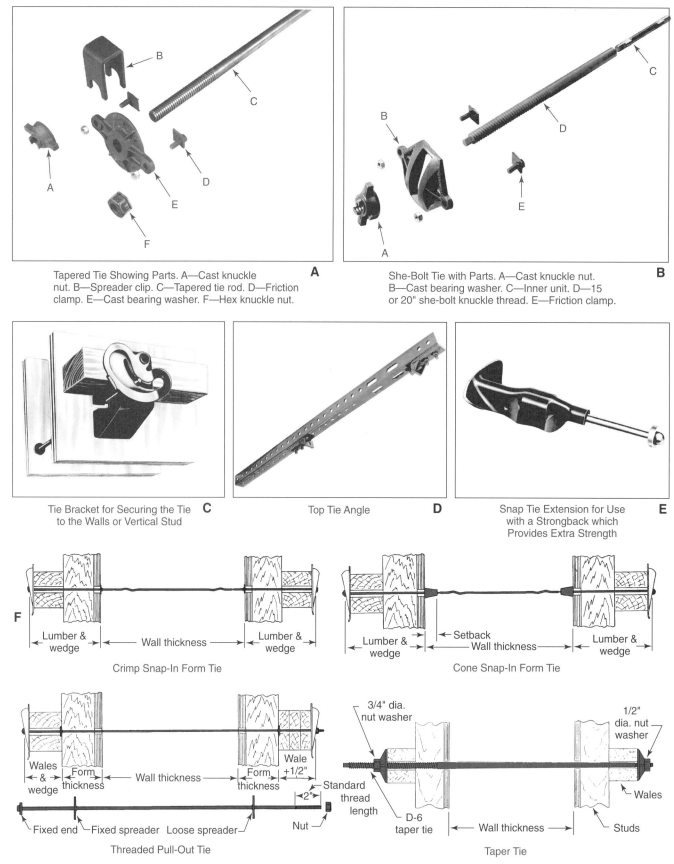

Tapered Tie Showing Parts. A—Cast knuckle nut. B—Spreader clip. C—Tapered tie rod. D—Friction clamp. E—Cast bearing washer. F—Hex knuckle nut.

**A**

She-Bolt Tie with Parts. A—Cast knuckle nut. B—Cast bearing washer. C—Inner unit. D—15 or 20" she-bolt knuckle thread. E—Friction clamp.

**B**

Tie Bracket for Securing the Tie to the Walls or Vertical Stud

**C**

Top Tie Angle

**D**

Snap Tie Extension for Use with a Strongback which Provides Extra Strength

**E**

**F**

Lumber & wedge — Wall thickness — Lumber & wedge

Crimp Snap-In Form Tie

Lumber & wedge — Setback — Wall thickness — Lumber & wedge

Cone Snap-In Form Tie

Wales & wedge — Form thickness — Wall thickness — Form thickness — Wale +1/2"

Fixed end — Fixed spreader — Loose spreader — Nut — Standard thread length — 2"

Threaded Pull-Out Tie

3/4" dia. nut washer — 1/2" dia. nut washer — D-6 taper tie — Wall thickness — Wales — Studs

Taper Tie

**Figure 15-11.** Top —Panel accessories. Bottom —Drawings show cross-sectional view of tie assemblies attached to the form. (Superior Concrete Accessories, Inc., Symons Corporation, and Dayton Sure-Grip and Shore Company)

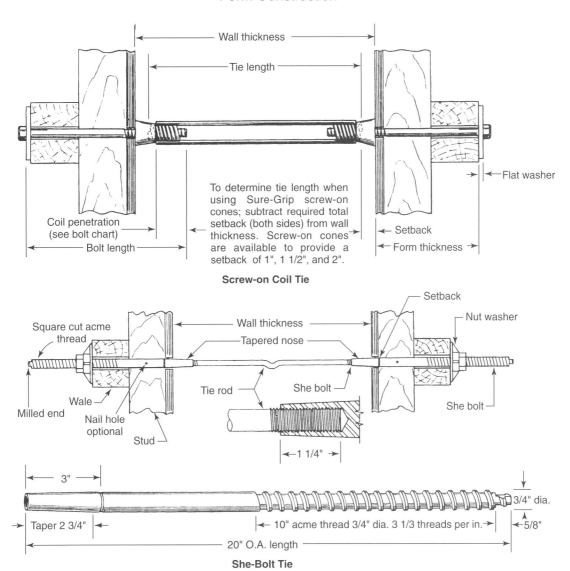

**Screw-on Coil Tie**

To determine tie length when using Sure-Grip screw-on cones; subtract required total setback (both sides) from wall thickness. Screw-on cones are available to provide a setback of 1", 1 1/2", and 2".

**She-Bolt Tie**

**Figure 15-12.** Threaded ties such as these can be disconnected. Protruding portions can be removed from the wall. (Dayton Sure-Grip and Shore Company)

**Figure 15-13.** Concrete form tie.

**Figure 15-14.** Modular form components are produced in several sizes: 2' × 2', 2' × 4', 4' × 2', 4' × 4', 8' × 2' and 8' × 4'. Plywood facing (3/4") is attached to a steel frame. (Symons Corporation)

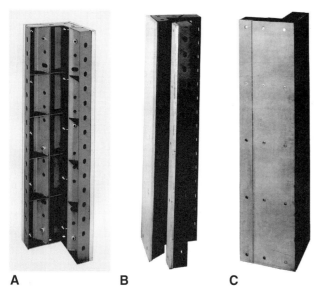

A          B          C

**Figure 15-15.** Strippable inside corner assemblies that provide a means of releasing forms from inside captive conditions such as counterfort walls, box conduits, and vertical, square and rectangular shafts. (Symons Corporation)

**Figure 15-17.** This clay soil will make an acceptable footing form.

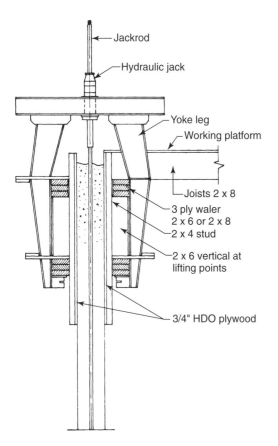

**Figure 15-16.** Section through a typical slip form. (American Plywood Association)

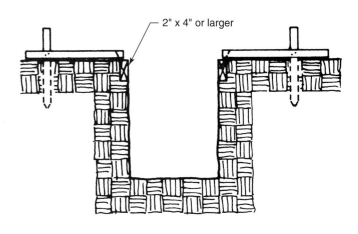

**Figure 15-18.** Earth footing forms.

Isolated footings for piers and columns are formed with bottomless boxes usually made with construction lumber. See Figure 15-21. The width of the boards should be the same as the thickness of the footing.

Tapered footing forms are built like a hopper. Two sides must have the exact dimensions of the footing while the other two are slightly larger so that cleats may be attached along their edges. See Figure 15-22. Side and end sections are held together with these 2 × 4 cleats. The form will try to rise as the concrete is poured so it must be securely anchored.

## Wall forms

Wall forms can be built-in-place or prefabricated. The type used depends on the complexity of the walls, need to reuse, or preference.

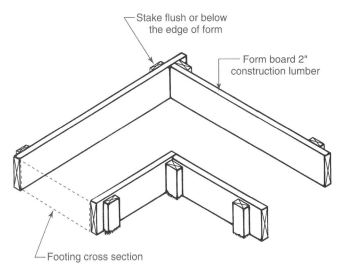

**Figure 15-19.** Footing forms made of 2" construction lumber.

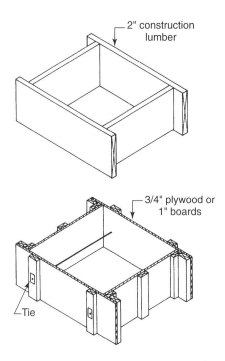

**Figure 15-21.** Isolated footing forms made of 2" construction lumber and 3/4" plywood.

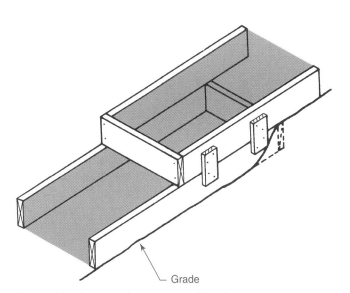

**Figure 15-20.** Form for a stepped footing.

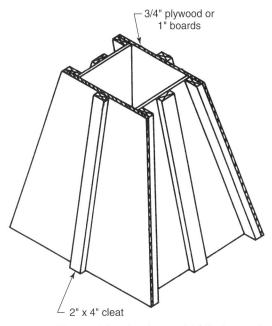

**Figure 15-22.** Tapered footing form of 3/4" plywood or 1" boards. This form will be held in place by stakes driven into the ground.

To build in-place forms, first attach a sole plate to the footing. Use concrete nails, power driven nails, or stakes. See Figure 15-23. The sole plate provides a place to attach the studs. It should be set back from the wall line the thickness of the sheathing. Be sure that it is properly located and straight because this will determine the position of the wall.

Then cut studs to desired length and toenail them to the top of the plate as in Figure 15-24. In a form for a typical basement wall, spacing of the studs is generally 24" with 3/4" plywood or 1" boards used for sheathing. Stud spacing should be decreased as the height of the wall increases.

Sheathing is nailed to the studs when they are in place. The first sheet should be leveled so that remaining sheets will be level. Holes for ties can be drilled as the sheathing is being attached. Placement of the ties determines the location of the wales. See Figure 15-25.

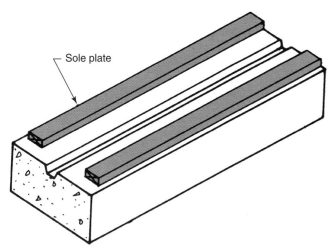

**Figure 15-23.** First step in form building is to attach sole plate to footing with power-driven nails or concrete nails.

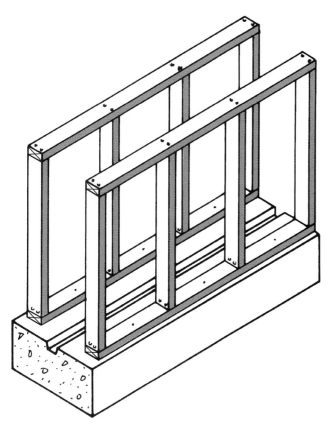

**Figure 15-24.** Second step in building wall forms is to toe-nail wall studs to sole plate.

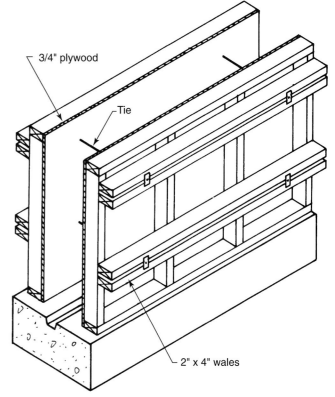

**Figure 15-25.** Completed section of a wall form with sheathing, ties, and wales in place.

Wales are attached to the outside of the studs using nails, clips, or wires. Either single or double wales can be used.

One end of the braces is attached to the wales. The other end is anchored in the ground or some other solid support.

If the wall is too narrow or deep for a person to work inside, the inside wall can be built in a horizontal position and tilted into place. Also, any reinforcing steel should be placed in position before the inside form is placed.

More and more prefab forms are being used. They can be assembled on the ground and then positioned. Bracing is required for them just as for the built-in-place forms.

Corners of forms are weak and must be given special attention. The forms must be tied together firmly to prevent concrete leakage. Corner brackets or tie rods can be used to give the needed strength. See Figure 15-26. Some forms have a series of metal eyes fastened to the sheathing panels that allow a metal rod to secure them together. Another method of tightening the corners is to lap the wales at the corner and secure them. Wooden wedges can be driven behind the wales to close any openings.

## Wall openings

Any openings in the wall, such as for windows, can be formed using bucks. **Bucks** are wood or steel frames set in the form between the inner and outer form to make an opening in the wall. See Figure 15-27.

A typical buck design consists of an outside frame reinforced by 2 × 4s placed on edge with horizontal cross braces. Either 3/8" plywood, 1" boards, or 2" construction lumber is used for the construction of bucks. Thinner material requires more bracing. If a recess is required to receive the frame, a recess strip is nailed to the outside of the buck's frame. If a wood nailing strip is required in the concrete, a wedge-shaped strip of wood is attached to the frame.

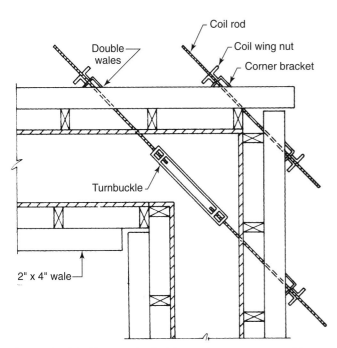

**Figure 15-26.** Form units held together at corner with corner brackets and tie rods.

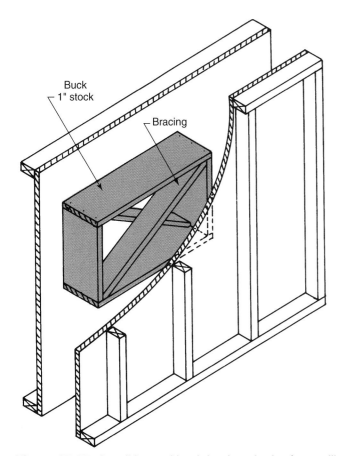

**Figure 15-27.** A well braced buck in place in the form will cast a void for a window or other opening through the wall. They are braced to prevent distortion.

Bucks for arches are generally made from two pieces of plywood that are fastened at the top of a typical rectangular buck. The arched form is enclosed with 1/4" plywood or heavier material as needed.

Some metal window and door frames can be set in the forms and serve as bucks. In this instance, they must be carefully plumbed, aligned, and well-braced to maintain proper placement.

Small openings for vents, ducts, and large diameter pipes can be required. They can be formed using small wood box frames that have been built to the proper size of the openings. Rigid foam plastic blocks and fiber or sheet metal sleeves can also be used.

### Slab forms

Forms for drives, walks, patios, or other flatwork are either wood or metal. Two-inch lumber is commonly used with wood or steel stakes for bracing.

For slabs 4" thick, 2 × 4 lumber is recommended. See Figure 15-28. One-inch lumber can be used, but stakes must be much closer together to keep the form straight. A 5" slab can be formed with 2 × 4's, but 2 × 6 materials are preferred. See Figure 15-28. Slabs that are 6" thick require 2 × 6 lumber.

Form material for slabs should be smooth, straight, and free of knot holes and other surface imperfections. Remember that *dressed* (machined smooth) 1" lumber is actually 3/4" thick and 2" lumber is 1 1/2" thick. Widths are usually 1/2" less than the nominal size. It is important to know the actual size of materials when building forms.

On large projects, metal forms are generally used to form slabs. They save time, stand rough handling, and can be used many more times than wood.

Wood stakes are made from 1 × 2, 1 × 4, 2 × 2, or 2 × 4 material. See Figure 15-29. They are spaced about 4' apart for 2" thick formwork. A spacing of about 2' is recommended for 1" thick formwork. Steel stakes are available for use with wood or metal forms. They are easier to drive and much stronger. Even though they cost more than wood stakes, they can be cheaper over a period of time due to their long life.

Stakes should be driven slightly below the top of the forms to aid in screeding and finishing the concrete slab. Drive the stakes straight and plumb so the form will be true. Use double-headed (form) nails to attach the stake to the form. Drive the nail through the stake into the form.

### Forming curves

Horizontal curves are easy to form if plywood, 1" boards, hardboard, or sheet metal is used. Curves with a short radius can be formed with 1/4" or 1/2" plywood with the face grain vertical. If heavier lumber is desired or a sharper bend is necessary, saw kerfing will help. Wet lumber will bend more easily than dry. Figure 15-30 shows some details for forming horizontal curves.

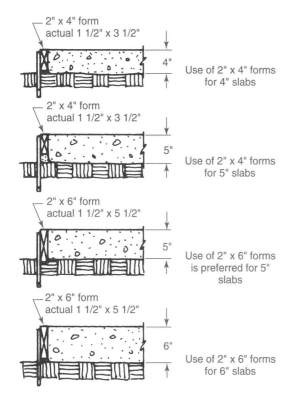

**Figure 15-28.** Recommended form lumber size for typical slab thicknesses.

**Figure 15-29.** Slab form being attached to stake using form nails. Note the nail has a double head for easy removal.

Long gentle curves may be formed using 2" thick material. Staking must be very secure because it must support a heavier load. It is a good idea to use extra stakes for all types of curves.

The proper grade can be maintained by using a mason's line or builder's level. The ***builder's level*** is an accurate spirit level combined with a telescope on a circular base. It sits on a tripod. When using the level, the grade is determined at a given point and a reading is taken on a

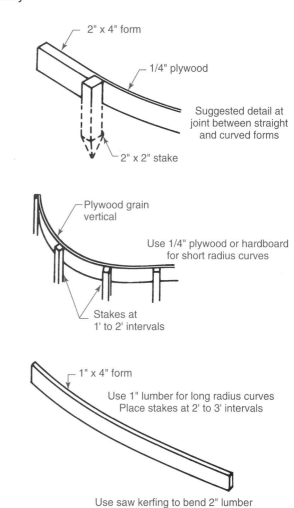

**Figure 15-30.** Forming horizontal curves using plywood.

rod or rule. Then the rod or rule is moved along the form line and measurements are made on the stakes indicating proper grade. When enough stakes have been sighted, the forms can be attached at the height indicated.

## Permanent forms

Divider strips in a patio are an example of permanent forms. They are left in place for decorative purposes and/or to serve as control joints. These forms should be made from 1 × 4 or 2 × 4 cypress, redwood, or cedar that has been treated with wood preservative.

The strips should be anchored to the concrete with 16d galvanized nails. Drive them through the board from alternate sides at 16" intervals. Outside forms would be nailed from only one side.

Masking tape can be used on the top edge of divider strips to protect the wood from abrasion and stain while placing the concrete. It is easily removed when the slab has set.

Another type of permanent form is a hollow tube. It is used to cast a void in concrete slabs, beams, or other applications. See Figures 15-31 and 15-32. The tubes are generally a fiber material and are produced in sizes from 2 1/4" to 36" in diameter. Lengths vary to meet specifications.

## Construction joints

A large slab or wall is often poured in sections that require a construction joint. Construction joints are formed by placing a bulkhead in the form. A **bulkhead** is a piece of material that prevents the concrete from moving past a certain point in the form. See Figure 15-33.

**Figure 15-32.** Hollow tube fiber forms used to cast voids in slabs, walls, piers, or other construction. (Sonoco Products Company)

Figure 15-34 shows a method of forming steps built directly on sloping ground. The sides of the form are usually made from 2 × 6 or 2 × 8 lumber. The riser forms are 1" boards. The beveled bottom edge allows troweling of the step surface right up to the riser.

**Figure 15-31.** Hollow tube form used to cast cylindrical column.

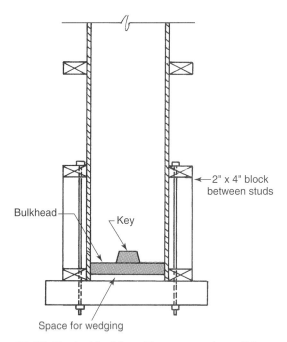

**Figure 15-33.** Typical bulkhead in a concrete wall form.

## Forms for steps

Two basic types of concrete steps can be constructed. One type is built directly on the slope of the ground. The other type is supported at the top and bottom with an open space under the steps.

Figure 15-35 shows one forming method for self-supporting steps. These steps must be reinforced with steel and must rest on a foundation. This form must be securely braced to hold the weight.

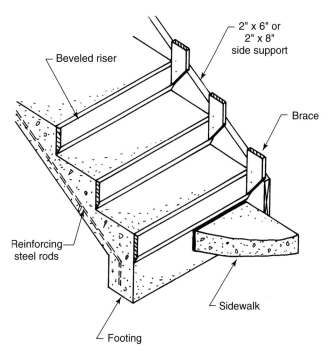

**Figure 15-34.** Forming technique for steps cast on sloping terrain.

### Column forms

Concrete columns reinforced with steel are commonly cast in square, rectangular, and round shapes. Forms for the square and rectangular columns are easily constructed from wood. See Figure 15-36. However, round column forms are usually metal, fiberboard, or paper. See Figure 15-37. Round prefabricated forms are produced for various lengths and diameters. Some forms are reusable. However, forms made of fiberboard and paper are not reusable, but have some advantages. They are economical, lightweight, and the paper surfaces produce a very

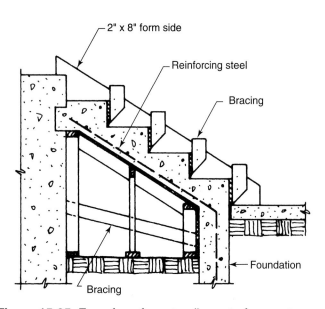

**Figure 15-35.** Form for a free-standing set of concrete steps.

smooth concrete surface with no seams. See Figure 15-37. Fiberboard forms often leave a visible seam. The form can be removed from cured concrete by sawing almost through the form material and then making the final cut with a sharp knife blade.

### Supporting steel reinforcing in forms

The amount, type, and spacing of steel reinforcing required in a slab, wall, or other concrete structure will depend on many factors. Frequently, residential construction requires no reinforcing at all if not in a seismic (earthquake) risk zone. The concern in form construction, however, is how to maintain the proper spacing between the forms and the rebars. Common methods include spacer blocks, plastic snap-on devices, and wood strips. See Figure 15-38. The spacer blocks and snap-on devices remain in the concrete, but the wood strips are removed as the concrete is placed.

One approach to supporting horizontal steel reinforcing bars in a concrete wall form is to place the rebar on the tie rods. Vertical rebar, if required, can be attached to the horizontal bars. This method maintains proper placement of the rebar and involves no special devices.

### Centering for arches

**Arch centering** is a structure used to support the masonry while the arch is being constructed. It is usually made from wood. The ribs are 2" lumber and the lagging is usually 3/4" × 2" strips of wood. See Figure 15-39. The lagging should be cut 1" shorter than the width of the masonry wall so that it will not interfere with the mason's line. The ribs are cut to the shape of the arch. The centering must be supported adequately for the weight it is to bear.

## Form Maintenance

Form maintenance is important because it reduces the cost by extending the life of a form. Many forms are damaged during stripping, which means to remove them from cured concrete. Do not use metal pry bars. These will damage the forms — especially wood forms. Use wood wedges and tap lightly if necessary.

Forms should be inspected, cleaned, repaired, and lightly oiled after they are removed. Use a stiff fiber brush on wood forms. A wire brush will raise the grain and roughen the surface. Mill oil of 100 or higher viscosity and pale in color is generally used for oiling forms. A liberal amount of oil should be applied a few days before the plywood is used to reduce sticking. Wipe to a thin film just before using the form. Forms should be solid stacked or stacked in small packages with faces together for future use. Panels should be protected from the sun and rain.

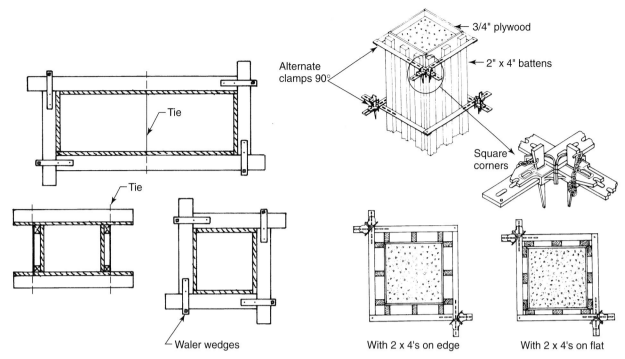

**Figure 15-36.** Square and rectangular column forms.

**Figure 15-37.** Top —Column form made of fiber paper. The form is only used once because it is destroyed by the removal process. Bottom —Concrete column after paper form has been removed.

# Insulated Concrete Wall Forms

An alternative to building a poured concrete wall that uses conventional forming materials and techniques is to use stay-in-place forms. See Figure 15-40. Several new products have become available recently that eliminate the use of traditional forms, speed the construction process, and produce an insulated wall. Five of these products are Lite Form™, Keeva Wall ™, Greenblock™, ConForm™, and Polysteel Forms™. As builders become more familiar with these and other similar products, the traditional removable type of forms may become a thing of the past in residential and some light commercial construction.

## Lite-Form

**Lite-Form** is an insulated wall form system that uses panels of plastic foam insulation as formwork. See Figure 15-41. It separates the panels with plastic ties so that the space between the panels can be filled with concrete. The insulation panels remain in place and become part of the wall. This results in a concrete wall that is insulated on both sides.

**Figure 15-38.** A spacer block is being used to support the horizontal rebar at the proper height in the form.

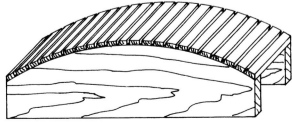

Jack Arch Center

Segmental Arch Center

**Figure 15-39.** Centering for jack and segmental arches.

**Figure 15-40.** The forms to be used to cast the foundation of this structure are made from rigid foam plastic modular units. (Lite-Form, Inc.)

**Figure 15-41.** This building technique allows the insulation panels to remain in place and become a part of the wall which results in an insulated wall. (Lite-Form, Inc.)

## Advantages of Lite-Form

The following are advantages of Lite-Form:

1. Thermal performance of the wall is increased. The fully insulated wall reduces thermal breaks and achieves an R-value of about 20.
2. Durability of the wall is increased because expansion and contraction of the wall is reduced and hence cracking is reduced.
3. Construction of the concrete wall requires less time. The light weight of the forms permits factory or on-site assembly and skilled labor is not required for installation. Costly stripping is not required since the forms remain in place. Furring strips are not required as drywall or paneling can be attached directly to the form boards or plastic ties.
4. The construction season can be extended in cold climates. The insulation permits the concrete to be cast in cold weather.

## Disadvantages of Lite-Form

The following are disadvantages of Lite-Form:
1. Extra care must be taken during the casting process. If the concrete is cast too rapidly, the forms can blow out causing damage and waste.
2. A pump truck with a hydraulic boom is suggested to place the concrete properly in the forms.
3. Foundations built with foam forms cost more.

## Installation/application

Form sections are built from 8" high by 8' long strips of polystyrene. Special form ties and corner ties are required. See Figure 15-42. The wall form is built course by course until the desired height is reached. A typical footing is generally used. The completed form is braced and reinforced before filling with concrete.

**Figure 15-42.** The patented form ties which are the trademark of this forming system are clearly visible as the wall forms are being put into place. (Lite-Form, Inc.)

Exterior siding or interior paneling can be screwed into the plastic ties. Any polystyrene that is above the grade and exposed should be covered with a trowel-applied protective coating. Electrical and plumbing lines can be installed in the 2" thick insulation. Conduit may be required by some codes.

Additional information about Lite-Form™ can be obtained by contacting:
Lite-Form Incorporated
P.O. Box 774
Sioux City, IA 51102

## *Keeva Wall*

Keeva Wall is essentially an insulated concrete post-and-beam form system. The basic material is expanded polystyrene (EPS) blocks. Each block has a row of 5" diameter holes on 8" centers running down the center. See Figure 15-43. When the blocks are stacked, the holes line up so that a 5" diameter void is created from the footing to the top plate. Bond beam blocks are used at floors and top of the wall. When vertical holes and horizontal channels

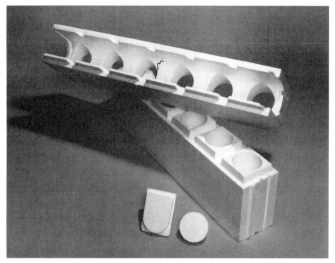

**Figure 15-43.** Keeva Wall blocks are 8" × 12" × 48". They are stacked course by course, joined together with interlocking tabs, are self-aligning, and require no bracing at the corners. (Keeva International, Inc.)

are filled with concrete and reinforcing steel, the result is a concrete post-and-beam structure that is encased in expanded polystyrene foam.

## Advantages of Keeva Wall

The following are advantages of Keeva Wall:
1. The system is easy to install and uses one-third less concrete than typical construction.
2. Keeva blocks interlock and are self-aligning, and no bracing is necessary at the corners.
3. As many of the cells can be filled with concrete as needed to resist lateral pressure. For example, if all cells are filled for a below grade application, the wall will have a 5" diameter post every 8" o.c.
4. The Keeva wall has an overall R-value of 24.

## Disadvantages of Keeva Wall

The following are disadvantages of Keeva Wall:
1. Extra care must be taken during the casting process. If the concrete is cast too rapidly, the form can blow out.
2. A pump truck with a hydraulic boom is suggested to place the concrete properly in the form.
3. Some time is needed to learn how to use this product.
4. Interior and exterior surface treatment can be somewhat more limited than with regular cast concrete.

## Installation/application

Keeva Wall blocks are 8" × 12" × 48". They are stacked course by course and joined together with interlocking tabs. They are self-aligning and no bracing is necessary at the corners. Floors are hung on ledgers bolted to the concrete bond beams. The anchor bolts are positioned during the pour. Later, 2 × 6 spacer blocks are put into cutouts so as not to crush the foam when the ledger is attached. Pressure treated 2 × 6s are bolted to the top of the wall with standard 1/2" anchor bolts.

The windows and doors are laid out after all the blocks have been stacked. Openings are cut after the concrete is poured. Reinforcing steel is placed around the openings as the blocks are stacked. Friction plugs are used to prevent concrete from entering the door and window opening areas.

Additional information about Keeva Wall blocks can be obtained by contacting:
Keeva International Inc.
1854 N. Acacia
Mesa, AZ 85213

## Greenblock

The Greenblock system is a foam concrete forming system that uses metric foam foundation units to form walls and foundations. The panel units are connected by polypropylene webs that hold the panels 5 1/2" apart. The webs are embedded in the foam every 5" o.c. and provide a dense fastening surface for wall treatments.

The blocks are 2 1/2" on the weather side and 2" on the interior side. A series of knobs and grooves provide for alignment and attachment. Foam end caps and blockouts are available. Notches in the webs eliminate the need to tie horizontal rebar.

### Advantages of Greenblock

The following are advantages of Greenblock:
1. The wall formed with Greenblock is energy efficient and reduces thermal breaks.
2. Construction of the concrete wall requires less time than conventional forming methods.
3. The construction season can be extended in cold weather.
4. The wall can be finished with most external building materials.

### Disadvantages of Greenblock

The following are disadvantages of Greenblock:
1. The construction is expensive—more than the conventional concrete or wood construction.
2. A pump truck with a hydraulic boom is necessary to place the concrete properly.
3. Extra care must be taken during the casting process to prevent blowout.

### Installation/application

Greenblocks are stackable units that consist of two 10" tall by 39" long panels held 5 1/2" apart with polypropylene webs. Dovetail grooves on the inside face of the panels bond the concrete and also lock in the foam end caps and blockouts.

Reinforcing steel connects the wall to the footing every 15" along the perimeter. A 2 × 4 kick plate is used to locate and anchor the first course of blocks. Laying the

blocks is generally begun at the corners working toward the middle of the wall. Vertical joints are staggered. If blocks don't meet on a module, they can be cut with a handsaw. Special top blocks taper out at the top to create a shelf 9 1/2" wide that can be used to support brick veneer or a beam flange.

The dense polypropylene flanges provide a handy anchor for most any exterior or interior wall finish. Dry wall, for example, can be screwed directly to the flanges.

Additional information about Greenblocks can be obtained by contacting:
Greenblock EPS Building System
P.O. Box 749
Woodland Park, CO 80866

## ConForm Polystyrene Block

ConForm Polystyrene Block is a type of insulated concrete wall form that uses interlocking blocks of plastic foam insulation. They are stackable so that the hollow cores can be filled with concrete. See Figure 15-44. ConForm is molded from a flame retardant EPS (expanded polystyrene) that possesses both thermal and acoustical properties.

### Advantages of ConForm Polystyrene Blocks

The following are some advantages of ConForm Polystyrene Blocks:
1. Thermal performance of the wall is increased. The fully insulated wall reduces thermal breaks and achieves an increased R-value.
2. Durability of the wall is increased because expansion and contraction of the wall is reduced and hence cracking is reduced.

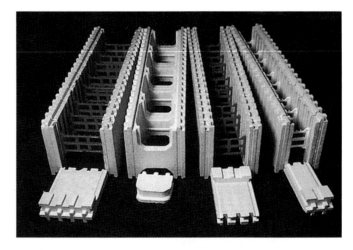

**Figure 15-44.** ConForm Polystyrene Block ™ is one type of insulated concrete wall form. The interlocking blocks of plastic foam insulation are stackable and provide the form for concrete. (American ConForm Industries, Inc.)

3. Construction of the concrete wall requires less time.
4. The construction season can be extended in cold climates.
5. The wall can be finished with most external building finishes.

## Disadvantages of ConForm Polystyrene Blocks

The following are some disadvantages of ConForm Polystyrene Blocks:

1. Extra care must be taken during the casting process to prevent blowout.
2. A pump truck with a hydraulic boom is necessary to place the concrete properly.
3. Foundations built with foam blocks cost more.

## Installation/application

ConForm Polystyrene Blocks are 10" × 10" × 40" and weigh 2 lbs. each. They are stacked course by course and joined together with an interlocking tongue and groove. See Figure 15-45. Unskilled labor can be used to stack the blocks so that the hollow cores are stacked over vertical reinforcing rods placed according to the local code requirements. Bracing is required as specified by the manufacturer. Electrical and plumbing lines can be installed within the thickness of the block shell.

Additional information about ConForm Polystyrene Blocks can be obtained by contacting:
American ConForm Industrial, Inc.
1820 South Santa Fe Street
Santa Ana, CA 92705

## Polysteel Forms

Polysteel Forms are a new and innovative expanded polystyrene (EPS) and steel form for the construction of super-insulated concrete walls. See Figures 15-46 and 15-47. Its core design creates a reinforced concrete wall

**Figure 15-46.** Polysteel Forms™ are made in two sizes. One size results in a 6" core of concrete for 8' deep basements, above grade residential walls, and light commercial construction. The other results in an 8" core of concrete for basements under two stories, basements with more than 8' of backfill, and large commercial projects, etc. (American Polysteel Forms)

**Figure 15-45.** ConForm products can be used to cost-effectively create almost any type of wall. (American ConForm Industries, Inc.)

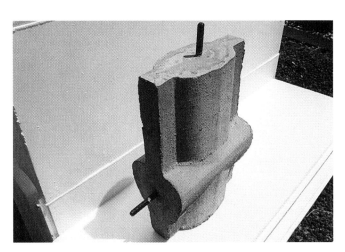

**Figure 15-47.** Polysteel Forms™ provide vertical columns of concrete at 12" o.c. and horizontal beams at 16" o.c. (American Polysteel Forms)

based on traditional post-and-beam construction and stay-in-place expanded polystyrene insulation. It provides for solid attachment for wall coverings in addition to a superior R-value rating.

Polysteel Forms are held together with five attachment and reinforcement strips, each 1" × 10" galvanized steel. Each form is 48" long by 16" high and is available in 6" and 8" core sizes. Vertical cores are located on 12" centers. Each form weighs 6 lb. and creates the same wall area as 140 lb. of concrete block. Walls constructed of Polysteel Forms provide a shear wall strength of up to 3500 pounds per linear foot according to the manufacturer.

## Advantages of Polysteel Forms

The following are advantages of Polysteel Forms:
1. Concrete can be poured at temperatures as low as −15°F (-9°C).
2. Requires 25% less concrete than conventional forming methods, while providing up to 50% stronger concrete per square inch.
3. Provides a superb building system for earthquake zones and hurricane prone areas.
4. Creates a super-insulated concrete wall which reportedly reduces heating and cooling costs by 50% to 80%.
5. Fire insurance rates are generally lower than for conventional, wood-frame buildings.

## Disadvantages of Polysteel Forms

The following are disadvantages of Polysteel Forms:
1. Costs more than conventional construction.

**Figure 15-48.** A pump truck with a hydraulic boom is required to place the concrete using this forming system. (American Polysteel Forms)

2. A pump truck with a hydraulic boom is necessary to place the concrete properly. See Figure 15-48.
3. Extra care must be taken during the casting process to prevent blowout.

**Figure 15-49.** Typical corner bracing. In this case, corner braces are attached to the Polysteel foundation forms that have already been filled with concrete and cured for several days. (American Polysteel Forms)

4. Proper bracing is necessary to produce a quality product. See Figure 15-49.

## Installation/application

Polysteel forms are stackable units that use a tongue and groove design to hold the units together during wall construction. See Figure 15-50. Forms can be placed on wet footings or cured concrete. Vertical reinforcing steel (#4) is placed at 2' intervals, or as required by the local building code, in the footing to coincide with the vertical cells in the forms. Vertical reinforcing is extended upward through the wall as the forms are positioned. See Figure 15-51. Form guides (2 × 4s) generally serve as guide rib-

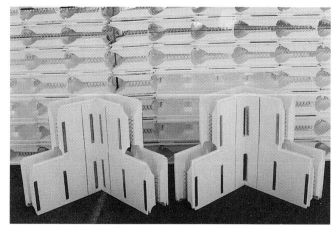

**Figure 15-50.** Factory pre-cut corner sets are color coded to ensure that the Polysteel™ furring strips and the concrete cores line up properly. Use all red lined forms in one corner and all black lined forms in the next corner. (American Polysteel Forms)

**Figure 15-51.** #4 (1/2" diameter) horizontal steel rebar is wired to the #4 vertical rebar using rebar ties and a special tool called a pigtail. (American Polysteel Forms)

**Figure 15-52.** Polysteel Forms™ enabled the homeowner to have three beautifully arched garage doors while supporting the roof load above. (American Polysteel Forms)

bons creating a channel for the forms to prevent movement on the footing.

Corner braces must be properly installed at corners and along the walls to support the assembly and prevent floating. Door frames (bucks) and window frames are also braced on both sides to be sure they remain plumb.

The forms are first placed at a corner of the wall using factory pre-cut, mitered corner forms. Form units can be cut to fill in non-standard lengths. These joints must be braced and taped with silver duct tape. Vertical joints are staggered to form a running bond so that the cores line up to develop a "post" of concrete. Intersecting walls can be formed by removing a section of the sidewall form to receive the intersecting wall form. These corners should also be braced. Horizontal reinforcing is installed along the wall as frequently as required by the application. Additional steel is required over openings. Curved walls, arches, and other unique shapes can be formed using Polysteel Forms. See Figure 15-52. Figure 15-53 shows a residence under construction using Polysteel forms and the completed structure.

The manufacturer of Polysteel Forms provides a detailed *User Manual* that describes each step in the process. Additional information can be obtained by contacting:

American Polysteel Forms
Berrenberg Enterprises, Inc.
5150-F Edith, NE
Albuquerque, New Mexico 87107

A

B

**Figure 15-53.** A —Polysteel Forms™ used to form the walls of this complex residential structure. B —The completed home. (American Polysteel Forms)

# REVIEW QUESTIONS
# CHAPTER 15 ▬▬▬▬▬▬▬

Write all answers on a separate sheet of paper. Do not write in this book.

1. Name three materials commonly used for concrete form construction.
2. The weight of a cubic foot of *average* concrete is about _____ lb.
3. Exterior plywood, a popular form material, is made with _____ glue.
4. A specially designed plywood for form construction is called _____.
5. Steel frame forms are generally stronger than wood frame forms. True or False?
6. Traditional concrete forms generally have five elements. Name them.
7. What is the function of wales?
8. The most common size of a prefab form module is _____.
9. _____ forms are used for casting very large structures of great height and long, low walls.
10. When may earth be used as a footing form material?
11. Stepped footings are used mainly for entrance concrete steps. True or False?
12. Wall forms rest on the _____.
13. Stud spacing for a typical basement wall form is about _____".
14. _____ are wood or steel frames set in the form between the inner and outer form to make a void in the wall.
15. What size form lumber is ordinarily used for a 4" thick slab?
16. The actual thickness of 1" lumber is _____".
17. Long, gentle curves may be formed with _____" thick lumber.
18. Give two examples of permanent forms (forms that are left in place).
19. A(n) _____ is a piece of material that prevents the concrete from moving past a certain point in the form.
20. *Centering* is a type of form used to support a(n) _____ during construction.
21. _____ can be used to coat forms to reduce sticking.
22. List three proprietary insulated concrete wall forms that are an alternative to conventional forming methods.
23. List three advantages of insulated concrete wall forms over conventional construction.
24. List three disadvantages of insulated concrete wall forms over conventional construction.

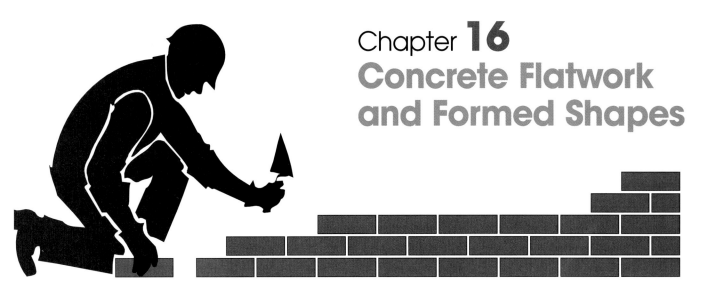

# Chapter **16**
# Concrete Flatwork and Formed Shapes

Concrete can be cast in practically any shape so long as a form can be built to contain it while it is in the plastic state. See Figure 16-1. This is one reason concrete is so valuable as a building material. However, most concrete is formed into slabs on grade, cast in place as walls or floors, or used in the manufacture of precast units (beams, panels, etc.). This chapter will cover the practical aspects of using concrete in these applications.

## Placing and Finishing Concrete

Concrete is moved about for placing by many methods. Some of the most popular methods include chutes, push buggies, buckets handled by cranes, and pumping through pipes. The method used should not restrict the consistency of the concrete. Consistency

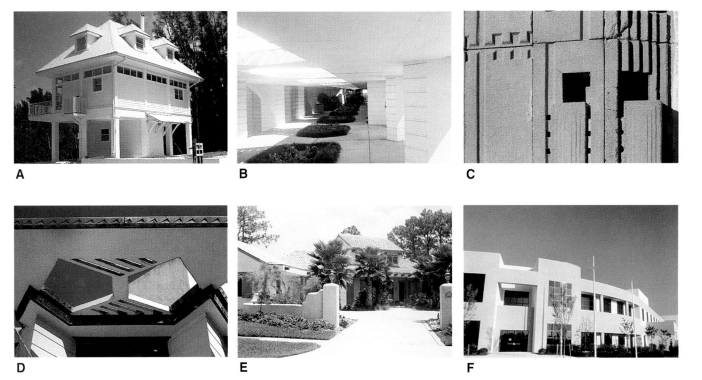

**Figure 16-1.** A—This modern beach cottage utilizes cast concrete piers and floor system. B—Concrete is the building material in this modern structure. C—Colored concrete cast in complex shapes to simulate stone. D—Geometric shapes sculpted in concrete form this roof overhang. E—Concrete was used for the fence and driveway. F—A modern tilt-up panel building. (Tilt-up Concrete Association)

A
B
C

D
E
F

should be governed by the placing conditions and the application. In other words, if the conditions permit the use of a stiff mix, the equipment should be designed to handle it.

## Preparation

Subgrade should be properly prepared before the concrete arrives. **Subgrade** is ground on which concrete is poured. Forms should be in place and level. Subgrades should be smooth and moist. See Figure 16-2.

**Trade Tip.** Moistening the subgrade prevents rapid loss of water from the concrete when pavements, floors, and similar flatwork are being placed. This is especially important in hot weather.

**Figure 16-2.** This worker is checking the subgrade for smoothness and proper height. (Portland Cement Association)

Forms should be tight, clean, and securely braced. Poorly constructed forms will sag and leak. Forms should be constructed from materials that will give the desired texture to the finished concrete.

Treating the forms with oil or other preparations will make form removal easier after the concrete has hardened. Wooden forms exposed to the sun for long periods should be saturated (soaked) thoroughly with water to tighten the joints.

Reinforcing steel should be clean and free of loose rust and scale. All hardened mortar should be removed from the steel before placing the concrete.

## Placing concrete

Once poured, concrete should be as near as possible to its intended location. For example, it should not be placed in large quantities in one place and allowed to run or be worked over a long distance in the form. Segregation (separating) of ingredients and sloping work planes result from this practice. It should be avoided. Generally, concrete should be placed in horizontal layers having uniform thickness.

If the form is deep, the concrete should be compacted after each layer is placed. Layers are usually 6" to 12" thick in reinforced concrete and up to 18" for nonreinforced applications.

Concrete should not be allowed to drop freely more than 3' or 4'. Drop chutes of rubber or metal can be used when placing concrete in thin vertical sections.

## Slab construction

Placement of concrete in slab construction should be started at the most distant point of the work so that each batch may be dumped against the previously placed concrete, not away from it. See Figure 16-3. Care should be taken to prevent stone pockets (areas of excessive large aggregate) from occurring. If this happens, some of the aggregate can be moved to areas where there is more cement paste to surround them.

Concrete placement in walls should begin at either end and progress toward the center. The same order should be used for each successive layer.

Concrete should be placed around the perimeter (outer edges) first in large flat open areas. Whatever method of placement is used, do not allow water to collect at the ends and corners of forms.

Compacting the concrete is always necessary. Work the mix with a spade or rod to be sure all spaces are filled and air pockets are worked out. This is called puddling, spading, or rodding. Mechanical vibrators can be used either in the concrete or on the forms.

This process should help to eliminate stone and air pockets, and consolidate each layer with those previously placed. It also will bring fine material to the faces and top for proper finishing.

Mechanical vibration does not make the concrete stronger, but it does permit the use of a stiffer mix that will be stronger than a wet mix. Too much vibration will cause segregation of particles. Judgment must be used to determine the proper amount. Indicators of sufficient vibration are the appearance of a line of mortar along the forms and by the sinking of the coarse aggregate into the mortar.

## Placing on hardened concrete

When fresh concrete is placed on hardened concrete, it is important to produce a good bond and a watertight

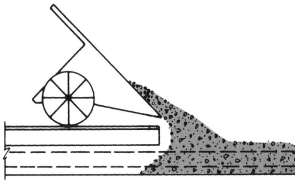

Proper placement of concrete slab

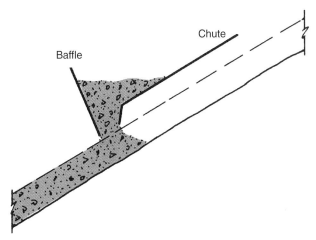

Proper placement of concrete on a
sloping surface

**Figure 16-3.** Proper method of dumping concrete when pouring a slab.

joint. To assure this result, the hardened concrete should be reasonably level, rough, clean, and moist. If some coarse aggregate are left exposed, it will aid in bonding. All loose or soft mortar should be removed from the top surface of the hardened concrete before placing the fresh batch.

For floors that require two courses of concrete, the top of the first course (lower level) can be broomed with a steel or stiff bristle broom just as it sets. The surface should be heavily scored and cleaned before the grout coat and top course are placed. The grout coat is a mixture of Portland cement and water. It usually has the consistency of thick paint and is scrubbed into the surface of the slab just before the top course is placed.

If old concrete is to receive a new topping, it must be thoroughly roughened and cleaned of dust and loose particles, grease, oil, or other materials. The surface might best be chipped with pneumatic tools or sand blasted to expose sound concrete. Hardened concrete must be moistened thoroughly before new concrete is placed on it, but no pools of water should be left standing on the surface of the existing concrete.

Where concrete is to be placed on hardened concrete or rock, a layer of mortar, 1/2" to 1" thick, is placed on the hard surface. This mortar provides a cushion for the new concrete and prevents stone pockets and aids in securing a tight joint. The mortar is generally made of the same materials as the concrete, but without the coarse aggregate. It should have a slump of less than 6".

## Pneumatic application of concrete

*Pneumatically applied concrete,* frequently called **shotcrete,** is a mixture of Portland cement, aggregate, and water shot into place by means of compressed air. Aggregate up to 3/4" can be used with some equipment. Shotcrete can be applied by either the wet mix or dry mix process.

In the dry mix process, the cement and aggregates are mixed in a relatively dry condition. This mix is pumped through a hose to a nozzle where water is added. At least 45 psi air pressure and a nozzle velocity of about 400 feet per second is used to force the dry materials through the hose. Water pressure at the nozzle should be at least 15 psi higher than the air pressure. In the wet mix process, the concrete is premixed before it is applied pneumatically.

Water content can be kept to a minimum using the pneumatic process. Therefore, high-strength, durable concrete may be obtained. The key element in this process is the person controlling the nozzle. In both processes the worker directs the nozzle and controls the thickness of the concrete layer and the angle of application. Pneumatic application can be used for new construction and repair work in difficult locations. It also works well where relatively thin sections and large areas are involved.

### Finishing concrete slabs

Concrete slabs can be finished several ways. It depends on the effect desired and the use of the product. Some surfaces can be left rough, others broomed, floated, or troweled. Still other surfaces can be textured, colored, or have exposed aggregate.

### Screeding

Screeding is usually the first finishing operation after the concrete is placed in the forms. It is performed with a screed.

*Screeding* is the process of striking off the excess concrete to bring the top surface to the proper grade or elevation. See Figure 16-4. The edge of the screed is either straight or curved, depending on the surface requirements. The screed rests on the top of the form and is moved across the concrete with a sawing motion. It is advanced forward slightly with each movement.

An excess of concrete should be carried along in front of the screed to fill low places as the tool is moved forward. But if too much concrete is allowed to build up in front of the screed, it may tend to leave hollows behind it.

In normal concrete, the dry materials used are heavier than water. They will begin to sink or settle to the

**Figure 16-4.** This concrete is being screeded to strike off the excess.

bottom of a plastic concrete mixture shortly after placement. This settling action causes bleeding, which is when excess water rises to the surface. Bleeding does not usually occur with air-entrained concrete.

 **Warning!** It is very important that the first operations of placing, screeding, and darbying be performed before any bleeding takes place. If any finishing operation is performed on the surface while the bled water is present, serious scaling, dusting, or crazing can result.

The high and low spots can be eliminated and large aggregate embedded using a darby or bull float. See Figure 16-5. This operation should follow immediately after screeding to prevent bleeding. Some surface finishes may not need any further finishing, but most will require edging and jointing, floating, troweling, or brooming. Most will require one or more of these operations.

**Edging and jointing.** If edging is necessary, this could be the next operation. See Figure 16-6. Edging provides a rounded edge or radius to prevent chipping or damage to the edge. The edger is run back and forth until the desired finish is obtained. Care should be taken to cover all coarse aggregate and not to leave too deep a depression on the top of the slab. This indentation could be difficult to remove during subsequent finishing operations.

As soon as edging has been completed, the slab is jointed or grooved. See Figure 16-7. The bit (cutting edge) of the jointing tool cuts a groove in the slab that is called a control or contraction joint. Any cracking due to shrinkage caused by drying out or temperature change will occur at the joint. These cracks are not noticeable when controlled.

**Figure 16-5.** A bull float is used to eliminate high and low spots and to embed large aggregate in the concrete.

**Figure 16-6.** Edging produces a radius on the edge of the slab that prevents chipping. A trowel may be inserted between form and concrete to provide a track for the edging operation shown here. (Stanley Goldblatt)

**Figure 16-7.** Jointing a slab helps control any cracking due to shrinkage caused by drying out or temperature change. (Stanley Goldblatt)

The joint weakens the slab and induces cracking at that location rather than some other place.

In sidewalk and driveway construction, the tooled joints are generally spaced at intervals equal to the width of the slab, but not more than 20' apart. They should be perpendicular (at right angles) to the edge of the slab. The groove is usually made with a 3/4" bit.

Use a straightedge as a guide when making the groove. A 1 × 8 or 1 × 10 board will be ideal. Be sure the board is straight.

Large concrete surfaces can be jointed by cutting grooves with a power saw using an abrasive or diamond blade. When grooves are cut rather than jointed, the operation should be performed 4 to 12 hours after the slab has been finished. The cutting must be done before random shrinkage cracks develop, but after the concrete is hard enough not to be torn or damaged by the blade.

**Floating.** After concrete has been edged and jointed, it should be allowed to harden enough to support a person and leave only a slight foot imprint. Floating should not begin until the water sheen has disappeared. When all bled water and water sheen has left the surface, the concrete has started to stiffen. The surface is floated with wood or metal floats, or with a finishing machine using float blades. See Figure 16-8. Aluminum or magnesium floats work better especially on air-entrained concrete.

Metal floats reduce the amount of work required by the finisher. Drag is reduced and the float slides more readily over the surface. A wood float tends to stick to the surface and produces a tearing action. The light metal float also forms a smoother surface texture than the wood float. There are three reasons why concrete is floated:

**Figure 16-8.** Floating a concrete surface with a wood float. (Stanley Goldblatt)

1. To embed aggregate particles just beneath the surface.
2. To remove slight imperfections, waves, and voids.
3. To compact the concrete at the surface in preparation for other finishing operations.

Be sure not to overwork the concrete. This will bring excess water and fine aggregate material to the surface that will result in surface defects.

Floating can be done to provide a coarse texture as the final finish. If this is done, then a second floating may be necessary after the concrete has partially hardened.

**Troweling.** When a smooth, dense surface is desired, steel troweling is performed after floating. Frequently the cement mason will float and trowel an area before moving their knee boards. Troweling produces a smooth, hard surface. See Figure 16-9.

**Figure 16-9.** Troweling produces a smooth, hard surface on the concrete. It can be done by hand (left) or with a troweling machine (right).

During the first troweling, whether by hand or power, the trowel blade must be kept as flat against the surface as possible. If the blade is held at too great of an angle, then a washboard effect will result.

A new trowel is not recommended for the first troweling. An old trowel that is *broken in* can be worked quite flat without the edges digging into the surface.

As the concrete progressively hardens it can be troweled several times to obtain a very smooth and hard surface. Usually smaller size trowels are used for successive applications so that sufficient pressure can be exerted for proper finishing. If necessary, tooled joints and edges can be rerun after troweling to maintain uniformity.

**Brooming.** Steel troweling produces a very smooth surface and is often slippery when wet. Sidewalks, driveways, and other outside flatwork frequently are broomed or brushed to produce a slightly roughened surface. The broomed surface is made by drawing a soft-bristled push broom over the surface of the slab after steel troweling. See Figure 16-10. The concrete must be hard enough so that brooming will not damage the edges of tooled joints. Coarser textures for steep slopes can be produced by using a stiffer bristled broom. Brooming is usually perpendicular (at right angles) to the traffic to provide the most resistance.

**Figure 16-10.** Brooming produces a slightly roughened surface that reduces the danger of slipping on a smooth troweled surface.

## Form removal

Generally, it is best to leave the forms in place as long as possible for better curing. Sometimes, however, it is desirable to remove the forms as soon as possible. In either case, leave them in place until the concrete is strong enough to support its own weight and any other loads that may immediately be placed upon it. The concrete should be hard enough to resist damage from form removal.

Usually, the side forms of relatively thick sections can be removed in 12 to 24 hours after placing. Testing the concrete to determine hardness is better than relying on an arbitrary age for form removal. The age-strength relationship should be determined from representative samples of

concrete used in the structure and cured under job conditions. Figure 16-11 shows the times required to attain certain strength under average conditions for air-entrained concrete. It should be stressed that these are averages and strength is affected by materials used, temperature, and other conditions.

| Strength PSI | Age | |
|---|---|---|
| | Type I or Normal Cement | Type III or High-Early-Strength Cement |
| 500 | 24 hours | 12 hours |
| 750 | 1 1/2 days | 18 hours |
| 1500 | 3 1/2 days | 1 1/2 days |
| 2000 | 5 1/2 days | 2 1/2 days |

**Figure 16-11.** This chart shows the age-strength relationship of air-entrained concrete that must be considered when removing forms.

If the forms are tight and require wedging, only wooden wedges should be used. Do not place a pinch bar or other metal tools against the concrete to wedge forms loose. Start removing forms some distance from a projection. This relieves pressure against projecting corners and reduces the likelihood of breaking off the edges.

### Finishing air-entrained concrete

Air-entrained concrete has a slightly different consistency than regular concrete. This requires a little change in finishing operations. Since air-entrained concrete contains many tiny air bubbles that hold all the materials in concrete in suspension, it requires less mixing water than standard concrete (concrete that is not air-entrained). It also bleeds less and is the reason for different finishing operations.

There is no waiting for the evaporation of free water from the surface before floating and troweling. If floating is done by hand, an aluminum or magnesium float is essential. A wooden float will drag and increase the work necessary to finish the surface. If floating is done with a power finishing machine, there is practically no difference in the finishing procedure for air-entrained and standard concrete except that finishing can begin sooner with the air-entrained concrete.

Since most all horizontal surface defects and failures are due to finishing operations performed while bled water or excess surface moisture is present, better results are usually obtained with air-entrained concrete.

### Curing concrete

Curing a concrete slab is one of the most important operations in producing quality work. It is also one of the most often neglected operations. Even if the concrete is mixed, placed, and finished properly, poor quality work will result if proper curing operations are not followed.

Little or no moisture loss should be allowed during the early stages of hardening. If necessary, the concrete should be protected during this critical period. Newly placed concrete should be protected from the sun and not allowed to dry out too fast. This can be accomplished with damp burlap, canvas, or polyethylene film coverings. See Figure 16-12. The covering can be applied as soon as the surface is hard enough that it will not be marred. Keep the covering moist for at least three days.

**Figure 16-12.** Polyethylene film can be used to prevent newly placed concrete from drying out too fast. (Portland Cement Association)

Another method of curing is called ponding. *Ponding* is done by keeping an inch or so of water on the concrete surface usually by earth dikes around the edges of the slab.

Membrane curing compounds sprayed on the surfaces of the concrete are frequently used. See Figure 16-13. Uniform coverage is necessary and often two coats are required to provide adequate protection.

**Figure 16-13.** Membrane curing compounds can be sprayed on the concrete surface immediately after the concrete has had its final finishing operation (Portland Cement Association).

## Curing temperatures

The rate of chemical reaction between cement and water is affected by the temperature. Therefore, the temperature affects the rate of hardening, strength, and other qualities of the concrete.

**Cold weather construction.** The American Concrete Institute (ACI) maintains that if the air temperature averages less than 40°F (4°C), and if it is below 50°F (10°C), more than half of each day for three weeks in a row, this is defined as cold weather. In cold weather, concrete placement may require heated materials, a covering for the fresh concrete, or heated enclosure. See Figure 16-14. Under these conditions, concrete sets up more slowly, takes longer to finish, and gains strength more slowly. The hydration of the cement generates some heat, but this may not be enough. If the concrete freezes before it hardens, the damage done by freezing can reduce the final compressive strength by as much as 50%. Ice starts to form in plastic concrete when the concrete temperature approaches 27°F (– 3°C). The freezing point can be as low as 20°F (–7 °C) if there are admixtures in the mix. Ice requires more space than water and this expansion in wet concrete weakens the product by creating void spaces that disrupts the bond between cement paste and the aggregate.

**Figure 16-14.** Straw is used to protect newly placed concrete in cold weather. (Jack Klasey)

The temperature of concrete at the time of placing should generally be 50°F to 70°F (10°C to 21°C). The materials should never be heated to the point that the fresh concrete is above 70°F (21°C). This will reduce the strength.

Concrete should never be placed on a frozen subgrade. When subgrades thaw, uneven settling and cracking of the slab usually results. Snow and ice in the form takes up space intended for the concrete. Thawed subgrade should be recompacted before placing concrete to avoid uneven settlement and cracked concrete. Forms, reinforcing steel, and embedded fixtures should be free of ice when the concrete is placed. A thin layer of warm concrete should be placed on cold, hardened concrete when an upper layer is to be poured. The thick upper layer will shrink as it cools, and the lower layer will expand as it warms. Failure of the bond will result if care is not taken.

In cold weather, moisture for curing is still very important. Keep the concrete moist, especially near heating units. Maintain the temperature of normal concrete at 70°F (21°C) for three days or at 50°F (10°C) for five days. Keep the temperature of high early-strength concrete at 70°F (21°C) for two days or at 50°F (10°C) for three days. Do not allow the concrete to freeze for the next four days.

It is estimated that for every 10°F drop in concrete temperature, set time increases by approximately one third. Hydration stops completely at 14°F, but it will resume when the temperature warms up.

Set time can be shortened and the early strength of the concrete increased by ordering it with extra cement, Type III cement, chemical accelerators, or a combination of the three. Type III cement is similar to ordinary Type I cement, but it is almost 50% stronger after 24 hours, because the particles are finer. Type III cement may not be readily available from smaller companies. Chemical accelerators will provide the same benefits as adding an extra 100 pounds of cement per cubic yard or using Type III cement. Accelerators (Type E and Type C) are available in chloride-based or nonchloride-based formulations. Calcium chloride can increase the potential for corrosion of steel reinforcing if water is present.

Air-entrained concrete is not as susceptible to damage caused by freezing and thawing as standard concrete (not air-entrained), because ice crystals form in the tiny air spaces in the mix. Air-entrained concrete, for this reason, is generally a good choice for exterior concreting that will be exposed to freezing and thawing.

**Cold weather protection.** Fresh concrete should be protected from freezing, and curing conditions maintained to assure adequate strength of the concrete. Two methods are generally used to achieve this: insulating the concrete, and providing a protective cover with heat.

The most common way to protect concrete against low temperatures is by insulating it. Unless the temperature is too low, the heat of hydration can provide enough heat to protect the concrete. However, 6" of straw held in place with polyethylene sheeting or tarps will provide extra protection. An alternative is to use insulating or curing blankets. Insulating blankets are easy to install and can be secured with weights or ties. Most blankets are made from fiberglass or closed-cell polypropylene insulation that is laminated to canvas or some other material. To be effective, insulating blankets must lie flat on the concrete surface and edges secured to prevent air movement under the blanket.

In very cold weather, a tent or other heated enclosure can be used to protect the concrete. Be sure to use vented heaters or electric heaters. Heaters that produce carbon dioxide can cause a soft, chalky layer to form on the surface of the concrete. Carbon monoxide is dangerous to workers.

**Hot weather construction.** There is a good reason not to place concrete in hot, dry, windy weather. The concrete sets much faster and may result in cracked or poorly finished slabs. The *American Concrete Institute (ACI)* states the following:

"If the initial setting time for a concrete mix is 2 1/2 hours at 60°F (16°C), that time is likely to be reduced to about an hour or less at 95°F (35°C)."

Concrete loses its slump faster in hot weather and can become unworkable before a large load can be placed. In addition, moisture evaporates rapidly from the slab surface that reduces finishing time. Plastic shrinkage cracking is a potential problem for any placement done in hot, dry, windy weather. These cracks can appear after the slab has been placed, screeded, and bull-floated. The only thing that can be done in this situation is to continue sealing the cracks by troweling.

To prevent rapid drying in extremely hot weather, avoid high temperatures in fresh concrete. Temperature of the mixing materials can be cooled by using chilled water or ice. The ice should be melted by the time the concrete leaves the mixer. When concreting in hot weather, you should get an early start while it is cooler, break big placements into smaller sections if possible, and shade the operation if practical. Some contractors use additives like set retarders and superplasticizers to reduce the problems of rapid set or slump loss. These are not complete solutions, however, because superplasticizers can make a stiff mix flow more easily, but the effect wears off suddenly—sometimes before the finishing process is completed. The use of set retarders requires experienced cement finishers, because unexpected results can happen using these chemicals.

Subgrades should be saturated sometime in advance and sprinkled just ahead of placing the concrete. Wood forms should be treated or wetted thoroughly. Placing should not be delayed and it should be screeded and darbied immediately after placing. Covers, such as burlap, which are kept constantly wet, should be placed over the concrete as soon as it is darbied. When the surface is ready for final finishing, a small section should be uncovered immediately ahead of the finishers and recovered as soon as possible.

The purpose of curing is to maintain conditions in the setting concrete that encourages complete hydration. In other words, the purpose is to keep the water in the concrete from evaporating. Moisture must be present for at least seven days. Longer than seven days is desirable, because the curing process can continue for 28 days. The longer the curing process, the stronger, harder, and more durable the finished product will be.

Several techniques are used to trap moisture in fresh concrete. The most common method is to spray on a liquid curing compound that forms a thin film over the slab. Some contractors use ponding; sprinkling or fogging; covering the concrete with sheets of plastic; or applying wet sand, hay, or burlap to the surface after the finishing is completed. All of these methods work, but each has its advantages and disadvantages depending upon the specific situation.

Air-entrained concrete requires special expertise, especially during hot weather. It develops a rubberlike surface in hot weather if finishing is delayed. This concrete is then very hard, if not impossible, to surface smoothly.

# Joints in Concrete

Three basic types of joints are frequently used in concrete construction. These joints are as follows:

1. **Isolation joints.** Isolation joints (sometimes called expansion joints) are used to separate different parts of a structure to permit both vertical and horizontal movement. This type of joint is used around the perimeter of a floating slab floor and around columns and machine foundations. See Figure 16-15.

2. **Control joints.** Control joints provide for movement in the same plane as the slab or wall is positioned. They are used to compensate for

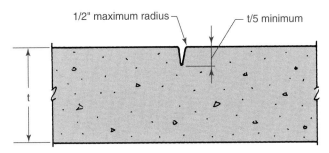

Hand Tooled Control Joint

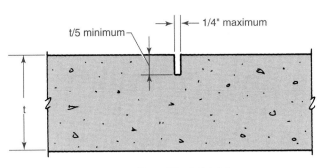

Sawed Control Joint

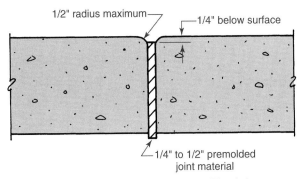

**Figure 16-15.** Detail of an isolation joint. The joint material can be flush in areas where no safety hazard from tripping exists. An example is a floating slab against a building wall.

contraction caused by drying shrinkage. Control joints should be constructed in such a way that they permit the transfer of loads perpendicular to the plane of the slab or wall. Figure 16-16 shows three types of control joints. If control joints are not used in slabs or lightly reinforced walls, random cracks will occur due to drying shrinkage. Control joints are sometimes called contraction or dummy joints.

3. **Construction joints.** Construction joints or bonded joints provide for no movement across the joint. They are only stopping places in the process of casting. Construction joints, however, may be made to perform as control joints. Figure 16-17 shows the details of a construction joint.

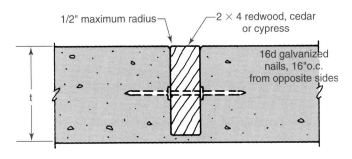

Wood Divider Strips

**Figure 16-16.** These types of control joints are used in sidewalks, drives, and patios.

## *Decorative and special finishes*

A variety of patterns, textures, and decorative finishes can be built into concrete during construction. Color can be added to the concrete. Aggregates can be exposed. Textured forms can be used. Concrete can be ground to produce a polished appearance. Geometric patterns can be scored or stamped into the concrete to resemble stone, brick, or tile. Divider strips can be used to form interesting patterns. The possibilities are unlimited.

## Colored concrete

Concrete can be colored using any of three methods: one-course, two-course, and dry-shake. See Figure 16-18.

The first two methods are similar. In both, the concrete mix is colored by adding a mineral oxide pigment prepared especially for use in concrete. White Portland cement will produce brighter colors or light pastel shades when used with light-colored sand. Normal gray cement can be used for black or dark gray colors.

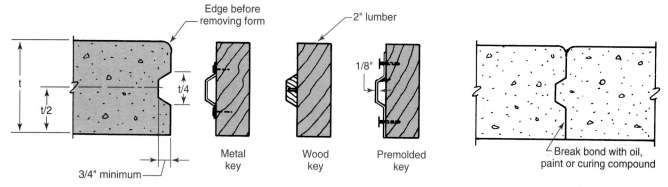

**Figure 16-17.** Details of a construction joint. The wood key can be made from a beveled 1" × 2" strip and is used for slabs 4" to 6" thick. The premolded key can be left in the slab permanently by tacking it lightly to the bulkhead.

**Figure 16-18.** This decorative exposed aggregate slab is further enhanced through the use of colored concrete to form a design. (Portland Cement Association)

All materials must be accurately controlled by weight to attain a uniform color in each batch. The amount of pigment used should never exceed 10% of the weight of the cement.

Color pigments should be mixed with the cement while dry. Use clean tools and a separate mixer to prevent streaking. When using the one-course method, uniform moistening of the subgrade is important for good color results.

The only difference between the one-course and two-course methods is that the two-course method uses a base coat of conventional concrete. The surface of the base coat is left rough to produce a good bond. The top coat may be placed as soon as the surface water disappears. The top coat of colored concrete is generally 1/2" to 1" thick.

In the dry-shake method, a commercially prepared dry color material is applied over the concrete surface after floating, edging, and grooving. Apply two coats of the dry-shake to the surface of the slab. Perform the finishing operations after each application of dry-shake. The color must be thoroughly worked into the concrete.

## Exposed aggregate

One of the most popular decorative concrete finishes is the exposed aggregate finish. It provides an unlimited color selection and a broad range of textures. See Figure 16-19. Not only are exposed aggregate finishes attractive, but they can be rugged, slip resistant, and immune to weather.

**Figure 16-19.** This exposed aggregate patio is striking in appearance and adds a factor of safety.
(Ideal Cement Company)

There are several ways to produce exposed aggregate finishes. One of the most common is called the *seeding* method. The procedure is to place, screed, and bull float or darby the concrete in the usual manner keeping the level of the surface about 3/8" to 1/2" lower than the forms. This space will be filled with the extra aggregate.

When these finishing operations have been completed, spread the aggregate uniformly with a shovel or by hand so that the entire surface is completely covered with a layer of stone. Next, embed the aggregate by tapping with a wood hand float or straightedge. Work the surface with the hand float until the surface is similar to that of a normal slab after floating.

Starting the next operation requires accurate timing. Usually you will wait until the slab can bear the weight of a worker on kneeboards with no indentations. Then brush the slab with a stiff nylon bristle broom to remove the extra mortar over the stones. Next, apply a fine spray of water along with brushing. If the aggregate becomes dislodged, stop the operation for a while. Continue washing and brushing until the water is clear and there is no noticeable cement film left on the aggregate. A surface retarder can be used for better control of the exposing operations, but is not necessary.

Another method of producing an exposed aggregate finish is to expose the stone in conventional concrete. No extra stone is added, but a high proportion of coarse to fine aggregate is necessary. The coarse aggregate should be uniform in size, bright in color, closely packed, and uniformly distributed. The slump of this concrete must be between 1" and 3".

A third method is to place a thin topping course of concrete containing special aggregates over a base of regular concrete. This technique is used for terrazzo construction.

Terrazzo toppings on outdoor slabs are generally 1/2" thick and contain marble, quartz, or granite chips. Random cracking is eliminated by using brass or plastic divider strips set in a bed of mortar. This type of terrazzo is called *rustic* or *washed terrazzo.*

## Textured finishes

Interesting decorative textures can be produced on vertical concrete surfaces by using textured form materials. See Figure 16-20. The variety of textures possible is almost endless. Figure 16-21 shows cast concrete textured to imitate stone. Special procedures may be required for these finishes.

Textured surfaces on slabs may be achieved by brooming, using mortar, dash coat, or rock salt, just to name a few. Brooming can be executed in a fine or coarse straight line pattern or a wavy texture. The procedure for brooming a surface was discussed earlier.

A travertine or keystone finish, as it is sometimes called, is produced by applying a dash coat of mortar over freshly leveled concrete. The dash coat is mixed to the consistency of thick paint and usually contains a yellow pigment. It is applied in a splotchy manner and with a dash brush. Numerous ridges and depressions are formed by this procedure.

After the coating has hardened slightly, the surface is troweled slightly to flatten the ridges and spread the mortar. The resulting finish is smooth on the high spots and coarse grained in the low areas. This effect looks like travertine marble from which it gets its name.

Another texture can be produced by scattering rock salt over the surface after hand floating or troweling. The salt is pressed into the surface so that the top of each grain is exposed. When the concrete has hardened, the surface is washed and brushed. The salt pellets will be dissolved leaving holes in the surface. Neither the rock salt nor travertine finish is recommended for areas that experience freezing weather. Water frozen in the recesses will ruin the surface.

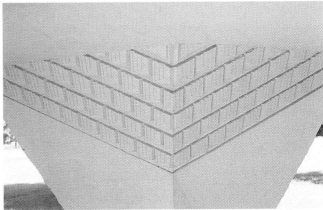

**Figure 16-20.** The rustic finish (top) was produced using rough boards. The pattern in the pier (bottom) was developed using thin strips attached to the form.

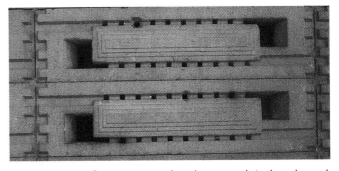

**Figure 16-21.** Cast concrete has been sculpted and used to imitate stone.

## Geometric patterns

A variety of geometric patterns can be stamped, sawed, or scored into a concrete surface to enhance the beauty of walks, drives, or patios. See Figure 16-22. Random flagstone or ashlar patterns are popular. They may be produced using a piece of 1/2" or 3/4" copper pipe bent into a flat S-shape to score the surface. Scoring must be done before the concrete becomes too hard to push the coarse aggregate aside. The best time is just after darbying or bull floating. After hand floating, a second scoring will be required to smooth the joints.

**Figure 16-22.** These colored concrete slabs have been scored to give the appearance of terra cotta tile. (Bomanite Corporation)

Other patterns such as stone, brick, or tile can be cut into partially set concrete with special stamping tools. See Figure 16-23. Color can also be added to create varying effects.

Still other patterns can be created using divider strips of wood, plastic, metal, or masonry units. These divider strips help to create interest. But they also aid in placing concrete, provide for combinations of various surface finishes, and greatly reduce random cracking. Wood divider strips should be made from pressure-treated lumber, redwood, or cypress.

## Nonslip and sparkling finishes

Nonslip finishes can be applied to surfaces that are frequently wet or that would be especially dangerous if slippery. Abrasive grains can be dry-shaken on the surface and lightly troweled. The two most widely used abrasive grains are silicon carbide and aluminum oxide. The silicon carbide grains are sparkling black in color and are also used to make *sparkling concrete*. Aluminum oxide is usually gray, brown, or white and does not sparkle. The grains should be spread uniformly over the surface. Use from 1/4 lb. to 1/2 lb. per sq. ft. of slab surface.

**Figure 16-23.** An imprint roller being used to simulate a stone pattern in concrete. (Stanley Goldblatt)

## Combination finishes

Concrete is a very versatile material. It can be used in many ways to create a beautiful walk, drive, or patio to complement the mood and style of any architectural design. See Figure 16-24.

**Figure 16-24.** The concrete driveway has been stamped and colored to complement the architectural design of this home.

Striking effects can be attained by combining several colors, textures, and patterns in concrete. Alternate areas of exposed aggregate and plain or colored concrete provide an exciting combination. Ribbons and borders of masonry or wood create a dramatic effect with plain concrete or exposed aggregate surfaces.

# Cast-in-Place Concrete Walls

Cast-in-place concrete permits structures of all shapes, sizes, and heights. Exterior wall surfaces can be rough or smooth, natural or colored. Some of the popular types of cast-in-place concrete walls include rustic, grid-patterned, colored aggregate, window walls, and sculptured.

## Rustic concrete walls

Rustic concrete walls can be produced by using rough form boards, bush hammering the surface, or casting vertical fins in the concrete surface. Forms made from rough-sawn boards have been used for years to produce a textured surface in concrete walls. See Figure 16-25.

**Figure 16-26.** The fins or ribs on this modern concrete structure add interest to the design.

**Figure 16-25.** This wall texture was the result of using rough sawn form boards. Note that the tie rod holes add to the overall texture effect.

Concrete will faithfully reproduce the wood texture and the rough board appearance tends to hide tie-rod holes and other imperfections. To assure uniformity of surface texture:
1. Use lumber of the same type throughout.
2. Or use a form coating to seal the surfaces of the boards. Bush hammering produces a coarse concrete texture. This is a method of mechanically breaking away the wall surface of hardened concrete to expose coarse aggregate. Interesting color variations and surface textures are produced by bush hammering.

Vertical fins or ribs can produce shadow effects on a concrete wall. See Figure 16-26. The fins can be smooth, sandblasted, or hammered. Inserts of wood, metal, or plastic can be used to create the ribs. Figure 16-27 shows a ribbed surface that has been hammered.

## Grid-patterned concrete walls

Well-planned joint patterns provide a low cost architectural treatment for cast-in-place walls or other structural elements. Construction joints can be inconspicuous or hidden by rustication strips. These strips are used to produce grooves in the concrete surface, which add something to the overall architectural effect and serve as control joints.

**Figure 16-27.** This concrete ribbed surface has been hammered to expose the aggregate. (American Plywood Association)

Removable architectural ties can be used to provide minimum-size, easily patched holes or the tie holes can be accentuated to reduce cost and enhance the appearance.

Figure 16-28 shows how the tie holes were formed in a grid pattern and left exposed. When form joints and tie holes are left unfinished, they are placed at predetermined locations and the pattern is repeated throughout the structure to create an architectural effect.

## Colored aggregate surfaces

One of the best methods of obtaining color in a cast-in-place concrete wall is through the use of exposed aggregate. A large percentage of coarse aggregate is used in the mix. The surface is sandblasted, bush hammered, or chisel-textured to expose the colored aggregate.

Another method of producing exposed aggregate cast-in-place walls is to preplace dry aggregate in the form and then grout under pressure with a cement-sand-water slurry. A hole near the bottom of the form is used for pumping in the grout and an external vibrator is used. The

**Figure 16-28.** Tie holes form a grid pattern in the structure and are left exposed as part of the surface texture.

aggregate is exposed by sandblasting two to seven days after grouting.

An alternate, patented method is to wrap wire mesh around the reinforcement and preplace aggregate between the mesh and the outside form. A special concrete mix is cast into the core of the mesh and dispensed to the outside.

### Cast-in-place window walls

Repetition of window openings makes an attractive design in large structures such as the one shown in Figure 16-29. White Portland cement concrete is well suited for window walls because the color is permanent. Any coarse aggregate can be used, but it should be reasonably uniform in color. White aggregate is preferred if maximum whiteness is desired.

**Figure 16-29.** This attractive design owes its overall effect to the use of white Portland cement. (ASG Corporation)

Reinforced plastic forms provide a concrete surface that is smooth and hard with few air voids and defects. The plastic forms are reinforced with fiberglass and are usually from 3/16" to 5/8" thick. Normal thickness of wall forms is 3/8". Column forms are usually 1/2" thick.

### Sculptured concrete walls

Sculptured concrete provides esthetic qualities. See Figure 16-30. A wide range of materials can be used to form decorative patterns in concrete walls. A few such materials are wood, corrugated metal sheets, plastic form liners, fiberglass sheets, formed plastic, and tempered fiberboard. Plaster waste molds can be used for fine sculpturing.

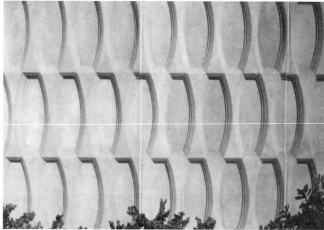

**Figure 16-30.** The fine sculptured details of these walls set them apart from the ordinary.

## Cast-in-Place Concrete Roof and Floor Systems

There are four basic cast-in-place concrete roof and floor structural systems. They are pan joists, waffles, flat plates, and flat slabs.

## Pan joist roof and floor system

Pan joist construction is a one-way structural system using a ribbed slab formed with pans. See Figure 16-31. This system is economical because the standard forming pans may be reused. Standard pan forms produce inside dimensions of 20" to 30" and depths from 6" to 20".

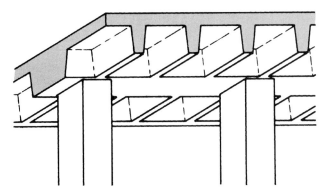

**Figure 16-31.** Pan joist roof construction is a one-way structural system using a ribbed slab formed with pans. Spans of up to 50' are common using this construction.

## Waffle roof and floor system

Waffle pans or forming domes are available in standard sizes but may be custom made for a particular job. Like pan joist construction, the forms can be reused. Figure 16-32 shows a structure using the waffle technique.

**Figure 16-32.** Waffle plate construction is used here to form the roof slab. Spans of up to 60' are possible.

Standard 30" by 30" square domes have a depth of 8", 10", 12", 14", 16", or 20". They have 3" flanges that provide for 6" wide joist ribs on 36" centers. Standard domes 19" × 19" square have a depth of 6", 8", 10", 12", or 14" and form 5" wide joist ribs on 24" centers using 2 1/2" flanges. Spans from 25' to 60' are possible using waffles.

## Flat plate roof and floor system

The main features of the flat plate system are minimum depth and architectural simplicity. See Figure 16-33. A flat plate is a two-way reinforced concrete framing system utilizing the simplest structural shape —a slab of uniform thickness.

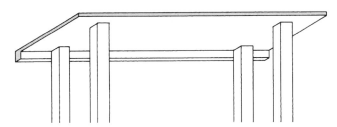

**Figure 16-33.** A flat plate roof is a two-way reinforced concrete framing system having a slab of uniform thickness.

The flat ceiling is economical to form and may be used for the finished ceiling without any additional treatment. Cantilevers are easily produced as well as other architectural projections. Slabs generally range from 5" to 14" thick and provide spans up to 35'. Flat plates provide a continuous solid ceiling with complete flexibility for locating partitions and mechanical equipment. Supporting columns need not be in a straight line. This adds flexibility. Electrical ducts and conduits may be embedded in the slab. The flat plate system is well suited for heavy loads such as roof parking.

## Flat slab roof and floor system

The flat slab system is designed for heavy roof loads with large open bays below. The difference between the flat-plate system and the flat slab is that the flat slab has a supporting panel in the area of each column for added support. See Figure 16-34.

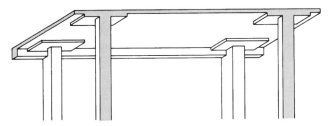

**Figure 16-34.** A flat slab roof is a two-way reinforced structural system that includes either drop panels or column capitals to carry heavier loads.

Flat slab thickness is usually 2.5% or 3% of the span and a minimum of 4" thick. The size of the supporting panel is about 33% of the span and 25% to 50% of the slab thickness. flat slabs are designed to span up to 40' with columns an equal distance apart. This system has all the advantages of the flat plate system and is stronger.

## Precast/prestressed concrete systems

With expansion of industrial processes in building construction has come greater use of modular layout, planning, and cost-conscious construction. Precast units for walls, floors, ceilings, and roofs can be mass produced at the factory or job site. Precast concrete units can be cast as tilt-up panels, standard-shaped concrete panels, or concrete window walls. A variety of prestressed panels and shapes are produced that are used in wall, roof, and floor systems.

## Tilt-up construction

*Tilt-up construction* is the process of casting concrete walls in a horizontal position on-site or other location and lifting them from the casting position or truck to their final location in the building. See Figure 16-35. Tilt-up construction is one of the fastest growing construction methods in the United States. This is mostly due to the reasonable cost, low maintenance, durability, and speed of construction. Tilt-up construction is especially suited for buildings greater than 10,000 sq. ft. with 20' or higher side walls that incorporate repetition in panel size and appearance. See Figure 16-36. These drawings are produced with AutoCAD, which is a Computer-Aided Drafting program. Each of the different aspects of the drawings are given a different color and are drawn on different layers. These layers can be turned on and off, so the drawing is not so busy when it is being produced. All of the layers that are to print can be turned on at the time of printing.

Thickness of the concrete panel is usually determined by a quantity called the *slenderness ratio*. This is the ratio of the unsupported panel height to the panel thickness. The ratio is generally 50, but a qualified engineer should make the calculation.

Panel connections to the footings, floors, roof, and between panels must be determined before construction. See Figure 16-37. Lifting loads are a major consideration, because they generally have the most influence on the panels. Reinforcement must be designed to accommodate these loads.

A drawing of each panel should be made showing the front and back as well as inserts and embeds. See Figure 16-38. Inserts provide the attachment points for lifting hardware and braces. Embeds are prefabricated steel plates with lugs that are cast into the panel to attach it to the footing, other panels, or roof system.

Spread footings are generally used for most tilt-up buildings, but pier footings can be used. See Figure 16-39. The footing must be straight and smooth if the panels are to fit properly. Footing size will be dependent on the soil bearing capacity, weight to be supported, and other factors specific to the site and building.

Tilt-up panels can be cast as individual panels or as larger slabs that are sawed into smaller panels after the concrete has been placed. Tilt-up panels have a grid of reinforcing steel embedded within the panel to provide the necessary strength. Standard Grade 40 or 60 bars are used. Plastic chairs to support re-bars are used instead of steel chairs to avoid rust on the panel face. Concrete must meet all mix specifications for the application.

Many panels are cast with a textured surface or pattern. The surface treatment must be planned carefully, but the results are impressive. Other finishes such as exposed aggregate, sandblasting, and brick or stone facings are commonly used with tilt-up panels to *dress up* the building.

**Figure 16-35.** Precast concrete panel units were used to construct this building.

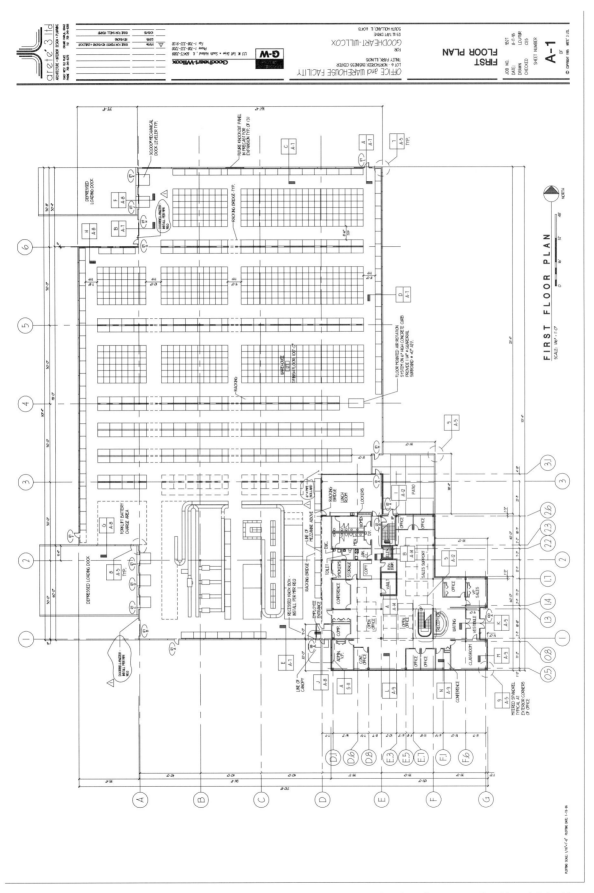

**Figure 16-36.** First floor plan of a warehouse and office building constructed using tilt-up panels. (Areté 3 Ltd.)

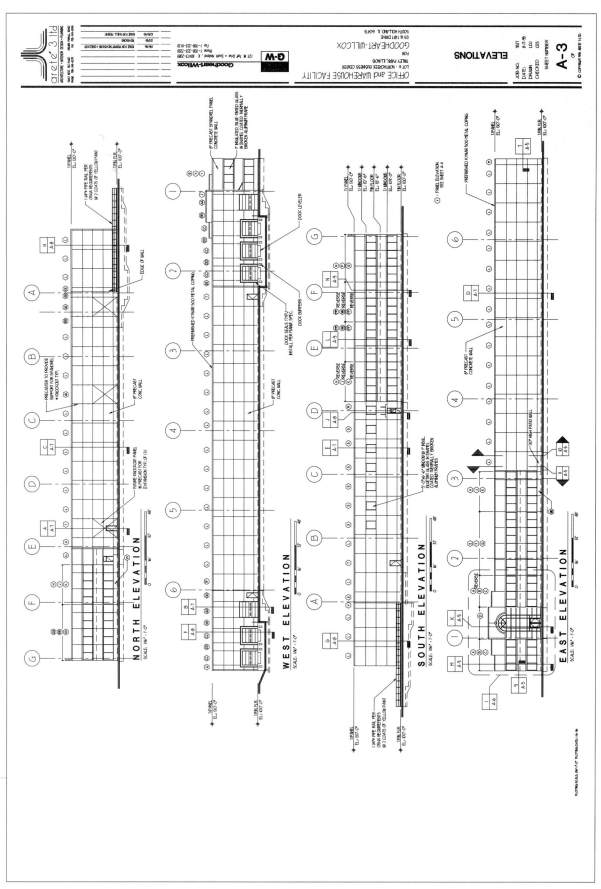

**Figure 16-36.** Continued. Exterior elevations of a warehouse and office building shown on the previous page. (Areté 3 Ltd.)

**Figure 16-37.** This metal connection will be used to attach a panel to the footing.

The erection process, should be well-planned for efficiency and safety. See Figure 16-40. Braces should be attached before the panels are lifted and not removed until the roof and decking are installed. Installing braces once the panels are in the vertical position is difficult and dangerous. The result of well-planned erection process is a spectacular finished structure.

## Prestressed panels

Standard-shaped concrete panels that are prestressed have brought a new concept to architectural precast concrete. These units are crack-free and highly resistant to deterioration. Some of the most common designs include *double-tee units*, *single-tee units*, and *hol-low-core panels*. Figure 16-41 shows a structure using double-tee roof units.

Double-tees are the most widely used prestressed, precast units. Widths are generally 4', 5', 6', 8', and 10'. The 4' width is most popular. Depths range from 6" to 36". Spans up to 60' are possible with the 4' wide unit. Greater spans are possible with wider units since they usually have a deeper web.

Single-tee units are generally used where long spans, beginning at about 30', are needed. They can be placed flange to flange as in Figure 16-42, or spread apart with concrete planks or cast-in-place concrete completing the enclosure. Flange widths range from 4' to 10' with depths from 12" to over 36" in 4" increments. Spans of 30' to 100' are common with single-tee units.

Cored slabs provide a flush ceiling with minimum depth required for the roof or floor system. See Figure 16-43. Slab thicknesses range from 4" to 12" in increments of 2". A variety of widths are available such as 1'-4", 1'-8", 2'-0", 3'-4", and 4'-0". All sizes are not available in a single core style. Lengths generally range from 12' to 40', but the length of a particular slab will depend on the thickness, width, and reinforcing. Generally, the thicker the slab, the longer the length available.

Precast concrete window walls may be cast as curtain walls or loadbearing walls. Forms or molds used to produce complicated designs are made from plastic, wood, or steel. These are custom units and must be designed by architects or engineers.

Typically, fiberglass molds are used for smooth concrete with sculptured mullions (wall members supporting a window). Precast window walls can be one-story or multistory. They can even be preglazed (glass installed) at the plant. Figure 16-44 shows a building with loadbearing wall panels.

Proper bracing of tilt up panels is important for both safety and efficient erection of the structure. The diagonal braces attached to the wall panels will not be removed until the roof decking is installed to tie the structure together.

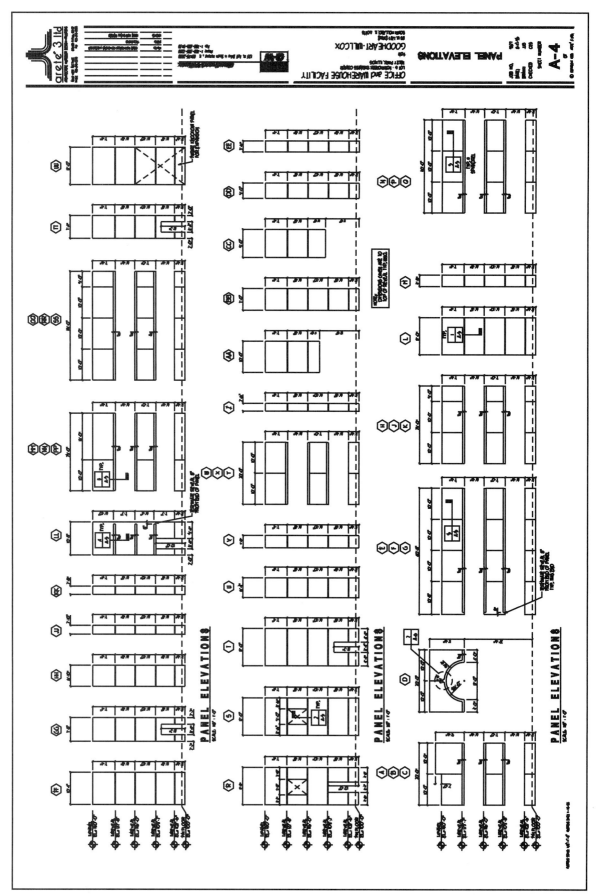

**Figure 16-38.** An elevation drawing of each panel type used in the construction of the building shown in Figure 16-36. (Areté 3 Ltd.)

**Figure 16-39.** Typical footing and foundation wall used in tilt-up construction.

**Figure 16-40.** Crane lifting panels in place. As more panels are erected, the final building structure becomes more identifiable.

**Figure 16-41.** Double-tee concrete units, as shown here, are the most widely used prestressed concrete product for medium range spans. Four foot wide double-tees are commonly used for spans up to 60'.

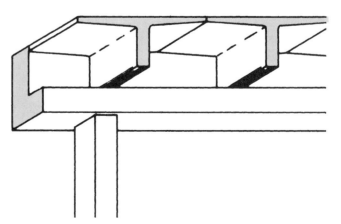

**Figure 16-42.** Single-tee precast, prestressed concrete units are generally used for very long spans. Typical spans are from 30' to 100'.

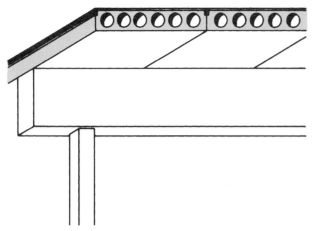

**Figure 16-43.** Hollow-core slabs are low cost and widely produced. Depths range from 4" to 12" and widths are available from 16" to 48". Spans possible with cored slabs are up to 50', but generally under 35'.

**Figure 16-44.** The precast concrete window walls in this building are load-bearing. (Kawneer Company, Inc.)

# REVIEW QUESTIONS
# CHAPTER 16 ▬▬▬

Write all answers on a separate sheet of paper. Do not write in this book.

1.  Why should the subgrade be moistened before placing concrete?
2.  Why are forms treated with oil or other preparation?
3.  When placing concrete, it should not be allowed to drop freely more than _____'.
4.  Air pockets can be worked out of plastic concrete by puddling, spading, _____ , and _____.
5.  What condition does too much vibration cause when removing air pockets by mechanical vibration?
6.  How is a good bond accomplished when placing fresh concrete on hardened concrete?
7.  Concrete that is shot into place by means of compressed air is called _____.
8.  The first finishing operation performed on concrete is usually _____.
9.  A radius may be put on the edge of a concrete slab using an _____.
10. What problems may result from performing finishing operations on a concrete slab while the bled water is present?
11. Name three reasons why concrete is floated.
12. Steel troweling produces a very smooth surface that is often slippery when wet. Sidewalks, driveways, and other outside flatwork frequently are _____ to produce a slightly roughened surface.
13. When may forms be removed?
14. Why is air-entrained concrete finished slightly different than normal concrete?
15. What is the most important consideration in curing concrete properly?

16. At what temperature does hydration stop completely?
17. What is the most common way to protect concrete against low temperatures during the curing process?
18. Why should concrete not be placed during hot, dry, windy weather?
19. List three things that can be done to reduce the effect of hot weather on placing and curing concrete.
20. Three basic types of joints are frequently used in concrete construction. What are they?
21. Why are isolation or expansion joints used?
22. Identify three decorative and special finishes that are used on concrete.
23. In addition to providing an attractive finish, what are three other attributes of exposed aggregate surfaces?
24. What kind of stone chips are generally used in terrazzo toppings?
25. What three methods are generally used to apply a geometric pattern to a concrete slab?
26. What are the four basic cast-in-place concrete roof and floor structural systems?
27. Pan joist construction is a one-way structural system using a ribbed slab formed with _____.
28. _____ construction is the process of casting concrete walls in a horizontal position on-site or other locations and lifting them from the casting position or truck to their final location in the building.
29. In a tilt-up concrete panel the thickness is usually determined by a quantity called the _____ ratio.
30. What type of footings are generally used for most tilt-up buildings?
31. Identify the three most common designs of prestressed panels.

Arc welding is used to join two precast panels at a corner. The welder is joining a piece of steel angle to embeds that were cast into the panels. Note the chain of a "come-along" winch being used to pull the panels together for welding. The lifting eye at right is threaded into an insert cast into the panel. After the eye is removed, a concrete finisher will cover the insert and smooth the surface.

# SECTION V
# SUCCEEDING ON THE JOB

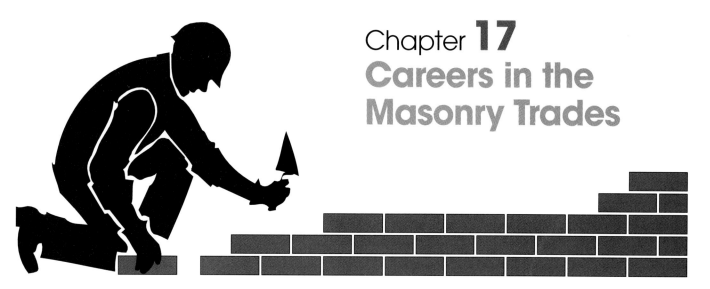

# Chapter 17
# Careers in the Masonry Trades

Masonry offers a rewarding career for people who have an interest in working with tools and materials. As masons, or bricklayers, they will be skilled workers. They will understand the basic principles and practices related to the construction industry. They will earn good pay and have an opportunity for advancement. What is more, they will play an important role in building homes, schools, and commercial structures. They can be proud of their skill and the fact that they can produce something that people need.

There are many benefits to being a professional mason. Your work is functional, long-lasting, and benefits others. You have a skill that is appreciated and can be used to earn a decent living over a long period of time. It is good, honest work. It is something that cannot be taken away from you by others. Each job is different, so it never becomes boring. You can be independent and creative while working outdoors in a healthy environment. Masonry work is not a high-stress occupation. See Figure 17-1.

**Figure 17-1.** These cement masons are placing concrete for a basement wall. Most of their work is performed outdoors. (Lite-Form, Inc.)

Physical qualifications are an important aspect of becoming a mason. One must have good health and strength, because the work is strenuous. A mason works outdoors much of the time. He or she must be able to lift heavy loads, bend, reach, and kneel. A mason must have good hand/eye coordination and balance. Good eyesight is very important. A mason must sometimes work on scaffolding high above the ground or in small, tight places such as an elevator shaft. In brief, a mason's job is very physical and requires a high degree of manual dexterity.

A mason must also have sharp mental skills. He or she must be able to handle basic math calculations such as adding, subtracting, multiplying, and dividing whole numbers as well as decimals and fractions. The mason must also be able to estimate quantities, lengths, weights, and volumes. To be successful, the mason must be able to handle language competently in written and spoken form. He or she must be able to read construction drawings and interpret them in terms of his or her duties. Performing as a mason is not just manual labor. It is a skilled occupation that requires study, concentration, and continued learning.

## What Masons Do

The term *masonry* has traditionally referred to the craft of building with brick. Today, it has a broader meaning. It includes two types of work. The first type consists of any construction bonded together with mortar. See Figure 17-2. The second type deals with cement masonry. See Figure 17-3. All workers who lay brick, block, tile, or stone are called masons. Cement masons specialize in concrete work such as slabs, footings, and foundations.

Masons use masonry units such as brick, block, or stone. With their tools and skills they form the units into buildings or other useful structures. This skill and knowledge must be learned and practiced if masons are to be successful in their trade.

**Figure 17-2.** A journeyman mason who knows the trade and takes pride in his workmanship.

**Figure 17-3.** A cement mason placing anchor bolts in a foundation.

Masons must learn to proportion the ingredients of mortar, figure trade-related problems, mix and spread mortar, read construction prints, handle different kinds of masonry units, and be willing to learn new techniques and procedures. See Figure 17-4. They must cooperate with other trades.

They will build exterior and interior walls, floors, patios, walks, columns, window and door openings, fireplaces, arches, and many other building elements. Modern masonry design will also require them to use masonry saws and other power-driven equipment. Masons must also use age-old tools such as the trowel, brick hammer, level, chisel, and mason's line. Masons are creative persons who can continue to learn more about the trade as they progress through the years.

Cement masons will learn how to work with concrete. More specifically, they will learn the proper methods of reinforcing, building forms, placing, and finishing concrete. As in any other type of masonry work, a high degree of skill is involved in quality cement mason's work.

**Figure 17-4.** A mason must know how to spread mortar properly.

Poured concrete construction industries employ many skilled workers as cement masons. Their job is to place concrete into forms where they will smooth and finish it. They will work on a wide range of construction projects from floors and slabs to roofs, sidewalks, highways, dams, airport runways, etc. See Figure 17-5.

# Advancement Opportunities ▬

The masonry trades recognize various levels of the skill. These levels are apprentice, journeyman, foreman, and superintendent.

## Apprenticeship

An **apprentice** is a person at least 17 years old and, preferably, not over 24 years old, who is under written agreement to work at and learn the trade. The apprenticeship agreement is made with a local apprenticeship and training committee acting as the agent of the contractor. See Figure 17-6. The agreement may also be directly made with a contractor approved by the local joint committee. See Figure 17-7. An apprenticeship applicant should have completed at least two years of high school, but preferably more.

The **U.S. Department of Labor, Bureau of Apprenticeship and Training** sets down the responsibilities of apprentices. These responsibilities are:

1. To perform diligently and faithfully the work of the trade and other pertinent duties as assigned by the contractor in accordance with the provisions of the standards.
2. To respect the property of the contractor and abide by the working rules and regulations of the contractor and the local joint committee.
3. To attend regularly and satisfactorily complete the required hours of instruction in subjects

**A**

**B**

**Figure 17-5.** A—Cement masons placing concrete at the construction site. B—Workers placing beam shear reinforcing for a major league baseball facility. (Wire Reinforcement Institute)

related to the trade, as provided under the local standards.

4. To maintain such records of work experience and training received on the job and in related instruction, as may be required by the local joint committee.

5. To develop safe working habits and conduct himself or herself in such manner as to assure personal safety and that of the safety of fellow workers.

6. To work for the contractor to whom assigned until the apprenticeship is completed, unless reassigned to another contractor or agreement is terminated by the local joint committee.

7. To conduct himself or herself at all times in a creditable, ethical, and moral manner, realizing that much time, money, and effort are spent to afford him or her an opportunity to become a skilled worker.

The first six months of employment after signing the apprenticeship agreement is a *trial period*. Before the end of the period, the committee will review the apprentice's ability and development. The normal term of apprenticeship is 4500 hours of employment or about three years. Apprenticeship is divided into six periods of advancement of six months each. The pay rate is a percentage of the journeyman rate. It is based on the period of advancement.

Each apprentice is required to take related instruction, outside of the job, for no less than 144 hours per year, during each year of the apprenticeship. Subjects include print reading, math, estimating, shop practice, and safety. Refer to the Reference Section of this text for a complete list.

A *Certificate of Completion of Apprenticeship* is awarded to a person who successfully completes the program. See Figure 17-8. The apprentice is now a *journeyman mason*.

## Journeyman

The *journeyman* is an experienced craftsperson who has successfully completed an apprenticeship in the trade. At this point, the worker is a free agent and can work for any contractor. The certificate will be recognized throughout the country.

## Foreman

A *foreman* is a journeyman who has the responsibility of supervising a group of workers. This job requires not only a high degree of knowledge about the craft, but also the ability to supervise people.

## Superintendent

The *superintendent* is generally a foreman who has been promoted because of outstanding performance. He or she is in charge of all the work in the field for the contractor. See Figure 17-9. This includes supervising the work of the foreman and making major decisions about the job under construction.

## Contractor

*Contractors* are responsible for the whole job. They are the top persons in charge. They organize people and work, prepare bids for jobs, inspect work, and run the business. The contractor must be knowledgeable about all phases of the business.

## Apprenticeship Agreement
Bureau of Apprenticeship and Training

## U.S. Department of Labor
Employment and Training Administration

**Warning: This agreement does not constitute a certification under Title 29, CFR, Part 5 for the employment of the apprentice on Federally financed or assisted construction projects. Current certifications must be obtained from the Bureau of Apprenticeship and Training or the recognized State Apprenticeship Agency shown below. (Item 22)**

OMB No. 1205-0223
Expires: 3/31/92

Privacy Act Statement: The information requested herein is used for apprenticeship program statistical purposes and will only be disclosed in accordance with the provisions of the Privacy Act. (Privacy Act of 1974) (P.L. 93-579).

The program sponsor and apprentice agree to the terms of Apprenticeship Standards incorporated as part of this Agreement. The sponsor will not discriminate in the selection and training of the apprentice in accordance with the Equal Opportunity Standards in Title 29 CFR Part 30.3, and Executive Order 11246. This agreement may be terminated by either of the parties, citing cause(s), with notification to the registration agency, in compliance with Title 29, CFR, Part 29.6.

**Part A: To be completed by sponsor**

1. Sponsor (Name and address) Program No. _____

2a. Trade (The work processes listed in the standards are part of this agreement)

2b. DOT symbol

3. Term (Hrs., Mos., Yrs.)

4. Probationary period (Hrs., Mos., Yrs.)

5. Credit for previous experience (Hrs., Mos., Yrs.)

6. Term remaining (Hrs., Mos., Yrs.)

7. Date apprenticeship begins (Indenture date)

8. Related instruction    a. Number of hours per year

b. Method
☐ Classroom
☐ Shop
☐ Correspondence

c. Source
☐ Voc. Ed.
☐ Sponsor
☐ Other

d. Apprentice wages for related instruction
☐ Will be paid
☐ Will not be paid

**9. Apprenticeship wages:** The apprentice schedule of pay shall be listed for each advancement period.

| | Period 1 | 2 | 3 | 4 | 5 | 6 | 7 | 8 | 9 | 10 |
|---|---|---|---|---|---|---|---|---|---|---|
| b. Term (Hrs., Mos., Yrs.) | | | | | | | | | | |
| c. Percent | | | | | | | | | | |

10a. Signature of committee (If applicable)        Date Signed

12. Name and address of sponsor designee to receive complaints (If applicable)

10b. Signature of committee (If applicable)        Date Signed

11. Signature of authorized representative (Employer/Sponsor)        Date Signed

**Part B: To be completed by apprentice**

13. Name (Last, first, middle), and address (No., Street, City, County, State, Zip Code)    Social Security number
| | | | | | | | | | |

17. Race/Ethnic Group (X one)
☐ White (Not Hispanic)
☐ Black (Not Hispanic)
☐ Hispanic
☐ Am. Indian or Alaska Native
☐ Asian or Pacific Islander
☐ Information not available
☐ Not elsewhere classified

18. Veteran Status
☐ Vietnam era veteran (8/15/64 to 5/7/75)
☐ Other veteran
☐ Non Veteran

C # _____

14. Date of birth (Mo, day, Yr)

15. Sex (X one)
☐ Male
☐ Female

16. Apprenticeship school linkage
☐ Yes
☐ No

19. Highest education level (X one)
☐ 8th grade or less
☐ 9th to 12th grade
☐ GED
☐ High School Graduate

20. Signature of apprentice        Date

21. Signature of parent/guardian (If minor)        Date

**Part C: To be completed by registration agency**

22. Registration agency and address

23. Signature (Registration agency)        Date registered

ETA 671
Rev. 8/91

**Figure 17-6.** Apprenticeship agreement between the apprentice and the local apprenticeship and training committee.

**U.S. DEPARTMENT OF LABOR • Employment and Training Administration**
Bureau of Apprenticeship and Training

**APPRENTICESHIP AGREEMENT BETWEEN APPRENTICE AND EMPLOYER**

CHECK APPROPRIATE BOX

☐ Vietnam-era Veteran ☐ Other Veteran ☐ Nonveteran

**PRIVACY ACT STATEMENT**

*The information requested herein is used for apprenticeship program statistical purposes and may not be otherwise disclosed without the express permission of the undersigned apprentice.*

Privacy Act of 1974 - P.L. 93-579

The employer and apprentice whose signatures appear below agree to these terms of apprenticeship.

The employer agrees to the nondiscriminatory selection and training of apprentices in accordance with the Equal Opportunity Standards stated in Section 30.3 of Title 29 Code of Federal Regulations, Part 30; and in accordance with the terms and conditions of the
*(Name of Apprenticeship Standards)* ..............................................................
which are made a part of this agreement.

The apprentice agrees to be diligent and faithful in learning the trade in accordance with this agreement.

*This AGREEMENT may be terminated by mutual consent of the parties, citing cause(s), with notification to the Registration Agency.*

NAME OF APPRENTICE *(Type or Print)*

SIGNATURE OF APPRENTICE

ADDRESS *(Number, Street, City, State, ZIP Code)*

SIGNATURE OF PARENT OR GUARDIAN

NAME OF EMPLOYER AND ADDRESS *(Company)*

SIGNATURE OF AUTHORIZED COMPANY OFFICIAL

APPROVED BY JOINT APPRENTICESHIP COMMITTEE

SIGNATURE OF CHAIRPERSON OR SECRETARY | DATE

REGISTERED BY *(Name of Registration Agency)*

TRAINING DATA

| Trade | Apprenticeship Term |
|---|---|
| Probationary Period | Credit for previous experience |
| Term remaining | Date apprenticeship begins |

**TO BE COMPLETED BY THE APPRENTICE**

DATE OF BIRTH *(Month, Day, Year)* ➤

SEX *(Check one)* ➤ ☐ Male ☐ Female

RACE/ETHNIC GROUP *(Check one)* ➤
☐ Caucasian/White
☐ Negro/Black
☐ Oriental
☐ American Indian
☐ Spanish American
☐ Information Not Available
☐ Not Elsewhere Classified

HIGHEST EDUCATION LEVEL *(Check one)* ➤
☐ 8th grade or less
☐ 9th grade or more
☐ 12th grade or more

SIGNATURE OF AUTHORIZED OFFICIAL | DATE *(Mo., Day, Yr.)*

**Figure 17-7.** Apprenticeship agreement between the apprentice and the employer.

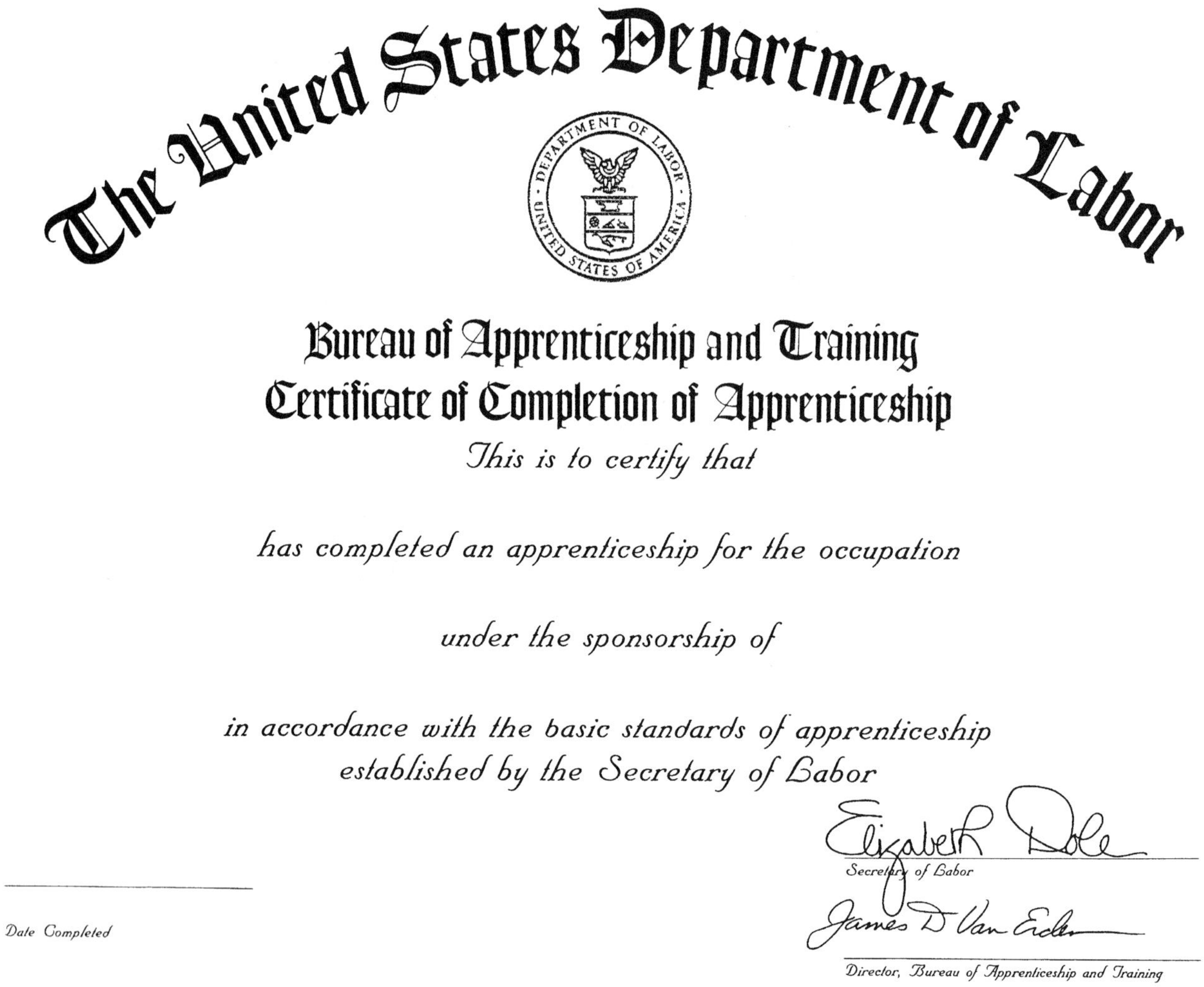

# The United States Department of Labor

## Bureau of Apprenticeship and Training
## Certificate of Completion of Apprenticeship

*This is to certify that*

*has completed an apprenticeship for the occupation*

*under the sponsorship of*

*in accordance with the basic standards of apprenticeship*
*established by the Secretary of Labor*

Elizabeth Dole
*Secretary of Labor*

James D Van Erden
*Director, Bureau of Apprenticeship and Training*

_____
*Date Completed*

**Figure 17-8.** The *Certificate of Completion of Apprenticeship* is awarded an apprentice upon completion of training.

**Figure 17-9.** The foreman and superintendent discuss progress of the job. (North Safety Products)

## REVIEW QUESTIONS
## CHAPTER 17

Write all answers on a separate sheet of paper. Do not write in this book.

1. Identify three physical qualifications that are important to masons.
2. What mental skills must a mason have?
3. There are two broad categories of masons. One type of mason works with masonry units that are bonded together with mortar. Describe what the other type of mason does.
4. In addition to using the tools, name five things that a mason must learn.
5. Name four basic tools that a mason uses.
6. In what three areas of work will cement masons be involved?

7. If you wish to become a mason, what is the beginning position?

8. Approximately how long does apprenticeship training last?

9. An apprenticeship applicant should have completed at least _____ years of high school, but preferably more.

10. A responsibility of apprentices is: to attend regularly and satisfactorily complete the required hours of instruction in subjects related to the trade, as provided under the local standards. True or False?

11. How many hours of related instruction must an apprentice successfully complete each year of apprenticeship training?

12. The first _____ months of employment after signing the apprenticeship agreement is a *trial period*.

13. When apprentices complete training, they receive a *Certificate of Completion of Apprenticeship* and become _____ masons.

14. After masons have gained experience on the job, they can be promoted to _____ if they can supervise people.

15. What is the title of the person who supervises a foreman on a construction job?

16. What is the responsibility of the superintendent?

17. Who is responsible for the whole job —considered the top person?

Cement finishers apply their skill to all types of structures. This worker prepares an expansion joint in a walkway so that the walk will not develop unsightly cracks. (The Associated General Contractors of America)

A mason who demonstrates the ability to direct the work of others may be promoted to the job of foreman, supervising a crew of workers like these cement masons.

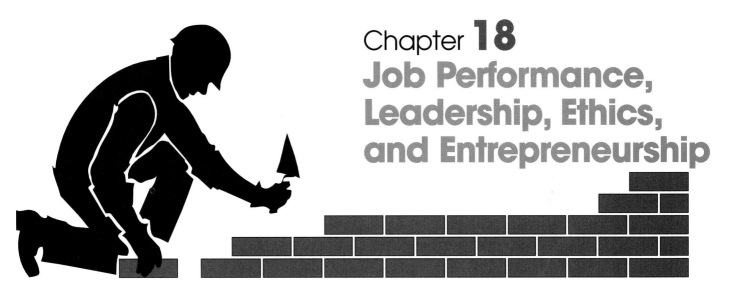

# Chapter **18**
# Job Performance, Leadership, Ethics, and Entrepreneurship

There are many factors that can affect your progress on the job. Special care must be taken to address these so that your position will not be in jeopardy. These factors include job performance, ethics, teamwork, leadership, and entrepreneurship.

## Job Performance

Job performance can be measured in several ways. Your general work habits, safety record on the job, and keeping current in your field are three of the primary areas that should be considered. This section will explore opportunities for improvement in these three areas.

### General work habits

General work habits play a large part in keeping your job as well as progressing on the job. You should always strive to improve your performance and the quality of your work. You might begin by asking yourself the following questions about your general work habits:

1. Do you observe safety rules? Safety is everyone's business. It involves not only your personal safety, but the safety of others. See Figure 18-1.
2. Do you keep the work area clean and organized? A cluttered workplace encourages accidents and slows down the work.
3. Does your nonwork-related conversation interfere with your work and the work of others? You are not being paid to tell jokes or catch up on gossip.
4. Do you observe company policies? Company policies are established to provide uniformity of response to foreseeable situations. They are vital to the smooth operation of the business.
5. Do you plan and execute your work efficiently? A professional comes to the job ready, to work, and has the necessary tools and attitude to perform the prescribed tasks.

**Figure 18-1.** Safety rules apply to everyone. It is your responsibility to work safely. (Trenwyth Industries, Inc.)

6. Do you always give your best effort to each job? One way to improve your performance is to practice good work habits. Always try to improve your skills and take pride in your work.

These are some of the questions that you should ask yourself about your general work habits. You can probably think of others. Paying attention to your work habits could result in increased pay and greater appreciation of you and your work. At the very least, it might preserve your job on the team. Eventually, your work habits and knowledge could help you be promoted.

### Safety on the job

Construction is a dangerous business. In fact, construction is cited by the **Occupational Safety and Health Administration (OSHA)** as the most dangerous occupation in the U.S. When all occupations are considered, an employee sustains an on-the-job injury every 18 seconds and a workman is killed every 47 minutes in this country. Job safety must, by necessity, be a primary concern for every employer and employee.

Statistics show that 5% of all accidents are caused by unsafe conditions; the other 95% are caused by unsafe actions. Unsafe workers will find a way to injure them-selves regardless of red warning signs, safety nets, guards, etc. The solution is to train workers to work safely. Working safely involves a thorough understanding of the tools, equipment, and materials you are working with. It also involves thinking safety. See Figure 18-2.

**Figure 18-2.** Is this worker thinking safety? Notice, he is supporting his weight on a thin piece of foam insulation.

Some of the techniques that can be employed to sharpen concern for safety on the job site include:
- ❖ Developing a set of company safety rules.
- ❖ Demonstrating safe work practices.
- ❖ Enforcing safe work practices.
- ❖ Discussing safety procedures every few weeks.
- ❖ Testing the knowledge of workers to be sure they understand the proper use of tools and machines.

A safety program isn't something you can just write up and distribute; it is an ongoing effort between a con-tractor, the crew leaders, and the workers on the site. When an accident does happen, everyone should partici-pate in a discussion about the particulars—how it hap-pened and what could have prevented it from happening.

Some residential construction workers like to work in shorts and tennis shoes, but this is not safe. See Figure 18-3. In a recent survey of small contractors by the *Journal of Light Construction*, several unsafe practices on the job site were detailed. No less than 25% of those surveyed used no respiration protection at all when doing general demolition. See Figure 18-4. Six percent used nothing when removing asbestos; 13% went without respiration protection when installing fiberglass insulation. Only 5% wore goggles or safety glasses all the time they were on the job site. When questioned about using saw guards, 13% did not use guards on portable circular saws, 37% did not use guards on power miter boxes, 57% did not use guards on radial arm saws, and 72% did not use guards on their table saws. A full 52% of the contractors reported that they used bounce-nailing when installing sheathing—a dangerous procedure. About one-third never use ear

protection on the job. Almost half said they weren't sure if their scaffolding would pass OSHA inspection. See Figure 18-5. Finally, half of the contractors reported that they never hold safety meetings or provide first aid instruction. More than 40% reported workers had experienced cuts and punctures, 18% reported an object in the eye, 18% reported sprains and muscle pulls, and 7% reported a back injury in the last three years. Clearly, safety is a problem in the residential construction industry.

**Figure 18-3.** Short pants and tennis shoes have no place on a construction site.

**Figure 18-4.** The proper respirator should always be worn when hazardous materials are airborne.

## Continuing education

It is imperative that people working in masonry con-tinue their education in order to become better informed about their field. When a person moves beyond the level of apprenticeship, the educational process should not stop. Workers must continue to update their skills, whether they are journeymen, foremen, superintendents, or contractors.

Changes in the masonry industry have evolved since the beginning and will continue in future years. This

**Figure 18-5.** An unsafe scaffold—it lacks the required safety rail.

creates a demand for craftsmen and contractors to institute changes in the way they think and build. It is important that managers keep their workers apprised of new products and systems and the reasons for carrying out the work in a particular way. See Figure 18-6.

**Figure 18-6.** These workers are using a state-of-the-art laser level to determine proper elevations.

One of the missions of the masonry industry should be to keep masons better informed on how to build and the reasons for doing the work in a given manner. Organizations at the national, regional, state, and local level provide an abundance of services for those who wish to continue their education. These services include technical seminars and training sessions, university classes, technical publications, directories, or construction information telephone lines.

Technical information from industry associations, manufacturers' literature, seminars, and ASTM Standards are available for workers in the masonry field who wish to continue to learn. To be successful, you must be alert, receptive to change, and have a desire to continue the educational process.

# Ethics

*Ethics* can be defined as, *"the rules or standards governing the conduct of the members of a profession."* Therefore, a brief discussion of ethics in the workplace is an appropriate subject.

## Work ethic

Your work ethic will be a very important consideration in keeping a job or receiving a promotion. Employers are very concerned about the attitude you bring to the workplace. Work ethic includes your enthusiasm for the work, your willingness to work late to meet deadlines, and whether you are on site and prepared to work at starting time. Other factors that have a bearing on your work ethic center upon how you relate to fellow workers. Ask yourself the following questions to shed light on your work ethic:

❖ Do you talk so much that others can't get their work done?
❖ Do you treat others with respect?
❖ Are you a gossip or critic of everyone else's work?
❖ Do you run to the boss with every little thing?
❖ Are you a constant complainer?
❖ Do you always have an excuse for poor work or work not finished?
❖ Are you willing to admit mistakes and learn from them?

These and many other related characteristics are part of your work ethic. If you expect to be happy at the workplace and progress to positions of more responsibility, then you will have to pay attention to your work ethic and personal behavior.

## Ethics code

Ethical practice has always been a concern of individuals and businesses who wish to be successful over the long term. Many organizations have recorded their goals as a means of at encouraging ethical practices. One such organization, the **National Association of Home Builders,** recently adopted a model ethics code. Following are the twelve objectives of their code of ethics:

1. To conduct business affairs with professionalism and skill.
2. To provide the best housing value possible.
3. To protect the consumer through the use of quality materials and construction practices backed by integrity and service.
4. To provide housing with high standards of safety, sanitation, and livability.
5. To meet all financial obligations in a responsible manner.
6. To comply with the spirit and letter of business contracts, and manage employees, subcontractors, and suppliers with fairness and honor.
7. To keep informed regarding public policies and other essential information which affect your business interests and those of the building industry as a whole.
8. To comply with the rules and regulations prescribed by law and government agencies for the health, safety, and welfare of the community.
9. To keep honesty as our guiding business policy.
10. To provide timely response to items covered under warranty.
11. To seek to resolve controversies through a non-litigation dispute resolution mechanism.
12. To support and abide by the decisions of the association in promoting and enforcing this Code of Ethics.

# Teamwork and Leadership

A mason with all of the technical knowledge and skill in the world is of little value to a contractor if they cannot get along with others and work as part of a team. See Figure 18-7. Building a structure demands cooperation among many people. Much time can be wasted because three people working on a single job think they are in charge. In actuality, different people are in charge at different times. Your performance should reflect this recognition.

The company is always looking for people who can *run a crew.* Some people have skill in running a large crew; others are capable of supervising a few other workers; still others shouldn't be managing anyone. If you wish to have the opportunity to manage, you should observe the traits in others that make them good managers. Practice these behaviors and you will likely become a manager. Remember that a manager is a leader. Others must follow your instructions and example.

The qualities of an effective leader have been written about and analyzed extensively. Most of the traits identified are common sense. For example, people with the ability to

**Figure 18-7.** This team is laying out the foundation location for a large commercial building.

lead can keep track of what is being built. They can foresee potential problems and can solve these problems as they arise. They know enough to make sure that the proper materials and equipment are on hand when they are needed. They can get along with all types of people and can communicate information clearly. They know who can handle each project and can get people to work hard and cooperate. They value their employees as people as well as members of the team. Most important, leaders lead by example. You must practice what you expect from others.

A few questions that you might ask yourself in the area of teamwork and leadership include the following:

1. How well do you receive instructions from others? If you can't accept instructions well from others, the chances are pretty that your instructions to others won't be accepted very well either.
2. Are you a team player? Team players are valued more in the workplace than solo artists. Remember that most jobs or tasks involve teamwork. Reflect on what it means to be a good team player.
3. Do you get into personality conflicts with other workers or the boss? Naturally, you can't completely ignore your personality traits, but they can be controlled. Don't see every situation as a crisis or a threat. Learn to take things in stride and place events in their total context. For example, put yourself in the other person's place and then ask how you would act.
4. Are you a good communicator? Many conflicts are the result of poor communication. Frequently, people take more offense at *how* something was said than at *what* was said. Think about what you wish to communicate and then put it into words that will be clearly understood. What is heard is frequently different from what is said.

5. Do you keep your foreman or supervisor informed about potential problems on the job? A little warning about a potential problem will be appreciated and help to establish you as a valued member of the team who is interested in saving the company money and producing the best quality job possible.

6. Do you have the ability to manage a crew? Others will most likely know the answer to this question even before you do. Seek their advice about improving your leadership skills. Study leadership styles or take a course in developing leadership skills. Learn from others. If you have what it takes to be a leader, it will be recognized.

As you progress to a job or position that involves managing others, you will be expected to show leadership. Consider the following attributes that successful leaders possess:

- Speak simply and directly
- Share the credit
- Have a vision
- Be positive
- Be optimistic and enthusiastic
- Avoid sarcasm
- Meet problems head on
- Check the small things
- Don't take yourself too seriously
- Be friendly, not a friend
- Control your emotions
- Always be truthful
- Treat everyone fairly
- Be constructive, don't criticize
- Always be a role model
- Have fun

# Entrepreneurship

Some day you may want to own or manage your own business. For this reason, you may want to think about the following considerations:

1. Every year many new businesses are started, but most fail. The reasons generally cited include lack of adequate financing, poor management of the enterprise, and lack of knowledge required.

2. There are positive and negative aspects of owning your own business. Some advantages include being your own boss, the chance to make more money, and the satisfaction of building a successful enterprise. Disadvantages include the enormous responsibility of making the right decisions, the long hours, the risk of failure, and the responsibility for the livelihood of others.

3. *Business opportunities* may not be opportunities at all. If it is too good to be true, it probably is. Ask questions such as:

- Is the location good?
- How much overhead will there be in the business (utilities, taxes, insurance)?
- What about the competition?
- How will you hire good employees?
- Can you manage a work force?
- What business skills and knowledge do you have?
- Can you get financing?
- What government regulations apply to your business?
- What type of business organization should you choose?
- What are your goals?
- How will you advertise your product or service?
- Where can you get good advice?
- What governmental agencies offer help?

4. Every business should have definite, well-defined goals. Goals provide direction and help in making decisions concerning your product or service to be offered for sale. Consider these questions:

- Will the product or service be conventional or unique?
- How will you gain the support and respect of your employees?
- What kind of image do you wish your company to convey to others?
- Who will purchase your product or service?
- How much should you charge?
- How will you reward productive workers?
- How will you protect your workers against injury on the job?
- What activities will help you meet your goals?
- How will you organize your business to maintain high efficiency and quality?
- Do you have the strength and determination to own and manage a business?
- What are the long-term consequences to yourself and your family?
- Are you prepared to risk failure?

Just having a good idea or quality product or service is not enough to have a successful business. It needs constant attention to detail and solid planning. Risk of failure is always present and may be brought about by forces beyond your control. However, the rewards of running a successful business generally outweigh the liabilities.

## Characteristics of entrepreneurs

A person should possess certain characteristics to be a successful entrepreneur. The following list summarizes the major characteristics:

1. **Good health.** Long hours and much physical labor make heavy demands on the mason or masonry worker.

2. **Knowledgeable.** To make a profit, one must know all aspects of the trade or trade specialty. In addition, the person must understand the industry and the products being used in the trade.

3. **Good planner.** Running a successful business means that nothing is left to chance. One must be able to foresee difficulties as well as plan how to take advantage of opportunities.

4. **Willing to take calculated risks.** Once a plan has been conceived that takes into account events likely to occur, the person must have the courage to risk money and their future on making the plan work.

5. **Innovative.** One must be successful in finding ways to improve and produce better work and thus gain the confidence of customers.

6. **Responsible.** One must be willing to accept the consequences for decisions whether good or bad. This includes paying debts, keeping promises, and accepting the responsibility for mistakes of their employees.

7. **Goal oriented.** An entrepreneur likes to set goals and works hard to achieve them. It has been predicted that by the year 2000, there will be a large increase in persons 30 to 60 years old. The population in this group will demand many services. Housing is one of the areas predicted to be in great demand. Thus, a business designed to offer services such as home construction or the rehabilitation of homes would likely be successful.

# REVIEW QUESTIONS CHAPTER 18

Write all answers on a separate sheet of paper. Do not write in this book.

1. Why is job site safety everyone's responsibility?
2. Why is nonwork-related conversation discouraged?
3. Whose responsibility is it to come to the job ready to work?
4. What are the benefits of good work habits?
5. How dangerous is the construction industry?
6. What is the primary cause of most accidents?
7. Name five techniques that can be employed to sharpen safety on the job.
8. Why is working in shorts and tennis shoes unsafe?
9. How can one keep up with changes in the masonry field?
10. Work ethic includes your enthusiasm for the work. True or False?
11. Why should a company or professional organization develop an ethics code?
12. Who should manage others?
13. Name five attributes that successful leaders possess.
14. Every year many new businesses are started, but most fail. True or False?
15. Name three reasons why most businesses fail.
16. Identify five characteristics of persons who are successful entrepreneurs.

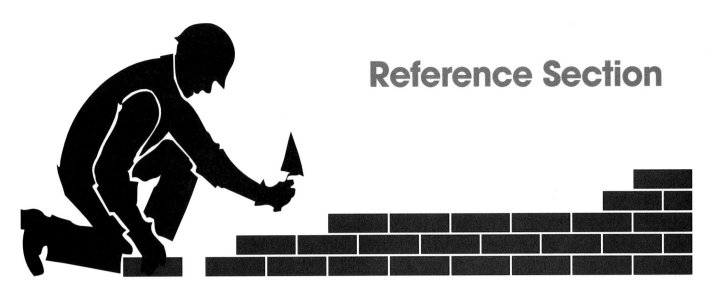

# Reference Section

## Mortar Proportions by Volume ▪

| Mortar Types for Classes of Construction | |
|---|---|
| **ASTM Mortar Type Designation** | **Construction Suitability** |
| M | Masonry subjected to high compressive loads, severe frost action, or high lateral loads from earth pressures, hurricane winds, or earthquakes. Structures below grade, manholes, and catch basins. |
| S | Structures requiring high flexural bond strength, but subject only to normal compressive loads. |
| N | General use in above grade masonry. Residential basement construction, interior walls and partitions. Concrete masonry veneers applied to frame construction. |
| O | Nonloadbearing walls and partitions. Solid load bearing masonry of allowable compressive strength not exceeding 100 psi. |
| K | Interior nonloadbearing partitions where low compressive and bond strengths are permitted by building codes. |

## Mortar Types for Classes of Construction ▬

| Mortar Needed for Concrete Masonry Units | |
|---|---|
| Nominal Height and Length of Units in Inches | Cubic Feet of Mortar per 100 Sq. Ft. |
| 8 × 16 | 6.0 |
| 8 × 12 | 7.0 |
| 5 × 12 | 8.5 |
| 4 × 16 | 9.5 |
| 2 1/4 × 8 | 14.0 |
| 4 × 8 | 12.0 |
| 5 × 8 | 11.0 |
| 2 × 12 | 15.0 |
| 2 × 16 | 15.0 |

## Metric Measure in Construction ▬▬▬▬

Conversion to the SI metric system can take two basic forms: *soft* conversion and *hard* conversion. The actual values of size, capacity, strength, etc., remain unchanged but are expressed in appropriate SI metric terms rather than U.S. Customary in soft conversion. Therefore a 15 acre plot of land would remain the same size but be soft converted to 6.07 ha (hectares). Hard conversion, however, involves an actual change in size, capacity, strength, etc., to arrive at a rational or more useful metric value. For example, a 10 lb. bag of sugar will become 4 kg (not 4.5359 kg, which is the equivalent of 10 lbs.). An actual change in the physical size of the product is usually required in hard conversion.

Units of linear measurement in the construction industry may be restricted to the meter (m) and the millimeter (mm). If required, the kilometer (km) may also be used. Other multiples and sub-multiples of the meter and millimeter should not be used in order to simplify calculations and eliminate unit symbols from drawings. Therefore on a drawing, whole number dimensions will always indicate millimeters, and decimalized numbers (to 3 decimal places) will always indicate meters. For example, the dimensions 200, 1000 and 10 000 are all millimeters, while 2.400, 6.500 and 10.000 are meters.

Even though SI metric units will be employed in the measurements used every day, the actual size of many materials will not change significantly and some will not change at all. The height of ceilings, doors and windows, heating capacities, and material sizes will appear much as they do now, but the way the sizes are described, communicated, and calculated will be very different.

To be able to use the new system of measurement you must establish new *recognition points* to aid in visualizing objects and understanding the effects such objects have. For example, a yard of concrete will automatically bring to mind various metal images such as: its physical size, how much it weights, the coverage it produces as a garage floor, etc. This kind of instinctive orientation and understanding of the quantities being considered must be developed again using metric measures.

Recognition points are already in common use in a limited sense. The mention of familiar 35 mm film can be used as a recognition point. Just remember that 35 mm film is about the same width as the thickness of an interior door. (35 mm= 1 3/8)

Plywood, wallboard, and other standard panel materials are now 4' × 8 '. When produced in SI dimensions they will be 1200 × 2400 mm which is slightly smaller than the present size. (1200 mm = 3' 11 1/4") Therefore, ceiling heights and the spacing of studs, joists, and trusses will be a little less.

Some other hard conversions which will affect the construction industry are shown in the chart below.

**Metric Measures in Construction**

| | |
|---|---|
| Spacing of wall studs | – 300, 400 and 600 mm (400 mm = 15 3/4") |
| Kitchen countertops | – 600 mm wide (600 mm = 1'-11 5/8") |
| | 900 mm high (900 mm = 2'-11 7/16") |
| Size of 2" × 4" stud | – 38 × 89 mm |
| Standard metric brick | – 67 × 100 x 200 mm (100 mm = 3 15/16") |
| Standard mortar joint | – 10 mm |
| Bag of cement | – 40 kg (40 kg = 88.2 lb.) |
| Temperature for pouring concrete | – Not less than 10° C nor |
| | more than 27° C |
| Insulation value | – R20 = RSI 3.5 |

**Areas and Volumes for Planes and Solids**

| | | |
|---|---|---|
| Parallelogram $$A = B \times H$$ | Circle $$A = \pi R^2$$ $$A = .7854 \times D^2$$ $$A = .0796 \times C^2$$ | Cylinder $$V = \pi R^2 \times H$$ |
| Trapezoid $$A = \frac{B + C}{2} \times H$$ | | Cone $$V = \frac{\pi R^2 \times H}{3}$$ |
| | Ellipse $$A = M \times m\ .7854$$ | |
| Triangle $$A = \frac{B \times H}{2}$$ | Rectangular Solids $$V = L \times W \times H$$ | Pyramids $$V = \text{Area of base} \times \frac{H}{3}$$ |
| Regular Polygon $$A = \frac{\text{Sum of sides (S)}}{2} \times R$$ | Prisms $$V = \text{Area of end} \times H$$ | Sphere $$V = \frac{1}{6} \times \pi D^3$$ |

# Size of Modular Bricks

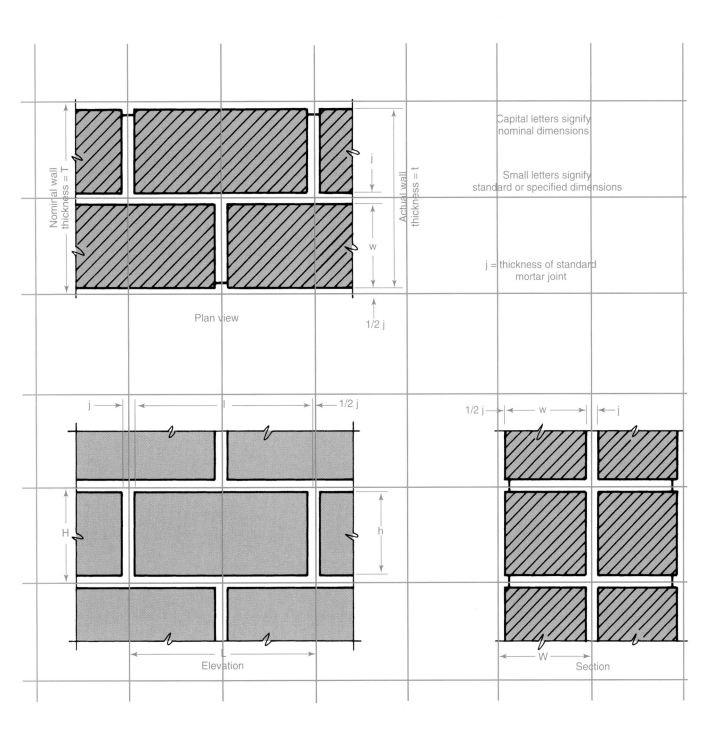

# Shapes and Sizes of Concrete Block

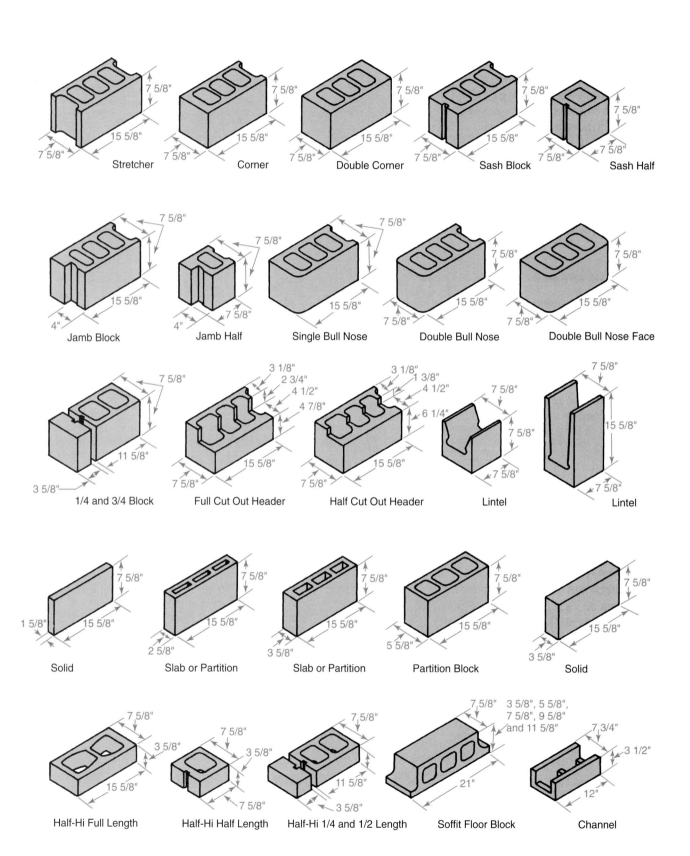

# Shapes and Sizes of Concrete Block (continued)

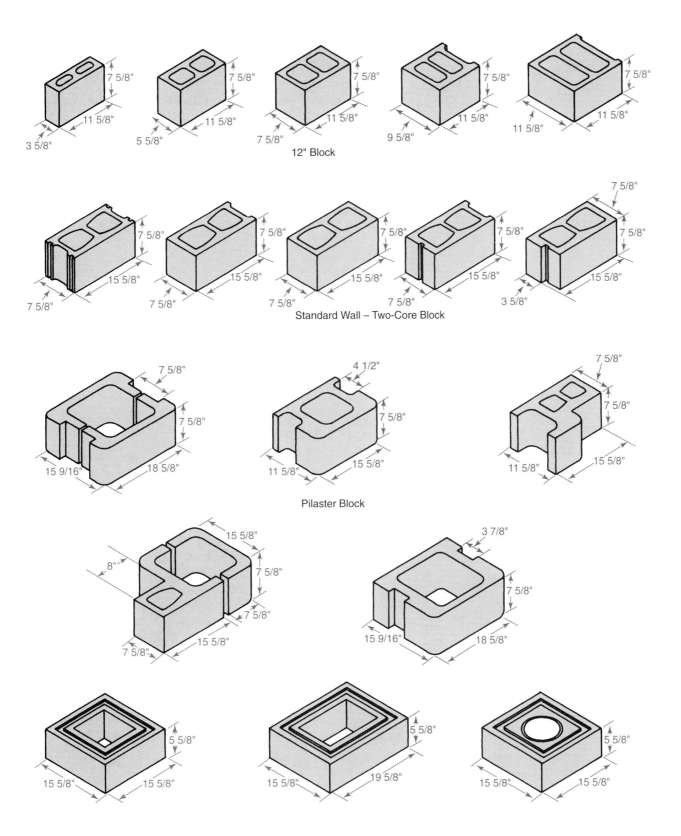

12" Block

Standard Wall – Two-Core Block

Pilaster Block

One-Piece Chimney Block

# Basic Shapes of Structural Clay Tile ▬▬▬▬▬▬

| Structural Clay Facing Tile Available Sizes | | |
|---|---|---|
| Series | Nominal Face Dimensions in Inches | Nominal Thickness in Inches |
| 6T | 5 1/3 × 12 | 2, 4, 6, 8 |
| 4D | 5 1/3 × 8 | 2, 4, 6, 8 |
| 4S | 2 2/3 × 8 | 2, 4 |
| 8W | 8 × 16 | 2, 4 |

| Nominal Modular Sizes* Face Dimensions | |
|---|---|
| Height by Length in Inches | Height by Length in Inches |
| 4 × 8 | 8 × 8 |
| 4 × 12 | 8 × 12 |
| 5 1/3 × 8 | 8 × 16 |
| 5 1/3 × 12 | 12 × 12 |

Thickness: All of the above are in nominal thicknesses of 4, 6 and 8 inches.

* Nominal sizes include the thickness of the standard mortar joint for all dimensions.

# Concrete Mixtures for Various Applications ▬▬▬▬▬

| Construction Application | Water/cement ratio gal. per bag | Consistency (amount of slump) | Maximum size of aggregate | Appropriate cement content bags per yd. | Probable 28th day strength (psi) |
|---|---|---|---|---|---|
| Footings | 7 | 4" to 6" | 1 1/2" | 5.0 | 2800 |
| 8" basement wall moderate ground water | 7 | 4" to 6" | 1 1/2" | 5.0 | 2800 |
| 8" basement wall severe ground water | 6 | 3" to 5" | 1 1/2" | 5.8 | 3500 |
| 10" basement wall moderate ground water | 7 | 4" to 6" | 2" | 4.7 | 2800 |
| 10" basement wall severe ground water | 6 | 3" to 5" | 2" | 5.5 | 3500 |
| Basement floor 4" thickness | 6 | 2" to 4" | 1" | 6.2 | 3500 |
| Floor slab on grade | 6 | 2" to 4" | 1" | 6.2 | 3500 |
| Stairs and steps | 6 | 1" to 4" | 1" | 6.2 | 3500 |
| Topping over concrete floor | 5 | 1" to 2" | 3/8" | 8.0 | 4350 |
| Sidewalks, patios, driveways, porches | 6 | 2" to 4" | 1" | 6.2 | 3500 |

# Welded Wire Reinforcement Stock Sizes ▬▬▬▬▬▬

| Type of Construction | Recommended Style | Remarks |
|---|---|---|
| Basement Floors | 6 × 6 – W1.4 × W1.4<br>6 × 6 – W2.1 × W2.1<br>6 × 6 – W2.9 × W2.9 | For small areas (15' maximum side dimension) use 6 × 6 – W1.4 × W1.4. As a rule of thumb, the larger the area or the poorer the subsoil, the heavier the gauge. |
| Driveways | 6 × 6 – W2.9 | Continuous reinforcement between 25' to 30' contraction joints |
| Foundation Slabs (Residential Only) | 6 × 6 – W1.4 × W1.4 | Use heavier gauge over poorly drained subsoil, or when maximum dimension is greater than 15'. |
| Garage Floors | 6 × 6 – W2.9 × W2.9 | Position at midpoint of 5" or 6" thick slab. |
| Patios and Terraces | 6 × 6 – W1.4 × W1.4 | Use 6 × 6 – W2.1 × W2.1 if subsoil is poorly drained. |
| Porch Floor<br>A. 6" thick slab up to 6' span<br>B. 6" thick slab up to 8' span | 6 × 6 – W2.9 × W2.9<br><br>4 × 4 – W4 × W4 | Position 1" from bottom form to resist tensile stresses. |
| Sidewalks | 6 × 6 – W1.4 × W1.4<br>6 × 6 – W2.1 × W2.1 | Use heavier gauge over poorly drained subsoil. Construct 25' to 30' slabs as for driveways. |
| Steps (Free Span) | 6 × 6 – W2.9 × W2.9 | Use heavier style if more than five risers. Position reinforcement 1" from bottom of form. |
| Steps (On Ground) | 6 × 6 – W2.1 × W2.1 | Use 6 × 6 – W2.9 × W2.9 for unstable subsoil. |

# Glass Block Layout Table ▬▬▬▬▬▬

## Glass Block Layout Table

| No. of Blocks | 6"<br>5 3/4 × 5 3/4 × 3 7/8 | 8"<br>7 3/4 × 7 3/4 × 3 7/8 | 12"<br>11 3/4 × 11 3/4 × 3 7/8 |
|---|---|---|---|
| 1 | 0' – 6" | 0' – 8" | 1' – 0" |
| 2 | 1' – 0" | 1' – 4" | 2' – 0" |
| 3 | 1' – 6" | 2' – 0" | 3' – 0" |
| 4 | 2' – 0" | 2' – 8" | 4' – 0" |
| 5 | 2' – 6" | 3' – 4" | 5' – 0" |
| 6 | 3' – 0" | 4' – 0" | 6' – 0" |
| 7 | 3' – 6" | 4' – 8" | 7' – 0" |
| 8 | 4' – 0" | 5' – 4" | 8' – 0" |
| 9 | 4' – 6" | 6' – 0" | 9' – 0" |
| 10 | 5' – 0" | 6' – 8" | 10' – 0" |
| 11 | 5' – 6" | 7' – 4" | 11' – 0" |
| 12 | 6' – 0" | 8' – 0" | 12' – 0" |
| 13 | 6' – 6" | 8' – 8" | 13' – 0" |
| 14 | 7' – 0" | 9' – 4" | 14' – 0" |
| 15 | 7' – 6" | 10' – 0" | 15' – 0" |
| 16 | 8' – 0" | 10' – 8" | 16' – 0" |
| 17 | 8' – 6" | 11' – 4" | 17' – 0" |
| 18 | 9' – 0" | 12' – 0" | 18' – 0" |
| 19 | 9' – 6" | 12' – 8" | 19' – 0" |
| 20 | 10' – 0" | 13' – 4" | 20' – 0" |
| 21 | 10' – 6" | 14' – 0" | 21' – 0" |
| 22 | 11' – 0" | 14' – 8" | 22' – 0" |
| 23 | 11' – 6" | 15' – 4" | 23' – 0" |
| 24 | 12' – 0" | 16' – 0" | 24' – 0" |
| 25 | 12' – 6" | 16' – 8" | 25' – 0" |

# Steel Reinforcing Bar Sizes

| Bar Number | Bar Diameter in Inches | Bar Area in sq. in. | Approximate Weight of 100 ft. |
|---|---|---|---|
| 2 | 1/4 | .05 | 17 |
| 3 | 3/8 | .11 | 38 |
| 4 | 1/2 | .20 | 67 |
| 5 | 5/8 | .31 | 104 |
| 6 | 3/4 | .44 | 150 |
| 7 | 7/8 | .60 | 204 |
| 8 | 1 | .79 | 267 |

# Specification for Reinforced Concrete Lintels with Stirrups

| Concrete Reinforced Lintels with Stirrups for Wall and Floor Loads | | | | | |
|---|---|---|---|---|---|
| Size of Lintel | | | Reinforcement | | Web Reinforcement No. 6 Gauge Wire Stirrups. Spacings From End of Lintel–Both Ends The Same |
| Height in. | Width in. | Clear Span of Lintel ft. | Top | Bottom | |
| 7 5/8 | 7 5/8 | 3 | None | 2 — 1/2" round | No stirrups required |
| 7 5/8 | 7 5/8 | 4 | None | 2 — 3/4" round | 3 stirrups, SP. :2, 3, 3 in. |
| 7 5/8 | 7 5/8 | 5 | 2 — 3/8" round | 2 — 7/8" round | 5 stirrups, SP. :2, 3, 3, 3, 3 in. |
| 7 5/8 | 7 5/8 | 6 | 2 — 1/2" round | 2 — 7/8" round | 6 stirrups, SP. :2, 3, 3, 3, 3, 3 in. |
| 7 5/8 | 7 5/8 | 7 | 2 — 1" round | 2 — 1" round | 9 stirrups, SP. :2, 3, 3, 3, 3, 3, 3, 3, 3 in. |

# Specification for Reinforced Concrete Lintels

| Concrete Reinforced One-Piece Lintels with Wall Load Only | | | | |
|---|---|---|---|---|
| Size of Lintel | | | Bottom Reinforcement | |
| Height in. | Height in. | Clear Span of Lintel ft. | Size Designation of Bars | Size of Bars |
| 5 3/4 | 7 5/8 | Up to 7 | No. 2 | 3/8" round deformed |
| 5 3/4 | 7 5/8 | 7 to 8 | No. 2 | 5/8" round deformed |
| 7 5/8 | 7 5/8 | Up to 8 | No. 2 | 3/8" round deformed |
| 7 5/8 | 7 5/8 | 8 to 9 | No. 2 | 1/2" round deformed |
| 7 5/8 | 7 5/8 | 9 to 10 | No. 2 | 5/8" round deformed |

| Concrete Reinforced Split Lintels with Wall Load Only | | | | |
|---|---|---|---|---|
| Size of Lintel | | | Bottom Reinforcement | |
| Height in. | Height in. | Clear Span of Lintel ft. | Size Designation of Bars | Size of Bars |
| 5 3/4 | 3 5/8 | Up to 7 | No. 1 | 3/8" round deformed |
| 5 3/4 | 3 5/8 | 7 to 8 | No. 1 | 5/8" round deformed |
| 7 5/8 | 3 5/8 | Up to 8 | No. 1 | 3/8" round deformed |
| 7 5/8 | 3 5/8 | 8 to 9 | No. 1 | 1/2" round deformed |
| 7 5/8 | 3 5/8 | 9 to 10 | No. 1 | 5/8" round deformed |

# Grade-Use Guide for Concrete Forms* ▬▬▬▬▬▬

| Use these terms when you specify plywood | Description | Typical Grade-trademarks | Veneer Grade | |
|---|---|---|---|---|
| | | | Faces | Inner Plies |
| B-B PLYFORM Class I & II** | Specifically manufactured for concrete forms. Many reuses. Smooth, solid surfaces. Edge-sealed. Mill-oiled unless otherwise specified. | B-B PLYFORM CLASS I EXTERIOR PS 1-66 DFPA | B | C |
| High Density Overlaid PLYFORM Class I & II** | Hard, semi-opaque resin-fiber overlay, heat-fused to panel faces. Smooth surface resists abrasion. Up to 200 reuses. Edge-sealed. Light oiling recommended between pours. | HDO - PLYFORM- I - EXT - DFPA - PS 1 -66. | B | C Plugged |
| Structural I PLYFORM** | Especially designed for engineered applications. All Group 1 species. Stronger and stiffer than PLYFORM Class I and II. Recommended for high pressures where face grain is paralleled to supports. Also available with High Density Overlay faces. | STRUCTURAL I B-B PLYFORM CLASS I EXTERIOR PS 1-66 DFPA | B | C or C Plugged |
| Special Overlays, proprietary panels and Medium Density Overlaid plywood specifically designed for concrete forming** | Produce a smooth uniform concrete surface. Generally mill treated with form release agent. Check with manufacturer for design specifications, proper use, and surface treatment recommendations for greatest number of reuses. | | | |

* Commonly available in 5/8" and 3/4" panel thicknesses (4' × 8' size).
** Check dealer for availability in your area.

# Specification for Reinforced Masonry Retaining Walls ▬▬▬▬▬

| Height of Wall | Width of Footing | Thickness of Footing | Dist. to Face of Wall | Size & Spacing of Vertical Rods in Wall | Size & Spacing of Horizontal Rods in Footing |
|---|---|---|---|---|---|
| 8" Walls | | | | | |
| 3' – 4" | 2' – 4" | 9" | 8" | 3/8" @ 32" | 3/8" @ 27" |
| 4' – 0" | 2' – 9" | 9" | 10" | 1/2" @ 32" | 3/8" @ 27" |
| 4" – 8" | 3' – 3" | 10" | 12" | 5/8" @ 32" | 3/8" @ 27" |
| 5' – 4" | 3' – 8" | 10" | 14" | 1/2" @ 16" | 1/2" @ 30" |
| 6' – 0" | 4' – 2" | 12" | 15" | 3/4" @ 24" | 1/2" @ 25" |
| 12" Walls | | | | | |
| 6' – 8" | 4' – 6" | 12" | 16" | 3/4" @ 24" | 1/2" @ 22" |
| 7' – 4" | 4' – 10" | 12" | 18" | 7/8" @ 32" | 5/8" @ 26" |
| 8" – 0" | 5' – 4" | 12" | 20" | 7/8" @ 24" | 5/8" @ 21" |
| 8' – 8" | 5' – 10" | 14" | 22" | 7/8" @ 16" | 3/4" @ 26" |
| 9' – 4" | 6' – 4" | 14" | 24" | 1" @ 8" | 3/4" @ 21" |

# U.S. Customary to Metric Conversions

## U. S. Customary to Metric Conversions

### Lengths
| | |
|---|---|
| 1 inch | = 2.540 centimeters |
| 1 foot | = 30.48 centimeters |
| 1 yard | = 91.44 centimeters or 0.9144 meters |
| 1 mile | = 1.609 kilometers |

### Areas
| | |
|---|---|
| 1 sq. in. | = 6.452 sq. centimeters |
| 1 sq. ft. | = 929.0 sq. centimeters or 0.0929 sq. meter |
| 1 sq. yd. | = 0.8361 sq. meter |

### Volumes
| | |
|---|---|
| 1 cu. In. | = 16.39 cu. centimeters |
| 1 cu. Ft. | = 0.02832 cu. meter |
| 1 cu. Yd. | = 0.7646 cu. meter |

### Weights
| | |
|---|---|
| 1 ounce | = 28.35 grams |
| 1 pound | = 453.6 grams or 0.4536 kilograms |
| 1 (short) ton | = 907.2 kilograms |

### Liquid Measurements
| | |
|---|---|
| 1 (fluid) ounce | = 0.0295 liter or 28.35 grams |
| 1 pint | = 473.2 cu. centimeters |
| 1 quart | = 0.9263 liter |
| 1 (US) gallon | = 3785 cu. centimeters or 3.785 liters |

### Power Measurements
| | |
|---|---|
| 1 horsepower | = 0.7457 kilowatt |

### Temperature Measurements
To convert degrees Fahrenheit to degrees Celsius, use the following formula: $C = 5/9 \times (F-32)$

# Metric to U.S. Customary Conversions

## Metric to U. S. Customary Conversions

### Lengths
| | |
|---|---|
| 1 millimeter (mm) | = 0.03937 in. |
| 1 centimeter (cm) | = 0.3937 in. |
| 1 meter (m) | = 3.281 ft. or 1.0937 yd. |
| 1 kilometer (km) | = 0.6214 miles |

### Areas
| | |
|---|---|
| 1 sq. millimeter | = 0.00155 sq. in. |
| 1 sq. centimeter | = 0.155 sq. in. |
| 1 sq. meter | = 10.76 sq. ft. or 1.196 sq. yd. |

### Volumes
| | |
|---|---|
| 1 cu. centimeter | = 0.06102 cu. In. |
| 1 cu. meter | = 35.31 cu. ft. or 1.308 cu. yd. |

### Weights
| | |
|---|---|
| 1 gram (G) | = 0.03527 oz. |
| 1 kilogram (kg) | = 2.205 lb. |
| 1 metric ton | = 2205 lb. |

### Liquid Measurements
| | |
|---|---|
| 1 cu. centimeter ($cm^3$) | = 0.06102 cu. in. |
| 1 liter (1000 $cm^3$) | = 1.057 quarts, 2.113 pints, or 61.02 cu. in. |

### Power Measurements
| | |
|---|---|
| 1 kilowatt (kw) | = 1.341 horsepower (hp) |

### Temperature Measurements
To convert degrees Celsius to degrees Fahrenheit, use the following formula: $F = (9/5 \times C) + 32$

# Resistivity to Heat Loss of Common Building Materials

| | Material | Resistivity | | Material | Resistivity |
|---|---|---|---|---|---|
| 4" | Concrete or stone | .32 | 1/2" | Plywood | .65 |
| 6" | Concrete or stone | .48 | 5/8" | Plywood | .80 |
| 8" | Concrete or stone | .64 | 3/4" | Plywood | .95 |
| 12" | Concrete or stone | .96 | 3/4" | Softwood sheathing or siding | .85 |
| 4" | Concrete block | .70 | | Composition floor covering | .08 |
| 8" | Concrete block | 1.10 | 1" | Mineral batt insulation | 3.50 |
| 12" | Concrete block | 1.25 | 2" | Mineral batt insulation | 7.00 |
| 4" | Common brick | .82 | 4" | Mineral batt insulation | 14.00 |
| 4" | Face brick | .45 | 2" | Glass fiber insulation | 7.00 |
| 4" | Structural clay tile | 1.10 | 4" | Glass fiber insulation | 14.00 |
| 8" | Structural clay tile | 1.90 | 1" | Loose fill insulation | 3.00 |
| 12" | Structural clay tile | 3.00 | 1/2" | Gypsum wallboard | .45 |
| 1" | Stucco | .20 | 1" | Expanded polystyrene, extruded | 4.00 |
| 15 lb | Building paper | .06 | 1" | Expanded polystyrene, molded beads | 3.85 |
| 3/8" | Sheet rock or plasterboard | .33 | | Single thickness glass | .88 |
| 1/2" | Sand plaster | .15 | | Glassweld insulation glass | 1.89 |
| 1/2" | Insulation plaster | .75 | | Single glass with storm window | 1.66 |
| 1/2" | Fiberboard ceiling tile | 1.20 | | Metal edge insulation glass | 1.85 |
| 1/2" | Fiberboard sheathing | 1.45 | 4" | Glass block | 2.13 |
| 3/4" | Fiberboard sheathing | 2.18 | 1 3/8" | Wood door | 1.92 |
| | Roll roofing | .15 | | Same with storm door | 3.12 |
| | Asphalt shingles | .16 | 1 3/4" | Wood door | 1.82 |
| | Wood shingles | .86 | | Same with storm door | 2.94 |
| | Tile or slate | .08 | | | |

# Quantities of Brick and Mortar Necessary to Construct Wall One Wythe in Thickness

| Nominal Size of Brick (In.) T  H  L | Number of Bricks per 100 Sq. Ft. | Cubic Feet of Mortar | | | |
|---|---|---|---|---|---|
| | | Per 100 Sq. Ft. | | Per 1000 Bricks | |
| | | 3/8" Joints | 1/2" Joints | 3/8" Joints | 1/2" Joints |
| **Modular Brick** | | | | | |
| 4 x 2 2/3 x 8 | 675 | 5.5 | 7.0 | 8.1 | 10.3 |
| 4 x 3 1/5 x 8 | 563 | 4.8 | 6.1 | 8.6 | 10.9 |
| 4 x 4 x 8 | 450 | 4.2 | 5.3 | 9.2 | 11.7 |
| 4 x 5 1/3 x 8 | 338 | 3.5 | 4.4 | 10.2 | 12.9 |
| 4 x 2 x 12 | 600 | 6.5 | 8.2 | 10.8 | 13.7 |
| 4 x 2 2/3 x 12 | 450 | 5.1 | 6.5 | 11.3 | 14.4 |
| 4 x 3 1/5 x 12 | 375 | 4.4 | 5.6 | 11.7 | 14.9 |
| 4 x 4 x 12 | 300 | 3.7 | 4.8 | 12.3 | 15.7 |
| 4 x 5 1/3 x 12 | 225 | 3.0 | 3.9 | 13.4 | 17.1 |
| 6 x 2 2/3 x 12 | 450 | 7.9 | 10.2 | 17.5 | 22.6 |
| 6 x 3 1/5 x 12 | 375 | 6.8 | 8.8 | 18.1 | 23.4 |
| 6 x 4 x 12 | 300 | 5.6 | 7.4 | 19.1 | 24.7 |
| **Nonmodular Brick** | | | | | |
| 3 3/4 x 2 1/4 x 8 | 655 | 5.8 | | 8.8 | |
| | 616 | | 7.2 | | 11.7 |
| 3 3/4 x 2 3/4 x 8 | 551 | 5.0 | | 9.1 | |
| | 522 | | 6.4 | | 12.2 |

# Correction Factors for Bonds with Full Headers

| Bond | Correction Factor |
|---|---|
| Full headers every fifth course only | 1/5 |
| Full headers every sixth course only | 1/6 |
| Full headers every seventh course only | 1/7 |
| English bond (Full headers every second course) | 1/2 |
| Flemish bond (alternate full headers and stretchers every course) | 1/3 |
| Flemish headers every sixth course | 1/18 |
| Flemish cross bond (Flemish headers every second course) | 1/6 |
| Double-stretcher, garden wall bond | 1/5 |
| Triple-stretcher, garden wall bond | 1/7 |

# Minimum Thickness and Maximum Span for Concrete Masonry Screen Walls

| Construction | Minimum Nominal Thickness, (t) | Maximum Distance between Lateral Supports (Height or Length, not Both) | | | |
|---|---|---|---|---|---|
| | | Nominal Thickness of Wall, inches | | | Other Nominal Thicknesses (t) |
| Nonload-Bearing Reinforced:* Exterior Interior | 4" 4" | 10'-0" 16'-0" | 15'-0" 24'-0" | 20'-0" 32'-0" | 30 t 48 t |
| Nonreinforced: Exterior Interior | 4" 4" | 6'-8" 12'-0" | 10'-0" 18'-0" | 13'-4" 24'-0" | 20 t 36 t |
| Loadbearing: Reinforced* Nonreinforced | 6" 6" | Not recommended | 12'-6" 9'-0" | 16'-8" 12'-0" | 25 t 18 t |

\* Total steel area, including joint reinforcement, not less than 0.002 times the gross cross-sectional area of the wall, not more than two-thirds of which may be used on either vertical or horizontal direction.

# Average Depth of Frost Penetration in Inches

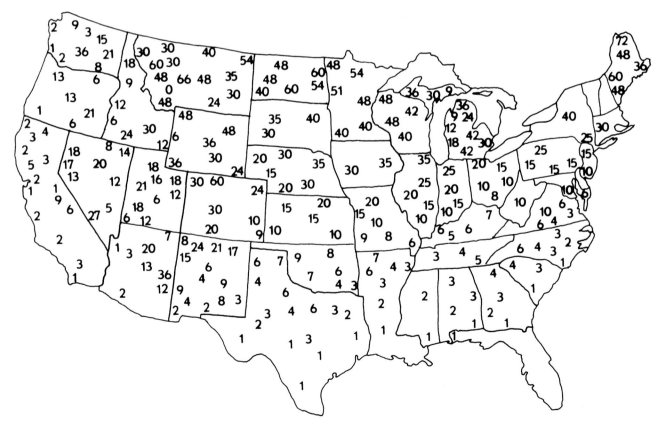

(U. S. Dept. of Commerce Weather Bureau)

# Vertical Coursing Table for Nonmodular Brick[1] ▬▬▬▬▬

| No. of Courses | Nominal Height (h) of Unit[2] | | | | |
|---|---|---|---|---|---|
| | 2" | 2²/₃" | 3¹/₅" | 4" | 5¹/₃" |
| 1 | 0' - 2" | 0' - 2¹¹/₁₆" | 0' - 3³/₁₆" | 0' - 4" | 0' - 5⁵/₁₆ |
| 2 | 0' - 4" | 0' - 5⁵/₁₆" | 0' - 6³/₈" | 0' - 8" | 0' - 10¹¹/₁₆" |
| 3 | 0' - 6" | 0' - 8" | 0' - 9⁵/₈" | 1' - 0" | 1' - 4" |
| 4 | 0' - 8" | 0' - 10¹¹/₁₆" | 1' - 0³/₁₆" | 1' - 4" | 1' - 9⁵/₁₆" |
| 5 | 0' - 10" | 1' - 1⁵/₁₆" | 1' - 4" | 1' - 8" | 2' - 2¹¹/₁₆" |
| 6 | 1' - 0" | 1' - 4" | 1' - 7³/₁₆" | 2' - 0" | 2' - 8" |
| 7 | 1' - 2" | 1' - 6¹¹/₁₆" | 1' - 10³/₈" | 2' - 4" | 3' - 1⁵/₁₆" |
| 8 | 1' - 4" | 1' - 9⁵/₁₆" | 2' - 1⁵/₈" | 2' - 8" | 3' - 6¹¹/₁₆" |
| 9 | 1' - 6" | 2' - 0" | 2' - 4¹³/₁₆" | 3' - 0" | 4' - 0" |
| 10 | 1' - 8" | 2' - 2¹¹/₁₆" | 2' - 8" | 3' - 4" | 4' - 5⁵/₁₆" |
| 11 | 1' - 10" | 2' - 5⁵/₁₆" | 2' - 11³/₁₆" | 3' - 8" | 4' - 10¹¹/₁₆" |
| 12 | 2' - 0" | 2' - 8" | 3' - 2³/₈" | 4' - 0" | 5' - 4" |
| 13 | 2' - 2" | 2' - 10¹¹/₁₆" | 3' - 5⁵/₈" | 4' - 4" | 5' - 9⁵/₁₆" |
| 14 | 2' - 4" | 3' - 1⁵/₁₆" | 3' - 8¹³/₁₆" | 4' - 8" | 6' - 2¹¹/₁₆" |
| 15 | 2' - 6" | 3' - 4" | 4' - 0" | 5' - 0" | 6' - 8" |
| 16 | 2' - 8" | 3' - 6¹¹/₁₆" | 4' - 3³/₁₆" | 5' - 4" | 7' - 1⁵/₁₆" |
| 17 | 2' - 10" | 3' - 9⁵/₁₆" | 4' - 6³/₈" | 5' - 8" | 7' - 6¹¹/₁₆" |
| 18 | 3' - 0" | 4' - 0" | 4 - 9⁵/₈" | 6' - 0" | 8' - 0" |
| 19 | 3' - 2" | 4' - 2¹¹/₁₆" | 5' - 0¹³/₁₆" | 6' - 4" | 8' - 5⁵/₁₆" |
| 20 | 3' - 4" | 4' - 5⁵/₁₆" | 5' - 4" | 6' - 8" | 8' - 10¹¹/₁₆" |
| 21 | 3' - 6" | 4' - 8" | 5' - 7³/₁₆" | 7' - 0" | 9' - 4" |
| 22 | 3' - 8" | 4' - 10¹¹/₁₆" | 5' - 10³/₈" | 7' - 4" | 9' - 9⁵/₁₆" |
| 23 | 3' - 10" | 5' - 1⁵/₁₆" | 6' - 1⁵/₈" | 7' - 8" | 10' - 2¹¹/₁₆" |
| 24 | 4' - 0" | 5' - 4" | 6' - 4¹³/₁₆" | 8' - 0" | 10' - 8" |
| 25 | 4' - 2" | 5' - 6¹¹/₁₆" | 6' - 8" | 8' - 4" | 11' - 1⁵/₁₆" |
| 26 | 4' - 4" | 5' - 9⁵/₁₆" | 6' - 11¹³/₁₆" | 8' - 8" | 11' - 6¹¹/₁₆" |
| 27 | 4' - 6" | 6' - 0" | 7' - 2³/₈" | 9' - 0" | 12' - 0" |
| 28 | 4' - 8" | 6' - 2¹¹/₁₆" | 7' - 5⁵/₈" | 9' - 4" | 12' - 5⁵/₁₆" |
| 29 | 4' - 10" | 6' - 5⁵/₁₆" | 7' - 8¹³/₁₆" | 9' - 8" | 12' - 10¹¹/₁₆" |
| 30 | 5' - 0" | 6' - 8" | 8' - 0" | 10' - 0" | 13' - 4" |
| 31 | 5' - 2" | 6' - 10¹¹/₁₆" | 8' - 3³/₁₆" | 10' - 4" | 13' - 9⁵/₁₆" |
| 32 | 5' - 4" | 7' - 1⁵/₁₆" | 8' - 6³/₈" | 10' - 8" | 14' - 2¹¹/₁₆" |
| 33 | 5' - 6" | 7' - 4" | 8' - 9⁵/₈" | 11' - 0" | 14' - 8" |
| 34 | 5' - 8" | 7' - 6¹¹/₁₆" | 9' - 0¹³/₁₆" | 11' - 4" | 15' - 1⁵/₁₆" |
| 35 | 5' - 10" | 7' - 9⁵/₁₆" | 9' - 4" | 11" - 8" | 15' - 6¹¹/₁₆" |
| 36 | 6' - 0" | 8' - 0" | 9' - 7³/₁₆" | 12' - 0" | 16' - 0" |
| 37 | 6' - 2" | 8' - 2¹¹/₁₆" | 9' - 10³/₈" | 12' - 4" | 16' - 5⁵/₁₆" |
| 38 | 6' - 4" | 8' - 5⁵/₁₆" | 10' - 1⁵/₈" | 12' - 8" | 16' - 10¹¹/₁₆" |
| 39 | 6' - 6" | 8' - 8" | 10' - 4¹³/₁₆" | 13' - 0" | 17' - 4" |
| 40 | 6' - 8" | 8' - 10¹¹/₁₆" | 10' - 8" | 13' - 4" | 17' - 9⁵/₁₆" |
| 41 | 6' - 10" | 9 - 1⁵/₁₆" | 10' - 1¹³/₁₆" | 13' - 8" | 18' - 2¹¹/₁₆" |
| 42 | 7' - 0" | 9' - 4" | 11' - 2³/₈" | 14' - 0" | 18' - 8" |
| 43 | 7' - 2" | 9' - 6¹¹/₁₆" | 11' - 5⁵/₈" | 14' - 4" | 19' - 1⁵/₁₆" |
| 44 | 7' - 4" | 9' - 9⁵/₁₆" | 11' - 8¹³/₁₆" | 14' - 8" | 19' - 6¹¹/₁₆" |
| 45 | 7' - 6" | 10' - 0" | 12' - 0" | 15' - 0" | 20' - 0" |
| 46 | 7' - 8" | 10' - 2¹¹/₁₆" | 12' - 3³/₁₆" | 15' - 4" | 20' - 5⁵/₁₆" |
| 47 | 7' - 10" | 10' - 5⁵/₁₆" | 12' - 6³/₈" | 15' - 8" | 20' - 10¹¹/₁₆" |
| 48 | 8' - 0" | 10' - 8" | 12' - 9⁵/₈" | 16' - 0" | 21' - 4" |
| 49 | 8' - 2" | 10' - 10¹¹/₁₆" | 13' - 0¹³/₁₆" | 16' - 4" | 21' - 9⁵/₁₆" |
| 50 | 8' - 4" | 11' - 1⁵/₁₆" | 13' - 4" | 16' - 8" | 22' - 2¹¹/₁₆" |
| 100 | 16' - 8" | 22' - 2¹¹/₁₆" | 26' - 8" | 33' - 4" | 44' - 5⁵/₁₆" |

[1] Brick positioned in wall as stretchers.
[2] For convenience in using table, nominal ¹/₃", ²/₃" and ¹/₅" heights of units have been changed to nearest ¹/₁₆". Vertical dimensions are from bottom of mortar joint to bottom of mortar joint.

# Vertical Coursing Table for Nonmodular Brick[1]

| No. of Courses | 2¼ in High Units | | 2⅝ in High Units | | 2¾ in High Units | |
|---|---|---|---|---|---|---|
| | 3/8" Joint | 1/2" Joint | 3/8" Joint | 1/2" Joint | 3/8" Joint | 1/2" Joint |
| 1 | 0' - 2⅝" | 0' - 2¾" | 0' - 3" | 0' - 3⅛" | 0' - 3⅛" | 0' - 3¼" |
| 2 | 0' - 5¼" | 0' - 5½" | 0' - 6" | 0' - 6¼" | 0' - 6¼" | 0' - 6½" |
| 3 | 0' - 7⅝" | 0' - 8¼" | 0' - 9" | 0' - 9⅜" | 0' - 9⅜" | 0' - 9¾" |
| 4 | 0' - 10½" | 0' - 11" | 1' - 0" | 1' - 0½" | 1' - 0½" | 1' - 1" |
| 5 | 1' - 1⅛" | 1' - 1¼" | 1' - 3" | 1' - 3⅝" | 1' - 3⅝" | 1' - 4¼" |
| 6 | 1' - 3¾" | 1' - 4½" | 1' - 6" | 1' - 6¾" | 1' - 6¾" | 1' - 7½" |
| 7 | 1' - 6⅜" | 1' - 7¼" | 1' - 9" | 1' - 9⅞" | 1' - 9⅞" | 1' - 10¾" |
| 8 | 1' - 9" | 1' - 10" | 2' - 0" | 2' - 1" | 2' - 1" | 2' - 2" |
| 9 | 1' - 11⅝" | 2' - 0¾" | 2' - 3" | 2' - 4⅛" | 2' - 4⅛" | 2' - 5¼" |
| 10 | 2' - 2¼" | 2' - 3½" | 2' - 6" | 2' - 7¼" | 2' - 7¼" | 2' - 8½" |
| 11 | 2' - 4⅞" | 2' - 6¼" | 2' - 9" | 2' - 10⅜" | 2' - 10⅜" | 2' - 11¾" |
| 12 | 2' - 7½" | 2' - 9" | 3' - 0" | 3' - 1½" | 3' - 1½" | 3' - 3" |
| 13 | 2' - 10⅛" | 2' - 11¾" | 3' - 3" | 3' - 4⅝" | 3' - 4⅝" | 3' - 6¼" |
| 14 | 3' - 0¾" | 3' - 2½" | 3' - 6" | 3' - 7¾" | 3' - 7¾" | 3' - 9½" |
| 15 | 3' - 3⅜" | 3' - 5¼" | 3' - 9" | 3' - 10⅞" | 3' - 10⅞" | 4' - 0¾" |
| 16 | 3' - 6" | 3' - 8" | 4' - 0" | 4' - 2" | 4' - 2" | 4' - 4" |
| 17 | 3' - 8⅝" | 3' - 10¾" | 4' - 3" | 4' - 5⅛" | 4' - 5⅛" | 4' - 7¼" |
| 18 | 3' - 11¼" | 4' - 1½" | 4' - 6" | 4' - 8¼" | 4' - 8¼" | 4' - 10½" |
| 19 | 4' - 1⅞" | 4' - 4¼" | 4' - 9" | 4' - 11⅜" | 4' - 11⅜" | 5' - 1¾" |
| 20 | 4' - 4½" | 4' - 7" | 5' - 0" | 5' - 2½" | 5' - 2½" | 5' - 5" |
| 21 | 4' - 7⅛" | 4' - 9¾" | 5' - 3" | 5' - 5⅝" | 5' - 5⅝" | 5' - 8¼" |
| 22 | 4' - 9¾" | 5' - 0½" | 5' - 6" | 5' - 8¾" | 5' - 8¾" | 5' - 11½" |
| 23 | 5' - 0⅜" | 5' - 3¼" | 5' - 9" | 5' - 11⅞" | 5' - 11⅞" | 6' - 2¾" |
| 24 | 5' - 3" | 5' - 6" | 6' - 0" | 6' - 3" | 6' - 3" | 6' - 6" |
| 25 | 5' - 5⅝" | 5' - 8¾" | 6' - 3" | 6' - 6⅛" | 6' - 6⅛" | 6' - 9¼" |
| 26 | 5' - 8¼" | 5' - 11½" | 6' - 6" | 6' - 9¼" | 6' - 9¼" | 7' - 0½" |
| 27 | 5' - 10⅞" | 6' - 2¼" | 6' - 9" | 7' - 0⅜" | 7' - 0⅜" | 7' - 3¾" |
| 28 | 6' - 1½" | 6' - 5" | 7' - 0" | 7' - 3½" | 7' - 3½" | 7' - 7" |
| 29 | 6' - 4⅛" | 6' - 7¾" | 7' - 3" | 7' - 6⅝" | 7' - 6⅝" | 7' - 10¼" |
| 30 | 6' - 6¾" | 6' - 1½" | 7' - 6" | 7' - 9¾" | 7' - 9¾" | 8' - 1½" |
| 31 | 6' - 9⅜" | 7' - 1¼" | 7' - 9" | 8' - 0⅞" | 8' - 0⅞" | 8' - 4¾" |
| 32 | 7' - 0" | 7' - 4" | 8' - 0" | 8' - 4" | 8' - 4" | 8' - 8" |
| 33 | 7' - 2⅝" | 7' - 6¾" | 8' - 3" | 8' - 7⅛" | 8' - 7⅛" | 8' - 11¼" |
| 34 | 7' - 5¼" | 7' - 9½" | 8' - 6" | 8' - 10¼" | 8' - 10¼" | 9' - 2½" |
| 35 | 7' - 7⅞" | 8' - 0¼" | 8' - 9" | 9' - 1⅜" | 9' - 1⅜" | 9' - 5¾" |
| 36 | 7' - 10½" | 8' - 3" | 9' - 0" | 9' - 4½" | 9' - 4½" | 9' - 9" |
| 37 | 8' - 1⅛" | 8' - 5¾" | 9' - 3" | 9' - 7⅝" | 9' - 7⅝" | 10' - 0¼" |
| 38 | 8' - 3¾" | 8' - 8½" | 9' - 6" | 9' - 10¾" | 9' - 10¾" | 10' - 3½" |
| 39 | 8' - 6⅜" | 8' - 11¼" | 9' - 9" | 10' - 1⅞" | 10' - 1⅞" | 10' - 6¾" |
| 40 | 8' - 9" | 9' - 2" | 10' - 0" | 10' - 5" | 10' - 5" | 10' - 10" |
| 41 | 8' - 11⅝" | 9' - 4¾" | 10' - 3" | 10' - 8⅛" | 10' - 8⅛" | 11' - 1¼" |
| 42 | 9' - 2¼" | 9' - 7½" | 10' - 6" | 10' - 11¼" | 10' - 11¼" | 11' - 4½" |
| 43 | 9' - 4⅞" | 9' - 10¼" | 10' - 9" | 11' - 2⅜" | 11' - 2⅜" | 11' - 7¾" |
| 44 | 9' - 7½" | 10' - 1" | 11' - 0" | 11' - 5½" | 11' - 5½" | 11' - 11" |
| 45 | 9' - 10⅛" | 10' - 3¾" | 11' - 3" | 11' - 8⅝" | 11' - 8⅝" | 12' - 2¼" |
| 46 | 10' - 0¾" | 10' - 6½" | 11' - 6" | 11' - 11¾" | 11' - 11¾" | 12' - 5½" |
| 47 | 10' - 3⅜" | 10' - 9¼" | 11' - 9" | 12' - 2⅞" | 12' - 2⅞" | 12' - 8¾" |
| 48 | 10' - 6" | 11' - 0" | 12' - 0" | 12' - 6" | 12' - 6" | 13' - 0" |
| 49 | 10' - 8⅝" | 11' - 2¾" | 12' - 3" | 12' - 9⅛" | 12' - 9⅛" | 13' - 3¼" |
| 50 | 10' - 11¼" | 11' - 5½" | 12' - 6" | 13' - 0¼" | 13' - 0¼" | 13' - 6½" |
| 100 | 21' - 10½" | 22' - 11" | 25' - 0" | 26' - 0½" | 26' - 0½" | 27' - 1" |

[1] Brick positioned in wall as stretchers. Vertical dimensions are from bottom of mortar joint to bottom of mortar joint.

# Weights and Measures

| Water Weight, lb. | Water U.S. gallons | Water Imp. gallons | Water Weight, lb. | Water U.S. gallons | Water Imp. gallons | Cement U.S. bags | Cement Can. bags | Cement Weight, lb. | Cement U.S. bags | Cement Can. bags | Cement Weight, lb. | W/C by weight | Can. w/c, gal./bag | U.S. w/c, gal./bag |
|---|---|---|---|---|---|---|---|---|---|---|---|---|---|---|
| 10 | 1.2 | 1.0 | 185 | 22.2 | 18.5 | 1.0 | 1.1 | 94 | 4.5 | 4.8 | 423 | .30 | 2.6 | 3.4 |
| 15 | 1.8 | 1.5 | 190 | 22.8 | 19.0 | 1.1 | 1.2 | 103 | 4.6 | 4.9 | 432 | .31 | 2.7 | 3.5 |
| 20 | 2.4 | 2.0 | 195 | 23.4 | 19.5 | 1.2 | 1.3 | 113 | 4.7 | 5.0 | 442 | .32 | 2.8 | 3.6 |
| 25 | 3.0 | 2.5 | 200 | 24.0 | 20.0 | 1.3 | 1.4 | 122 | 4.8 | 5.2 | 451 | .33 | 2.9 | 3.7 |
| 30 | 3.6 | 3.0 | 205 | 24.6 | 20.5 | 1.4 | 1.5 | 132 | 4.9 | 5.3 | 461 | .34 | 3.0 | 3.8 |
| 35 | 4.2 | 3.5 | 210 | 25.2 | 21.0 | 1.5 | 1.6 | 141 | 5.0 | 5.4 | 470 | .35 | 3.1 | 3.9 |
| 40 | 4.8 | 4.0 | 215 | 25.8 | 21.5 | 1.6 | 1.7 | 150 | 5.1 | 5.5 | 479 | .36 | 3.2 | 4.1 |
| 45 | 5.4 | 4.5 | 220 | 26.4 | 22.0 | 1.7 | 1.8 | 160 | 5.2 | 5.6 | 489 | .37 | 3.2 | 4.2 |
| 50 | 6.0 | 5.0 | 225 | 27.0 | 22.5 | 1.8 | 1.9 | 169 | 5.3 | 5.7 | 498 | .38 | 3.3 | 4.3 |
| 55 | 6.6 | 5.5 | 230 | 27.6 | 23.0 | 1.9 | 2.0 | 179 | 5.4 | 5.8 | 508 | .39 | 3.4 | 4.4 |
| 60 | 7.2 | 6.0 | 235 | 28.2 | 23.5 | 2.0 | 2.1 | 188 | 5.5 | 5.9 | 517 | .40 | 3.5 | 4.5 |
| 65 | 7.8 | 6.5 | 240 | 28.8 | 24.0 | 2.1 | 2.3 | 197 | 5.6 | 6.0 | 526 | .41 | 3.6 | 4.6 |
| 70 | 8.4 | 7.0 | 245 | 29.4 | 24.5 | 2.2 | 2.4 | 207 | 5.7 | 6.1 | 536 | .42 | 3.7 | 4.7 |
| 75 | 9.0 | 7.5 | 250 | 30.0 | 25.0 | 2.3 | 2.5 | 216 | 5.8 | 6.2 | 545 | .43 | 3.8 | 4.8 |
| 80 | 9.6 | 8.0 | 255 | 30.6 | 25.5 | 2.4 | 2.6 | 226 | 5.9 | 6.3 | 555 | .44 | 3.9 | 5.0 |
| 85 | 10.2 | 8.5 | 260 | 31.2 | 26.0 | 2.5 | 2.7 | 235 | 6.0 | 6.4 | 564 | .45 | 3.9 | 5.1 |
| 90 | 10.8 | 9.0 | 265 | 31.8 | 26.5 | 2.6 | 2.8 | 244 | 6.1 | 6.6 | 573 | .46 | 4.0 | 5.2 |
| 95 | 11.4 | 9.5 | 270 | 32.4 | 27.0 | 2.7 | 2.9 | 254 | 6.2 | 6.7 | 583 | .47 | 4.1 | 5.3 |
| 100 | 12.0 | 10.0 | 275 | 33.0 | 27.5 | 2.8 | 3.0 | 263 | 6.3 | 6.8 | 592 | .48 | 4.2 | 5.4 |
| 105 | 12.6 | 10.5 | 280 | 33.6 | 28.0 | 2.9 | 3.1 | 273 | 6.4 | 6.9 | 602 | .49 | 4.3 | 5.5 |
| 110 | 13.2 | 11.0 | 285 | 34.2 | 28.5 | 3.0 | 3.2 | 282 | 6.5 | 7.0 | 611 | .50 | 4.4 | 5.6 |
| 115 | 13.8 | 11.5 | 290 | 34.8 | 29.0 | 3.1 | 3.3 | 291 | 6.6 | 7.1 | 620 | .51 | 4.5 | 5.7 |
| 120 | 14.4 | 12.0 | 295 | 35.4 | 29.5 | 3.2 | 3.4 | 301 | 6.7 | 7.2 | 630 | .52 | 4.6 | 5.9 |
| 125 | 15.0 | 12.5 | 300 | 36.0 | 30.0 | 3.3 | 3.5 | 310 | 6.8 | 7.3 | 639 | .53 | 4.6 | 6.0 |
| 130 | 15.6 | 13.0 | 305 | 36.6 | 30.5 | 3.4 | 3.7 | 320 | 6.9 | 7.4 | 649 | .54 | 4.7 | 6.1 |
| 135 | 16.2 | 13.5 | 310 | 37.2 | 31.0 | 3.5 | 3.8 | 329 | 7.0 | 7.5 | 658 | .55 | 4.8 | 6.2 |
| 140 | 16.8 | 14.0 | 315 | 37.8 | 31.5 | 3.6 | 3.9 | 338 | 7.1 | 7.6 | 667 | .56 | 4.9 | 6.3 |
| 145 | 17.4 | 14.5 | 320 | 38.4 | 32.0 | 3.7 | 4.0 | 348 | 7.2 | 7.7 | 677 | .57 | 5.0 | 6.4 |
| 150 | 18.0 | 15.0 | 325 | 39.0 | 32.5 | 3.8 | 4.1 | 357 | 7.0 | 7.8 | 686 | .58 | 5.1 | 6.5 |
| 155 | 18.6 | 15.5 | 330 | 39.6 | 33.0 | 3.9 | 4.2 | 367 | 7.4 | 7.9 | 696 | .59 | 5.2 | 6.7 |
| 160 | 19.2 | 16.0 | 335 | 40.2 | 33.5 | 4.0 | 4.3 | 376 | 7.5 | 8.1 | 705 | .60 | 5.3 | 6.8 |
| 165 | 19.8 | 16.5 | 340 | 40.8 | 34.0 | 4.1 | 4.4 | 385 | 7.6 | 8.2 | 714 | .61 | 5.3 | 6.9 |
| 170 | 20.4 | 17.0 | 345 | 41.4 | 34.5 | 4.2 | 4.5 | 395 | 7.7 | 8.3 | 724 | .62 | 5.4 | 7.0 |
| 175 | 21.0 | 17.5 | 350 | 42.0 | 35.0 | 4.3 | 4.6 | 404 | 7.8 | 8.4 | 733 | .63 | 5.5 | 7.1 |
| 180 | 21.6 | 18.0 | 355 | 42.6 | 35.5 | 4.4 | 4.7 | 414 | 7.9 | 8.5 | 743 | .64 | 5.6 | 7.2 |

## Metric conversions

| To convert form: | To: | Multiply by: |
|---|---|---|
| U.S., gal. | cu. meters | 0.003785 |
| imp. gal | cu. meters | 0.004546 |
| U.S. gags | kilograms | 42.637 |
| Can. bags | kilograms | 39.689 |
| pounds | kilograms | 0.4536 |
| cu. ft. | cu. meters | .02832 |
| cu. yd. | cu. meters | 0.7646 |

## Equivalent weights and measures

| | | |
|---|---|---|
| 1 gal. of water, U.S. | = | 8.33 lb. |
| 1 gal. of water, imp. | = | 10.00 lb. |
| 1 cu. ft. of water | = | 62.40 lb. |
| 1 bag of cement, U.S. | = | 94.00 lb. |
| 1 bag of cement, Can. | = | 87.50 lbs |

Note: Larger quantities can be secured by addition of above figures

# Construction Details of Screen Walls

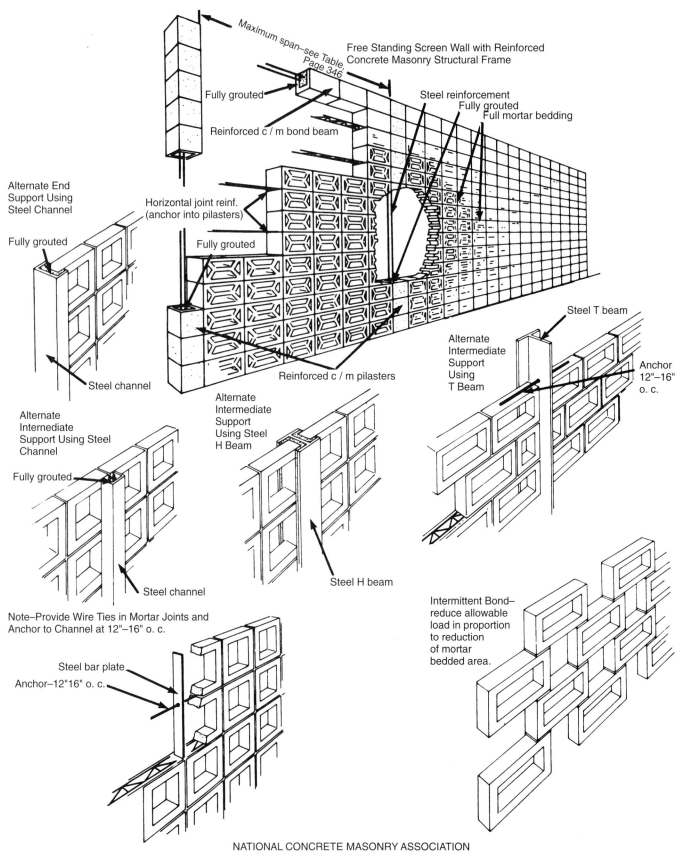

Maximum span—see Table, Page 346

Free Standing Screen Wall with Reinforced Concrete Masonry Structural Frame

Fully grouted

Reinforced c / m bond beam

Steel reinforcement
Fully grouted
Full mortar bedding

Alternate End Support Using Steel Channel

Fully grouted

Horizontal joint reinf. (anchor into pilasters)

Fully grouted

Steel channel

Reinforced c / m pilasters

Steel T beam

Alternate Intermediate Support Using T Beam

Anchor 12"–16" o. c.

Alternate Interrnediate Support Using Steel Channel

Fully grouted

Steel channel

Alternate Intermediate Support Using Steel H Beam

Steel H beam

Note—Provide Wire Ties in Mortar Joints and Anchor to Channel at 12"–16" o. c.

Steel bar plate

Anchor–12"16" o. c.

Intermittent Bond— reduce allowable load in proportion to reduction of mortar bedded area.

NATIONAL CONCRETE MASONRY ASSOCIATION
P. O. Box 9185, Rosslyn Station, Arlington, Virginai 22209

# Shapes and Sizes of Structural Clay Tile

Various Sizes and Shapes of Structural Clay Tile

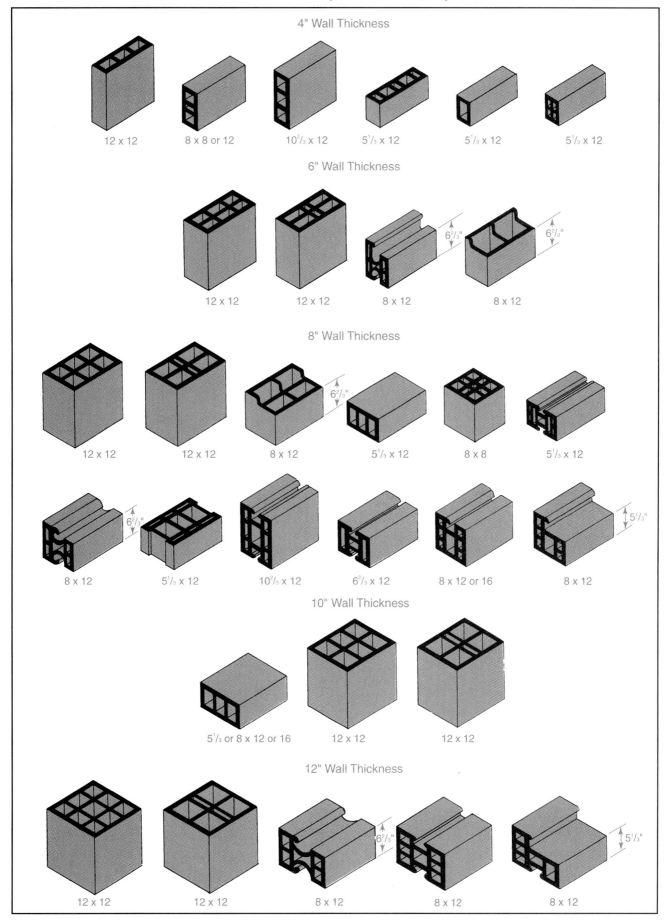

4" Wall Thickness

12 x 12     8 x 8 or 12     $10^2/_3$ x 12     $5^1/_3$ x 12     $5^1/_3$ x 12     $5^1/_3$ x 12

6" Wall Thickness

12 x 12     12 x 12     8 x 12 ($6^2/_3$")     8 x 12 ($6^2/_3$")

8" Wall Thickness

12 x 12     12 x 12     8 x 12 ($6^2/_3$")     $5^1/_3$ x 12     8 x 8     $5^1/_3$ x 12

8 x 12 ($6^2/_3$")     $5^1/_3$ x 12     $10^2/_3$ x 12     $6^2/_3$ x 12     8 x 12 or 16     8 x 12 ($5^1/_3$")

10" Wall Thickness

$5^1/_3$ or 8 x 12 or 16     12 x 12     12 x 12

12" Wall Thickness

12 x 12     12 x 12     8 x 12 ($6^2/_3$")     8 x 12     8 x 12 ($5^1/_3$")

# Most Widely Available Sizes of Structural Clay Tile

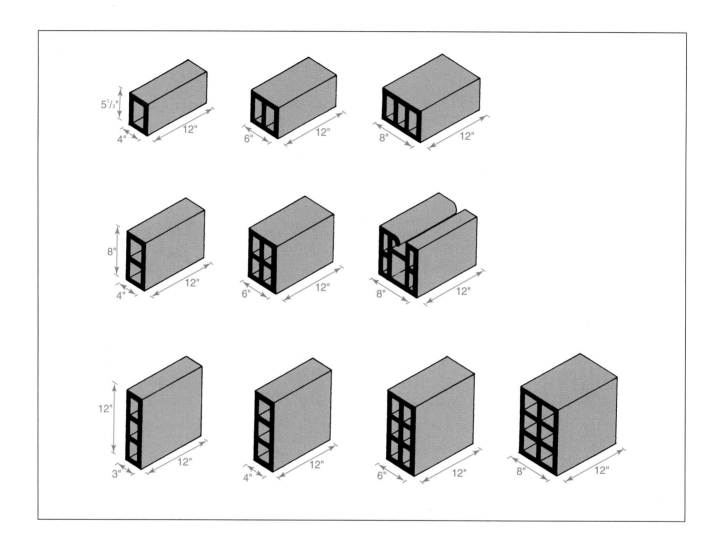

# Line Symbols

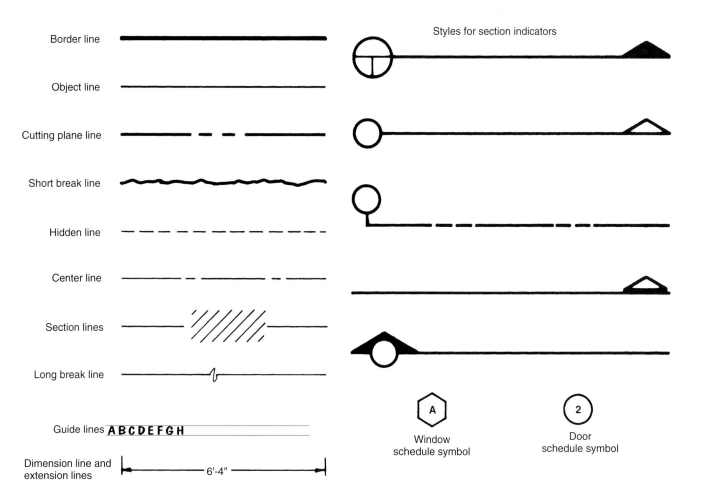

Border line

Object line

Cutting plane line

Short break line

Hidden line

Center line

Section lines

Long break line

Guide lines **ABCDEFGH**

Dimension line and extension lines — 6'-4" —

Styles for section indicators

Window schedule symbol

Door schedule symbol

# Building Material Symbols

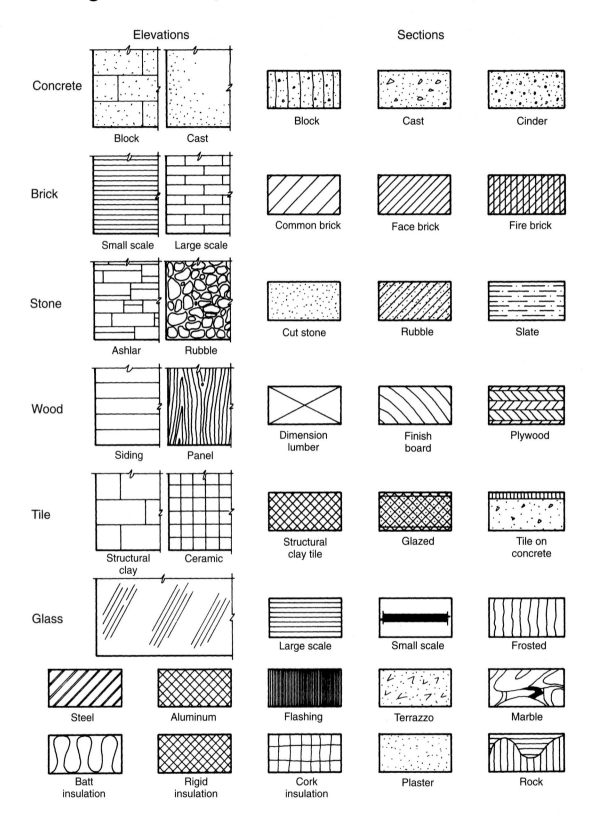

Elevations

Sections

Concrete — Block, Cast

Block, Cast, Cinder

Brick — Small scale, Large scale

Common brick, Face brick, Fire brick

Stone — Ashlar, Rubble

Cut stone, Rubble, Slate

Wood — Siding, Panel

Dimension lumber, Finish board, Plywood

Tile — Structural clay, Ceramic

Structural clay tile, Glazed, Tile on concrete

Glass — Large scale, Small scale, Frosted

Steel, Aluminum, Flashing, Terrazzo, Marble

Batt insulation, Rigid insulation, Cork insulation, Plaster, Rock

# Topographical Symbols

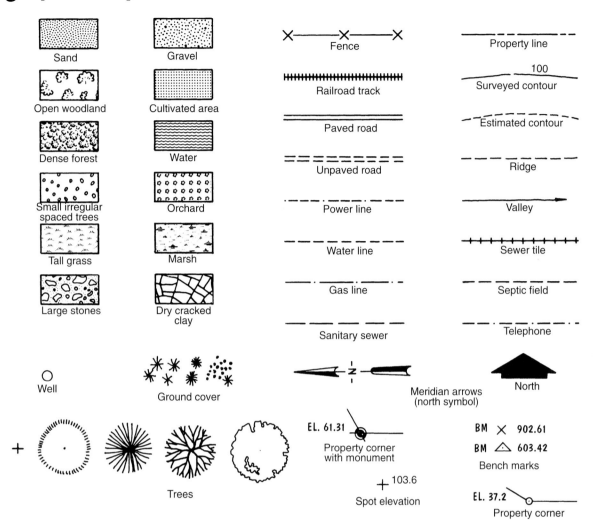

Sand

Gravel

Open woodland

Cultivated area

Dense forest

Water

Small irregular
spaced trees

Orchard

Tall grass

Marsh

Large stones

Dry cracked
clay

Fence

Property line

Railroad track

100
Surveyed contour

Paved road

Estimated contour

Unpaved road

Ridge

Power line

Valley

Water line

Sewer tile

Gas line

Septic field

Sanitary sewer

Telephone

Well

Ground cover

Meridian arrows
(north symbol)

North

Trees

EL. 61.31
Property corner
with monument

BM ✕ 902.61
BM △ 603.42
Bench marks

+ 103.6
Spot elevation

EL. 37.2
Property corner

# Plumbing Symbols

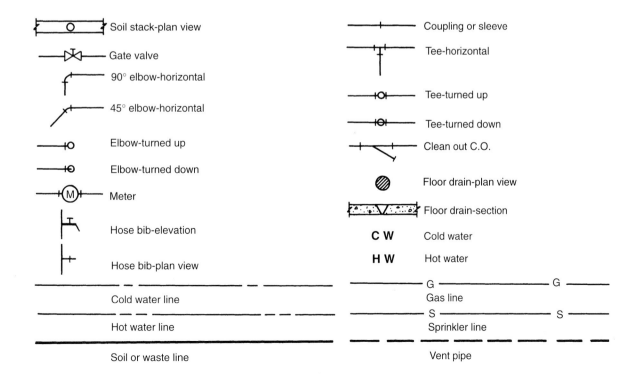

Soil stack-plan view

Gate valve

90° elbow-horizontal

45° elbow-horizontal

Elbow-turned up

Elbow-turned down

Meter

Hose bib-elevation

Hose bib-plan view

Cold water line

Hot water line

Soil or waste line

Coupling or sleeve

Tee-horizontal

Tee-turned up

Tee-turned down

Clean out C.O.

Floor drain-plan view

Floor drain-section

C W    Cold water

H W    Hot water

G ——— G
Gas line

S ——— S
Sprinkler line

Vent pipe

# Climate Control Symbols

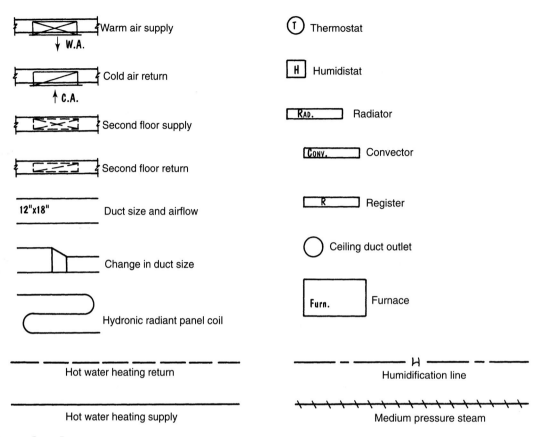

Warm air supply
↓ W.A.

Cold air return
↑ C.A.

Second floor supply

Second floor return

12"x18" Duct size and airflow

Change in duct size

Hydronic radiant panel coil

Hot water heating return

Hot water heating supply

T Thermostat

H Humidistat

RAD. Radiator

CONV. Convector

R Register

Ceiling duct outlet

Furn. Furnace

Humidification line

Medium pressure steam

# Electrical Symbols

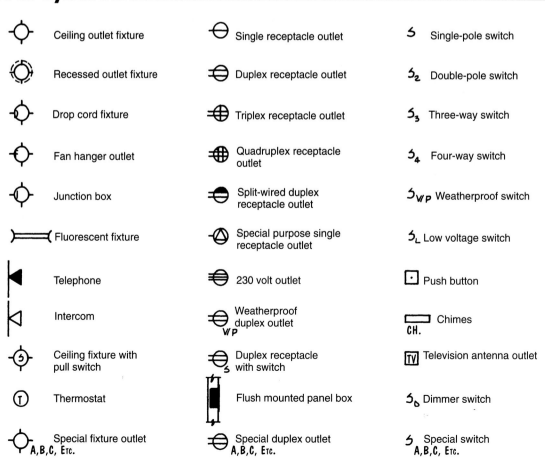

Ceiling outlet fixture

Recessed outlet fixture

Drop cord fixture

Fan hanger outlet

Junction box

Fluorescent fixture

Telephone

Intercom

Ceiling fixture with pull switch

T Thermostat

Special fixture outlet
A,B,C, ETC.

Single receptacle outlet

Duplex receptacle outlet

Triplex receptacle outlet

Quadruplex receptacle outlet

Split-wired duplex receptacle outlet

Special purpose single receptacle outlet

230 volt outlet

Weatherproof duplex outlet
W P

Duplex receptacle with switch

Flush mounted panel box

Special duplex outlet
A,B,C, ETC.

S Single-pole switch

S₂ Double-pole switch

S₃ Three-way switch

S₄ Four-way switch

S_WP Weatherproof switch

S_L Low voltage switch

Push button

Chimes
CH.

TV Television antenna outlet

S_b Dimmer switch

S Special switch
A,B,C, ETC.

# Patterns for Concrete Masonry

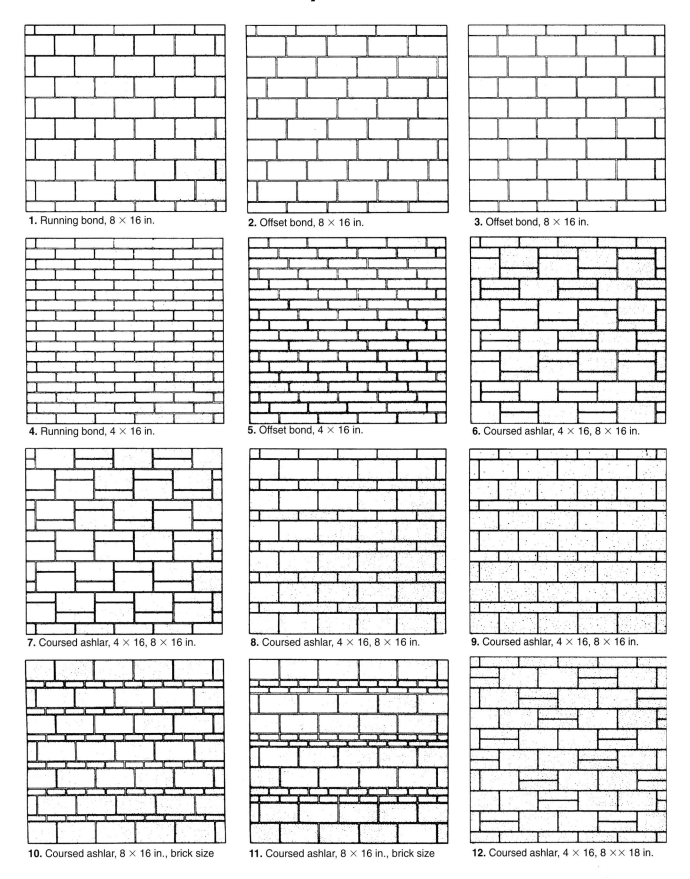

**1.** Running bond, 8 × 16 in.

**2.** Offset bond, 8 × 16 in.

**3.** Offset bond, 8 × 16 in.

**4.** Running bond, 4 × 16 in.

**5.** Offset bond, 4 × 16 in.

**6.** Coursed ashlar, 4 × 16, 8 × 16 in.

**7.** Coursed ashlar, 4 × 16, 8 × 16 in.

**8.** Coursed ashlar, 4 × 16, 8 × 16 in.

**9.** Coursed ashlar, 4 × 16, 8 × 16 in.

**10.** Coursed ashlar, 8 × 16 in., brick size

**11.** Coursed ashlar, 8 × 16 in., brick size

**12.** Coursed ashlar, 4 × 16, 8 × × 18 in.

# Patterns for Concrete Masonry (Continued)

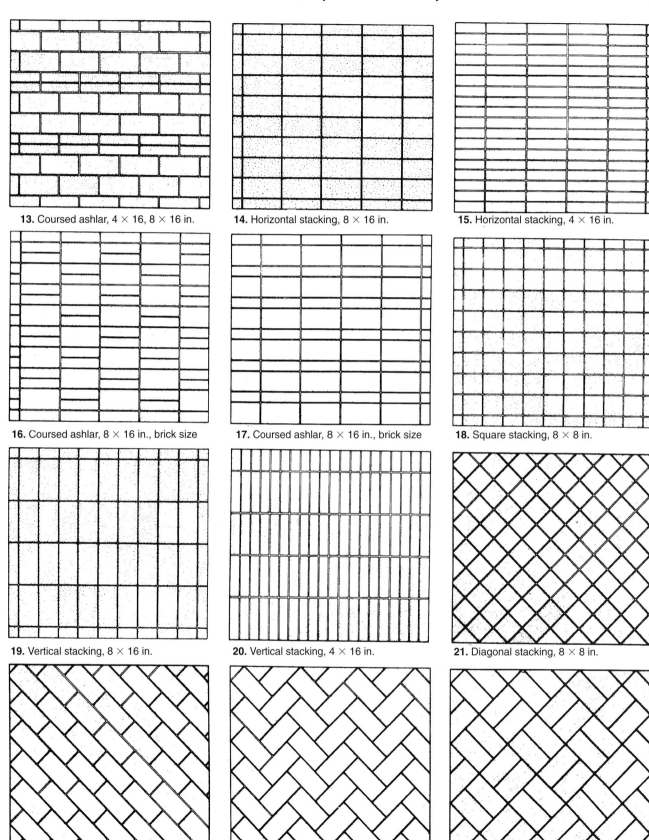

**13.** Coursed ashlar, 4 × 16, 8 × 16 in.

**14.** Horizontal stacking, 8 × 16 in.

**15.** Horizontal stacking, 4 × 16 in.

**16.** Coursed ashlar, 8 × 16 in., brick size

**17.** Coursed ashlar, 8 × 16 in., brick size

**18.** Square stacking, 8 × 8 in.

**19.** Vertical stacking, 8 × 16 in.

**20.** Vertical stacking, 4 × 16 in.

**21.** Diagonal stacking, 8 × 8 in.

**22.** Diagonal bond, 8 × 16 in.

**23.** Diagonal basket weave, 8 × 16 in.

**24.** Diagonal basket weave, 8 × 16 in.

# Patterns for Concrete Masonry (Continued)

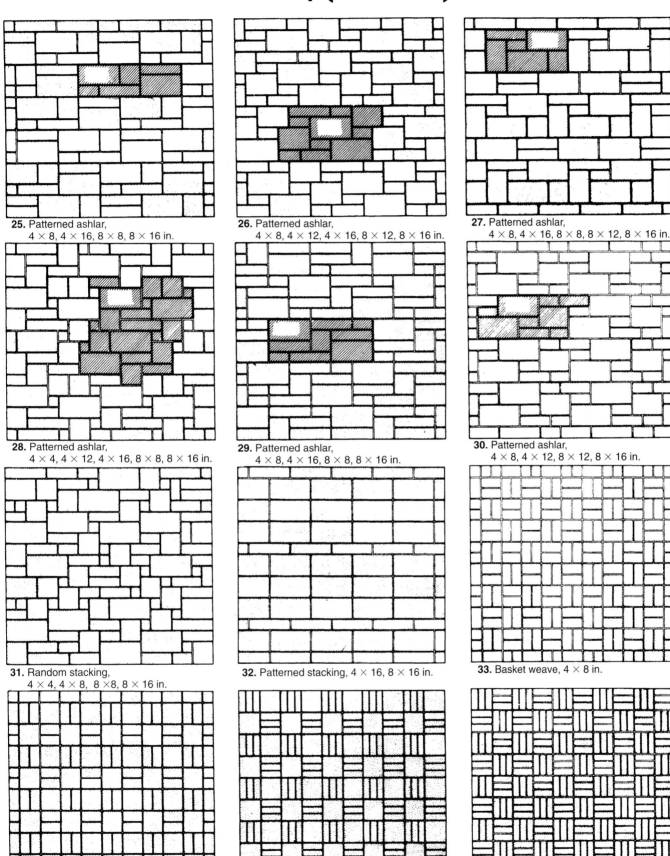

**25.** Patterned ashlar,
  4 × 8, 4 × 16, 8 × 8, 8 × 16 in.

**26.** Patterned ashlar,
  4 × 8, 4 × 12, 4 × 16, 8 × 12, 8 × 16 in.

**27.** Patterned ashlar,
  4 × 8, 4 × 16, 8 × 8, 8 × 12, 8 × 16 in.

**28.** Patterned ashlar,
  4 × 4, 4 × 12, 4 × 16, 8 × 8, 8 × 16 in.

**29.** Patterned ashlar,
  4 × 8, 4 × 16, 8 × 8, 8 × 16 in.

**30.** Patterned ashlar,
  4 × 8, 4 × 12, 8 × 12, 8 × 16 in.

**31.** Random stacking,
  4 × 4, 4 × 8, 8 ×8, 8 × 16 in.

**32.** Patterned stacking, 4 × 16, 8 × 16 in.

**33.** Basket weave, 4 × 8 in.

**34.** Basket weave, 4 × 8, 8 × 8 in.

**35.** Basket weave, 8 × 8 in., brick size

**36.** Basket weave, brick size

# Patterns for Concrete Masonry (Continued)

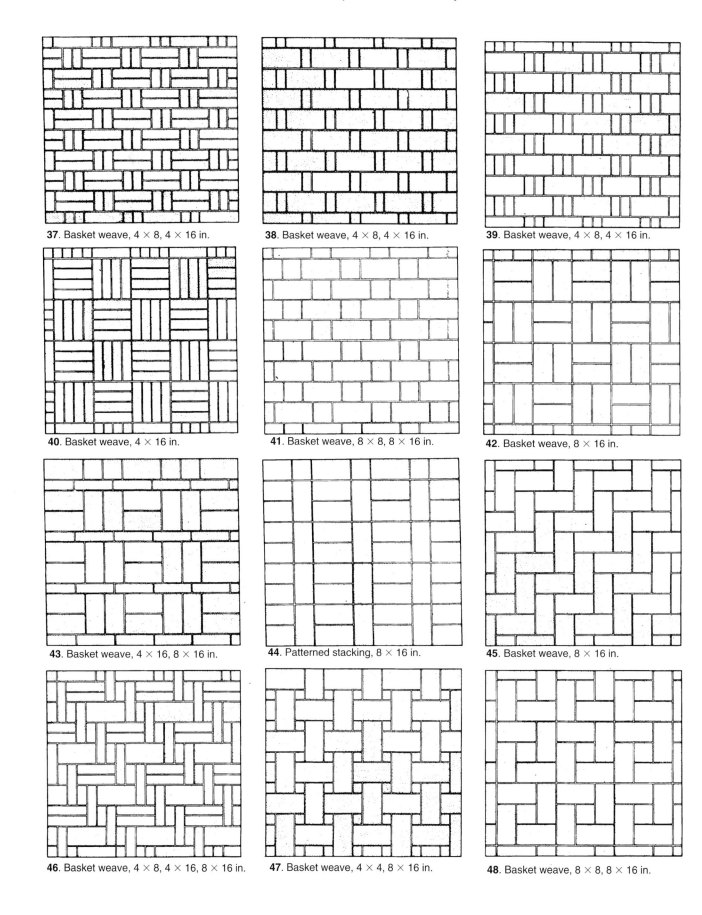

**37**. Basket weave, 4 × 8, 4 × 16 in.

**38**. Basket weave, 4 × 8, 4 × 16 in.

**39**. Basket weave, 4 × 8, 4 × 16 in.

**40**. Basket weave, 4 × 16 in.

**41**. Basket weave, 8 × 8, 8 × 16 in.

**42**. Basket weave, 8 × 16 in.

**43**. Basket weave, 4 × 16, 8 × 16 in.

**44**. Patterned stacking, 8 × 16 in.

**45**. Basket weave, 8 × 16 in.

**46**. Basket weave, 4 × 8, 4 × 16, 8 × 16 in.

**47**. Basket weave, 4 × 4, 8 × 16 in.

**48**. Basket weave, 8 × 8, 8 × 16 in.

# Specifications for Single-Face Fireplaces

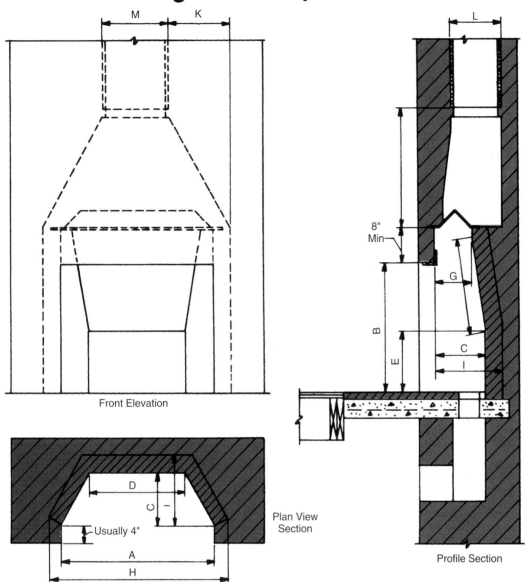

Front Elevation

Plan View Section

Profile Section

Usually 4"

8" Min

Design Data for Single-face Fireplaces

| Width | Hgt. | Depth | Back | Vert. Back | Slope Back | Throat | Width | Depth | Smoke Chamb. | Flue Lining Sizes | | | | |
|---|---|---|---|---|---|---|---|---|---|---|---|---|---|---|
| | | | | | | | | | | Rect. | | Rnd. | Modular | |
| A | B | C | D | E | F | G | H | I | J | K | L × M | K | K | L × M |
| 24 | 24 | 16 | 11 | 14 | 15 | 8 3/4 | 32 | 20 | 19 | 11 3/4 | 8 1/2 × 8 1/2 | 8 | 10 | 8 × 12 |
| 26 | 24 | 16 | 13 | 14 | 15 | 8 3/4 | 34 | 20 | 21 | 12 3/4 | 8 1/2 × 8 1/2 | 8 | 11 | 8 × 12 |
| 28 | 24 | 16 | 15 | 14 | 15 | 8 3/4 | 36 | 20 | 21 | 11 1/2 | 8 1/2 × 13 | 10 | 12 | 8 × 12 |
| 30 | 29 | 16 | 17 | 14 | 18 | 8 3/4 | 38 | 20 | 24 | 12 1/2 | 8 1/2 × 13 | 10 | 13 | 12 × 12 |
| 32 | 29 | 16 | 19 | 14 | 21 | 8 3/4 | 40 | 20 | 24 | 13 1/2 | 8 1/2 × 13 | 10 | 14 | 12 × 12 |
| 36 | 29 | 16 | 23 | 14 | 21 | 8 3/4 | 44 | 20 | 27 | 15 1/2 | 13 × 13 | 12 | 16 | 12 × 12 |
| 40 | 29 | 16 | 27 | 14 | 21 | 8 3/4 | 48 | 20 | 29 | 17 1/2 | 13 × 13 | 12 | 16 | 12 × 12 |
| 42 | 32 | 16 | 29 | 14 | 23 | 8 3/4 | 50 | 20 | 32 | 18 1/2 | 13 × 13 | 12 | 17 | 16 × 16 |
| 48 | 32 | 18 | 33 | 14 | 23 | 8 3/4 | 56 | 22 | 37 | 21 1/2 | 13 × 13 | 15 | 20 | 16 × 16 |
| 54 | 37 | 20 | 37 | 16 | 27 | 13 | 68 | 24 | 45 | 25 | 13 × 18 | 15 | 26 | 16 × 20 |
| 60 | 37 | 22 | 42 | 16 | 27 | 13 | 72 | 27 | 45 | 27 | 13 × 18 | 15 | 26 | 16 × 20 |
| 60 | 40 | 22 | 42 | 16 | 29 | 13 | 72 | 27 | 45 | 27 | 18 × 18 | 18 | 26 | 16 × 20 |
| 72 | 40 | 22 | 54 | 16 | 29 | 13 | 84 | 27 | 56 | 33 | 18 × 18 | 18 | 32 | 20 × 20 |
| 84 | 40 | 24 | 64 | 20 | 26 | 13 | 96 | 29 | 67 | 36 | 20 × 20 | 20 | 36 | 20 × 20 |
| 96 | 40 | 24 | 76 | 20 | 26 | 13 | 108 | 29 | 75 | 42 | 24 × 24 | 22 | 42 | 20 × 20 |

Dimensions are in inches.

Flue sizes for chimney height of at least 14' – 0"

# Wire Area, Diameter, and Mass ▬▬▬▬▬▬▬▬▬▬

## Metric Wire Area, Diameter & Mass With Equivalent U.S. Customary Units

| Metric Units* | | | | U.S. Customary Units** | | | | Gage Guide |
|---|---|---|---|---|---|---|---|---|
| Size ★ (mw=Plain) (mm²) | Area (mm²) | Diameter (mm) | Mass (kg/m) | Size ★ (w=Plain) (in²x100) | Area (in²) | Diameter (in) | Weight (lb/ft) | |
| MW200 | 200 | 16.0 | 1.57 | W31 | .310 | .628 | 1.054 | |
| MW130 | 130 | 12.9 | 1.02 | W20.2 | .202 | .507 | .687 | 7/0 |
| MW120 | 120 | 12.4 | .941 | W18.6 | .186 | .487 | .632 | 6/0 |
| MW100 | 100 | 11.3 | .784 | W15.5 | .155 | .444 | .527 | 5/0 |
| MW90 | 90 | 10.7 | .706 | W14.0 | .140 | .422 | .476 | |
| MW80 | 80 | 10.1 | .627 | W12.4 | .124 | .397 | .422 | 4/0 |
| MW70 | 70 | 9.4 | .549 | W10.9 | .109 | .373 | .371 | 3/0 |
| MW65 | 65 | 9.1 | .510 | W10.1 | .101 | .359 | .343 | |
| MW60 | 60 | 8.7 | .470 | W9.3 | .093 | .344 | .316 | 2/0 |
| MW55 | 55 | 8.4 | .431 | W8.5 | .085 | .329 | .289 | |
| MW50 | 50 | 8.0 | .392 | W7.8 | .078 | .314 | .263 | 1/0 |
| MW45 | 45 | 7.6 | .353 | W7.0 | .070 | .298 | .238 | |
| MW40 | 40 | 7.1 | .314 | W6.2 | .062 | .283 | .214 | 1 |
| MW35 | 35 | 6.7 | .274 | W5.4 | .054 | .262 | .184 | 2 |
| MW30 | 30 | 6.2 | .235 | W4.7 | .047 | .245 | .160 | 3 |
| MW26 | 26 | 5.7 | .204 | W4.0 | .040 | .226 | .136 | 4 |
| MW25 | 25 | 5.6 | .196 | W3.9 | .039 | .223 | .133 | |
| MW20 | 20 | 5.0 | .157 | W3.1 | .031 | .199 | .105 | |
| MW19 | 19 | 4.9 | .149 | W2.9 | .029 | .192 | .098 | 6 |
| MW15 | 15 | 4.4 | .118 | W2.3 | .023 | .171 | .078 | |
| MW13 | 13 | 4.1 | .102 | W2.0 | .020 | .160 | .068 | 8 |
| MW10 | 10 | 3.6 | .078 | W1.6 | .016 | .143 | .054 | |
| MW9 | 9 | 3.4 | .071 | W1.4 | .014 | .135 | .048 | 10 |

*Metric wire sizes can be specified in 1 mm² increments.

**U.S. customary sizes can be specified in .001 in² increments.

Note ∴ — For other available wire sizes, consult other WRI Publications or discuss with WWF manufacturers.

Note ★ — Wires may be deformed, use prefix MD or D, except where only MW or W is required by building codes (usually less than MW26 or W4).

# Load Table for Standard Steel Pipe Columns ▬▬▬▬▬▬▬▬

| Maximum Allowable Concentric Loads | | | | | | | | | | | |
|---|---|---|---|---|---|---|---|---|---|---|---|
| Nominal size in inches | Weight per ft. in pounds | Unbrace length in feet | | | | | | | | | |
| | | 6 | 7 | 8 | 9 | 10 | 11 | 12 | 14 | 16 | 18 |
| 3 | 7.58 | 38 | 36 | 34 | 31 | 28 | 25 | 22 | 16 | 12 | 10 |
| 3 1/2 | 9.11 | 48 | 46 | 44 | 41 | 38 | 35 | 32 | 25 | 19 | 15 |
| 4 | 10.79 | 59 | 57 | 54 | 52 | 49 | 46 | 43 | 36 | 29 | 23 |
| 5 | 14.62 | 83 | 81 | 78 | 76 | 73 | 71 | 68 | 61 | 55 | 47 |
| 6 | 18.97 | 110 | 108 | 106 | 103 | 101 | 98 | 95 | 89 | 82 | 75 |

Loads are in kips. 1 kip = 1,000 pounds

(American Institute of Steel Construction)

# Common Styles of Metric Welded Wire Reinforcement (WWR) with Equivalent U.S. Customary Units[3]

| | A_s (mm²/m) | Metric Styles (MW = Plain wire)[2] | Wt. (kg/m²) | Equivalent Inch-Pound Styles (W = Plain Wire)[2] | A_s (in²/ft) | Wt. (lbs/CSF) |
|---|---|---|---|---|---|---|
| A[1 & 4] | 88.9 | 102x102 - MW9xMW9 | 1.51 | 4x4 - W1.4xW1.4 | .042 | 31 |
| | 127.0 | 102x102 - MW13xMW13 | 2.15 | 4x4 - W2.0xW2.0 | .060 | 44 |
| | 184.2 | 102x102 - MW19xMW19 | 3.03 | 4x4 - W2.9xW2.9 | .087 | 62 |
| | 254.0 | 102x102 - MW26xMW26 | 4.30 | 4x4 - W4.0xW4.0 | .120 | 88 |
| | 59.3 | 152x152 - MW9xMW9 | 1.03 | 6x6 - W1.4xW1.4 | .028 | 21 |
| | 84.7 | 152x152 - MW13xMW13 | 1.46 | 6x6 - W2.0xW2.0 | .040 | 30 |
| | 122.8 | 152x152 - MW19xMW19 | 2.05 | 6x6 - W2.9xW2.9 | .058 | 42 |
| | 169.4 | 152x152 - MW26xMW26 | 2.83 | 6x6 - W4.0xW4.0 | .080 | 58 |
| B[1] | 196.9 | 102x102 - MW20xMW20 | 3.17 | 4x4 - W3.1xW3.1 | .093 | 65 |
| | 199.0 | 152x152 - MW30xMW30 | 3.32 | 6x6 - W4.7xW4.7 | .094 | 68 |
| | 199.0 | 305x305 - MW61xMW61 | 3.47 | 12x12 - W9.4xW9.4 | .094 | 71 |
| | 362.0 | 305x305 - MW110xMW110 | 6.25 | 12x12 - W17.1xW17.1 | .171 | 128 |
| C[1] | 342.9 | 152x152 - MW52xMW52 | 5.66 | 6x6 - W8.1xW8.1 | .162 | 116 |
| | 351.4 | 152x152 - MW54xMW54 | 5.81 | 6x6 - W8.3xW8.3 | .166 | 119 |
| | 192.6 | 305x305 - MW59xMW59 | 8.25 | 12x12 - W9.1xW9.1 | .091 | 69 |
| | 351.4 | 305x305 - MW107xMW107 | 9.72 | 12x12 - W16.6xW16.6 | .166 | 125 |
| D[1] | 186.3 | 152x152 - MW28xMW28 | 3.22 | 6x6 - W4.4xW4.4 | .088 | 63 |
| | 338.7 | 152x152 - MW52xMW52 | 5.61 | 6x6 - W8xW8 | .160 | 115 |
| | 186.3 | 305x305 - MW57xMW57 | 3.22 | 12x12 - W8.8xW8.8 | .088 | 66 |
| | 338.7 | 305x305 - MW103xMW103 | 5.61 | 12x12 - W16xW16 | .160 | 120 |
| E[1] | 177.8 | 152x152 - MW27xMW27 | 3.08 | 6x6 - W4.2xW4.2 | .084 | 60 |
| | 317.5 | 152x152 - MW48xMW48 | 5.52 | 6x6 - W7.5xW7.5 | .150 | 108 |
| | 175.7 | 305x305 - MW54xMW54 | 3.08 | 12x12 - W8.3xW8.3 | .083 | 63 |
| | 317.5 | 305x305 - MW97xMW97 | 5.52 | 12x12 - W15xW15 | .150 | 113 |

[1]Group A - Compares areas of WWR at $f_y$ = 60,000 psi with other reinforcing at $f_y$ = 60,000 psi
Group B - Compares areas of WWR at $f_y$ = 70,000 psi with other reinforcing at $f_y$ = 60,000 psi
Group C - Compares areas of WWR at $f_y$ = 72,500 psi with other reinforcing at $f_y$ = 60,000 psi
Group D - Compares areas of WWR at $f_y$ = 75,000 psi with other reinforcing at $f_y$ = 60,000 psi
Group E - Compares areas of WWR at $f_y$ = 80,000 psi with other reinforcing at $f_y$ = 60,000 psi

[2]Wires may also be deformed, use prefix MD or D, except where only MW or W is required by building codes (usually less than a MW26 or W4). Also wire sizes can be specified in 1mm² (metric) or .001 in.² (inch-pound) increments.
[3]For other available styles or wire sizes, consult other WRI publications or discuss with WWR manufacturers.
[4]Styles may be obtained in roll form.   Note: It is recommended that rolls be straightened and cut to size before placement.

# Abbreviations

| | | | | | | | |
|---|---|---|---|---|---|---|---|
| aggr | aggregate | frbk | firebrick | ptn | partition |
| alum. | aluminum | fdr. | fire door | d | penny |
| L | angle | fp | fireplace | perp | perpendicular |
| arch. | architectural | fprf | fireproof | pc | piece |
| asb. | asbestos | flge | flange | plas | plaster |
| asph | asphalt | flr | floor | pl | plate |
| AT. | asphalt tile | flg | flooring | PL GL | plate glass |
| avg | average | fluor | fluorescent | lb | pound, pounds |
| bbl | barrel, barrels | fl | flush | psi | pounds per square inch |
| basmt | basement | ' or ft | foot, feet | P/C | poured concrete |
| B | bathroom | ft lb | foot pound | prect | precast |
| BR | bedroom | ftg | footing | prefab | prefabricated |
| BM | bench mark | fdn | foundation | PC | pull chain |
| bev | beveled | fr | frame | PB | push button |
| BP | blueprint | gal. | gallon, gallons | r or rad | radius |
| BRKT | bracket | galv | galvanized | recp | receptacle |
| br | brass | GA | gage | rect | rectangle |
| BRK | brick | gl | glass | ref | refrigerator |
| Btu | British thermal unit | gr | grade | reg | register |
| bro | bronze | g | granite | reinf | reinforce |
| Bldg | building | grtg | grating | ret | return |
| Buz | buzzer | gyp | gypsum | RH | right-hand |
| CAB | cabinet | hdw | hardware | R | riser |
| CP | candlepower | hdwd | hardwood | rm | room |
| CI | cast iron | hgt | height | rgh | rough |
| clk | caulk | hor | horizontal | rnd | round |
| clg | ceiling | HB | hose bibb | rub. | rubber |
| cem | cement | hw | hot water | sad. | saddle |
| c or CL | center line | C | hundred | sch | schedule |
| cm | centimetre, centimetres | I | I beam | smls | seamless |
| cer | ceramic | " or in. | inches, inches | sec | second |
| chfr | chamfer | incl | include | " | second (angular measure) |
| chkd | checked | ID | inside diameter | shthg | sheathing |
| cin bl | cinder block | insul | insulation | ship | ship lap |
| cir | circular | int | interior | sh | shower |
| CIRC | circumference | jt | joint | sdg | siding |
| c − o | cleanout | j | junction | sc | sill cock |
| cir | clean | jb | junction box | S | sink |
| CL | closet | kg | kilogram | sl | slate |
| col | column | km | kilometre | sp | soil pipe |
| com | common | kw | kilowatt, kilowatts | S | south |
| con | concrete | kwh | kilowatt-hour | SE | southeast |
| conc b | concrete block | k | kip (1000 lb.) | SW | southwest |
| conc clg | concrete ceiling | K | kitchen | sq | square |
| conc fl | concrete floor | ldg | landing | sq ft | square foot, feet |
| cnd | conduit | lth | lath | sst | stainless steel |
| const | construction | Lau | laundry | Stp | standpipe |
| cop | copper | LT | laundry tray | stl | steel |
| cu | cubic | lav | lavatory | stiff. | stiffener |
| cu ft | cubic foot, feet | LDR | leader | stn | stone |
| cu in | cubic inch, inches | ld | left-hand | SP | sump pit |
| cu yd | cubic yard, yards | LH | left-hand | susp clg | suspending ceiling |
| DP | dampproofing | lgth | length | sw | switch |
| det | detail | lev | level | T | tee |
| diag | diagram | lt | light | tel | telephone |
| dia. | diameter | lwc | lightweight concrete | TC | terra cotta |
| dim. | dimensions | Ls | limestone | ter | terrazzo |
| DR | dining room | lin ft | linear feet | thermo | thermostat |
| d − h | double hung | L CL | linen closet | thk | thick, thickness |
| dn | down | lino | linoleum | thd | thread |
| ds | downspout | Ll | live load | T & G | tongue and groove |
| dr | drain | LR | living room | tr | tread |
| dwg | drawing | l | lumen | trnbkl | turnbuckle |
| drwn | drawn | mip | malleable iron pipe | unfin | unfinished |
| dw | dry well | mfr | manufacture | V | vent, ventilator |
| E | east | mr | marble | vent. | ventilation |
| elec | electric | MO | masonry opening | vert. | vertical |
| el | elevation | matl | material | vol | volume |
| elev | elevator | max | maximum | WV | wall vent |
| ent | entrance | mm | millimetre, millimetres | wp | waterproofing |
| equip. | equipment | min | minimum | w | watt, watts |
| est | estimate | (') | minute (angular measure) | whr | watt-hour |
| exc | excavate | moldg | molding | WS | weather stripping |
| exist. | existing | ni | nickle | wp | weatherproof |
| exp bt | expansion bolt | nom | nominal | whp | weep hole |
| exp jt | expansion joint | N | north | W | west |
| EXT | extension | No. | number | WF | wide flange |
| ext | exterior | oc | off center | wth | width |
| xh | extra heavy | opg | opening | wdw | window |
| fab | fabricate | oz | ounce, ounces | w/ | with |
| F | fahrenheit | out. | outlet | wf | wood frame |
| fig. | figure | OD | out diameter | wrt | wrought |
| fil | fillet | oa | overall | WI | wrought iron |
| fin. | finish | ovfl | overflow | yd | yard, yards |
| FAO | finish all over | ovhd | overhead | z | zinc |

# OUTLINE OF SUGGESTED RELATED INSTRUCTION COURSE
# FOR APPRENTICE BRICKLAYERS

The following is a suggested course outline of related instruction to supplement on-the-job training and the work processes recommended as necessary for a 4500-hour apprenticeship program.

## FIRST AND SECOND YEARS

### A. Masonry Material

1. Masonry Units

History, description, manufacture, classification, types, special units, structural characteristics, physical properties, color, texture, and uses for:
   a. Clay and shale brick
   b. Fire brick.
   c. Sandlime brick.
   d. Concrete masonry units.
   e. Tile (structural and facing).
   f. Stone (granite, limestone, sandstone, marble).
   g. Acid brick.
   h. Glass block.
   i. Terra cotta.

2. Mortar

Properties, description, uses, workability, water retentivity, bond, durability, and admixtures for:
   a. Hydrated lime.
   b. Cement lime.
   c. Cement mortar.
   d. Prepared masonry cement mortar.
   d. Special mortar for:
      Fire brick.
      Glass block.
      Acid brick.
      Stone (granite, marble, etc.).

3. Sand

Classification, description, selection, tests, types, and uses.

### B. Tools and Equipment

Use, care, operation, and safe practices for:
1. Brick trowel.
2. Brick hammer, blocking chisels, six-foot rules, levels, and jointing tools.
3. Story-pole and spacing rule.
4. Stone setting:
   a. Wood wedges.
   b. Setting tools.
   c. Caulking gun.
   d. Chain hoists.
   e. Cranes.
   f. Hangers.
5. Accessories:
   a. Wall ties.
   b. Expansion strips.
   c. Clip angles.
   d. Nailing blocks.
   e. Reinforced steel for grouted walls and lintels.
   g. Flashing materials
   h. Anchor bolts.
   i. Steel bearing plates.
6. Welding Equipment.

### C. Trade Arithmetic

1. Review of the fundamental operations of arithmetic, including:
   a. Fractions.
   b. Decimals.
   c. Conversions.
   d. Weights.
   e. Measures.
2. Reading the rule:
   a. Six-foot rule.
   b. Spacing rule.

### D. Plan, Blueprint Reading, and Trade Sketching

1. Fundamentals of plan and blueprint reading:
   a. Types of plans.
   b. Kinds of plans
   c. Conventions.
   d. Symbols.
   e. Scale representation.
   f. Dimensions.
2. Trade Sketching:
   a. Tools (types).
   b. Straight line sketching.
   c. Circles and arcs.
   d. Making a working sketch.

### E. Construction Details

1. Trade terms, motion study, bonds (structural and pattern), laying of units, points, etc., for:
   a. Walls.
   b. Footings.
   c. Pilasters, columns, and piers.
   d. Chasers.
   e. Recesses (corbelling).
   f. Chimneys and fireplaces.
2. Cleaning, caulking, and pointing.
3. Reinforced masonry lintels.

### F. Shop Practices

1. Spreading mortar.
2. Laying bricks to line – building inside and outside corners for a 4, 8, and 12-inch wall.
3. Layout and erect:
   a. Walls and corners with:
      Flemish and Dutch bond
      Tile backing.
      Pilasters and chase.
      A cavity.
      Reinforced grouted brick.
   b. Brick piers.
   c. Chimneys – single and double flues.
4. Setting sills, copings, and quoins.

### G. Safety

## THIRD YEAR

### A. Tools and Equipment

1. Use, care, operation, and safe practices for:
   a. Builder's level and transit.
   b. Frames, beams, lintels, and rods.
   c. Welding equipment.

### B. Blueprint Reading

1. Specifications.
2. Job layout.
3. Shop drawings.
4. Modular measure.

### C. Construction Details

1. Arch construction.
2. Modular masonry.
3. Firebox construction.
4. Layout of story poles and batter boards

### D. Estimating

1. Mortar.
2. Masonry units (modular and nonmodular types).
3. Concrete footings.

### E. Shop Practice

1. Layout and erect:
   a. Reinforced masonry lintels.
   b. Story pole and batter boards.
   c. Fireplaces (with and without steel fireplace forms).
   d. Project with glazed tile leads and panels.

e. Project with marble or granite setting, adhesive terra cotta, glass block
f. Modular wall.
g. Circular corner.

**F. Prefabricated Masonry Panels**
1. Layout.
2. Assembly
3. Welding and erection.
4. Installation.
5. Caulking, painting, and cleaning.

**G. Application of Insulating Materials for Masonry Walls**

**H. Safety**

## WORK EXPERIENCES

To acquire the necessary skills of the trade in its various categories, the apprentice shall (as near as possible) be provided with employment in the following categories and amounts:

| Type of Experience | Approximate Hours: |
|---|---|
| 1. Laying of bricks, including: | 2250 |
| a. Mixing mortar, cement, and patent mortar; spreading mortars, bonding and tying. | |
| b. Building footings and foundations. | |
| c. Plain exterior brickwork (straight wall work, backing up brickwork). | |
| d. Building arches, groins, columns, piers, and corners. | |
| e. Planning and building chimneys, fireplaces and flues, and floors and stairs. | |
| f. Building masonry panels. | |
| g. Welding. | |
| 2. Laying of stone, including: | 450 |
| a. Cutting and setting of rubblework or stonework. | |
| b. Setting of cut-stone trimmings. | |
| c. Butting ashlar. | |
| 3. Pointing, cleaning, and caulking, including: | 150 |
| a. Pointing brick and stone, cutting and raking joints. | |
| b. Cleaning stone, brick and tile (water, acid, sandblast). | |
| c. Caulking stone, brick, and glass block. | |
| 4. Installation of building units, such as: | 1275 |
| a. Tile block cutting and setting. | |
| b. Cutting, setting, and pointing of cement blocks, artificial stone, glass blocks, and cork and styrofoam panels. | |
| c. Blockarching. | |
| 5. Fireproofing, including: | 225 |
| a. Building party walls (partition tile, gypsum blocks, glazed tile). | |
| b. Standardized firebrick. | |
| c. Specialties. | |
| 6. Care and use of tools and equipment, including: | 150 |
| a. Trowels. | |
| b. Brickhammer. | |
| c. Plumb rule. | |
| d. Scaffolds. | |
| e. Cutting saws. | |
| f. Welding equipment. | |
| Total | 4500 |

All such work shall be performed under the supervision of a journeyman. Supervision should not be of such nature as to prevent the development of responsibility and initiative.

An employer who is to train apprentices shall meet the qualifying requirements as set forth in the local bargaining agreement, and shall be able to provide the necessary work experience for training.

# American Society for Testing and Materials (ASTM) Standards Used in the Masonry Industry

C 5    *Specification for Quicklime for Structural Purposes*

C 31   *Method of Making and Curing Concrete Test Specimens in the Field*

C 33   *Specification for Concrete Aggregates*

C 39   *Method of Test for Compressive Strength of Cylindrical Concrete Specimens*

C 55   *Specification for Concrete Building Brick*

C 90   *Specification for Hollow Load-Bearing Concrete Masonry Units*

C 91   *Specification for Masonry Cement*

C 94   *Specification for Ready-Mixed Concrete*

C 129  *Specification for Non-Load-Bearing Concrete Masonry Units*

C 139  *Specification for Concrete Masonry Units for Construction of Catch Basins and Manholes*

C 140  *Methods of Sampling and Testing Concrete Masonry Units*

C 143  *Test Method for Slump of Hydraulic Cement Concrete*

C 144  *Specification for Aggregate for Masonry Mortar*

C 145  *Specification for Solid Load-Bearing Concrete Masonry Units*

C 150  *Specification for Portland Cement*

C 207  *Specification for Hydrated Lime for Masonry Purposes*

C 236  *Test Method for Steady-State Thermal Performance of Building Assemblies by Means of a Guarded Hot Box*

C 270  *Specification for Mortar for Unit Masonry*

C 315  *Specification for Clay Flue Linings*

C 404  *Specification for Aggregates for Masonry Grout*

C 423  *Test Method for Sound Absorption and Sound Absorption Coefficients by the Reverberation Room Method*

C 426  *Test Method for Drying Shrinkage of Concrete Block*

C 476  *Specification for Grout for Masonry*

C 595  *Specification for Blended Hydraulic Cements*

C 617  *Practice for Capping Cylindrical Concrete Specimens*

C 631  *Specification for Bonding Compounds for Interior Plastering*

C 634  *Definitions of Terms Relating to Environmental Acoustics*

C 744  *Specification for Prefaced Concrete and Calcium Silicate Masonry Units*

C 780  *Method for Pre-Construction and Construction Evaluation of Mortars for Plain and Reinforced Unit Masonry*

C 847  *Specification for Metal Lath*

C 887  *Specification for Packaged, Dry, Combined Materials for Surface Bonding Mortar*

C 897  *Specification for Aggregate for Job-Mixed Portland Cement-Based Plasters*

C 901  *Specification for Prefabricated Masonry Panels*

C 920  *Specification for Elastometric Joint Sealants*

C 926  *Specification for Application of Portland Cement-Based Plaster*

C 932  *Specification for Surface-Applied Bonding Agents for Exterior Plastering*

C 933  *Specification for Welded Wire Lath*

C 936  *Specification for Solid Concrete Interlocking Paving Units*

C 946  *Practice for Construction of Dry-Stacked, Surface-Bonded Walls*

C 952  *Test Method for Bond Strength of Mortar to Masonry Units*

C 962  *Guide for Use of Elastomeric Joint Sealants*

C 1006 *Test Method for Splitting Tensile Strength of Masonry Units*

C 1012 *Test Method for Length Change of Hydraulic-Cement Mortars Exposed to a Sulfate Solution*

C 1019 *Method of Sampling and Testing Grout*

C 1032 *Specification for Woven Wire Plaster Base*

C 1038 *Test Method for Expansion of Portland Cement Mortar Bars Stored in Water*

C 1063 *Specification for Installation of Lathing and Furring for Portland Cement-Based Plaster*

C 1072 *Method for Measurement of Masonry Flexural Bond Strength*

C 1093 *Practice for the Accreditation of Testing Agencies for Unit Masonry*

C 1142 *Specification for Ready Mixed Mortar for Unit Masonry*

C 1148 *Test Method for Measuring the Drying Shrinkage of Masonry Mortar*

D 3665 *Practice for Random Sampling of Construction Materials*

D 4258 *Practice for Surface Cleaning Concrete for Coating*

D 4259 *Practice for Abrading Concrete*

D 4260 *Practice for Acid Etching Concrete*

D 4261 *Practice for Surface Cleaning Concrete Unit Masonry for Coating*

D 4262 *Test Method for pH of Chemically Cleaned or Etched Concrete Surfaces*

D 4263 *Test Method for Indicating Moisture in Concrete by the Plastic Sheet Method*

E 72   *Methods of Conducting Strength Tests of Panels for Building Construction*

E 90   *Method for Laboratory Measurement of Airborne Sound Transmission Loss of Building Partitions*

E 119  *Methods of Fire Tests of Building Construction and Materials*

E 336  *Method for Measurement of Airborne Sound Insulation in Buildings*

E 380  *Practice for Use of the International System of Units (SI) (the Modernized Metric System)*

E 413  *Classification for Determination of Sound Transmission Class*

E 447  *Test Methods for Compressive Strength of Masonry Prisms*

E 492  *Method of Laboratory Measurement of Impact Sound Transmission through Floor-Ceiling Assemblies Using the Tapping Machine*

E 597  *Practice for Determining a Single-Number Rating of Airborne Sound Isolation in Multi-Unit Building Specifications*

---

ASTM, 1916 Race Street, Philadelphia, PA 19103.

# Building Code Requirements for Masonry Structures ▬▬

The adoption of the Building Code Requirements for Masonry Structures, ACI 530-92/ASCE 5-92/TMS 402-92, represents the consolidation of several masonry codes into one consistent set of provisions. (National Concrete Masonry Association)

# Specifications for Masonry Structures

Construction requirements and quality assurance provisions in the code—Building Code Requirements for Masonry Structures—are included by reference to Specifications for Masonry Structures, ACI 530.1-92/ASCE 6-92/TMS 602-92. (National Concrete Masonry Association)

A curved form for concrete curbing often requires complex and sturdy bracing to resist the pressure of the concrete when it is placed. (Jack Klasey)

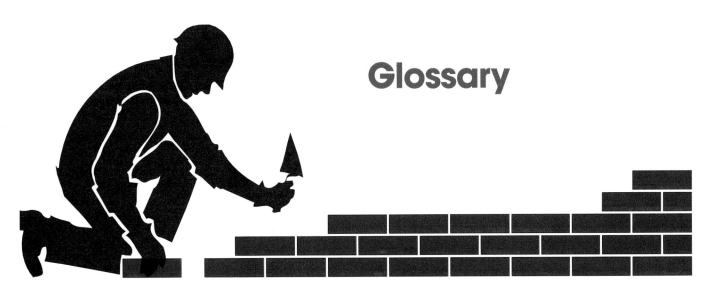

# Glossary

## A

**Absorbent:** Capable of taking in water or moisture.

**Absorption:** The weight of water a brick or tile unit absorbs when immersed in either cold or boiling water for a stated length of time. It is expressed as a percentage of the weight of the dry unit.

**Absorption rate:** The weight of water absorbed when a brick is partially immersed for one minute, usually expressed in either grams or ounces per minute. Also called *suction* or *initial rate of absorption*.

**Abutment:** **1.** A supporting wall carrying the end of a bridge or span and sustaining the pressure of the earth next to it. **2.** A skewback and the masonry which supports it.

**Accelerator:** A material that speeds hardening of concrete or mortar.

**Acid-resistant brick:** Brick suitable for use in contact with chemicals, usually used along with acid-resistant mortars.

**Adhesion-type ceramic veneer:** The inner sections of ceramic veneer held in place by adhesion of mortar to unit and backing. No metal anchors are required.

**Admixtures:** Materials added to mortar as water-repellent or coloring agents, or to retard or hasten setting.

**Adobe brick:** Large roughly-molded, sun-dried clay brick of varying size.

**Aggregate:** Inert particles that are mixed with Portland cement and water to form concrete.

**Air-entraining agent:** A material used to trap air in mortar to improve its workability and durability.

**Air space:** A cavity or space in the wall, or between building materials.

**All stretcher bond:** Bond showing only stretchers on the face of the wall, each stretcher divided evenly over the stretchers below it.

**Alumina:** A mineral contained in clay used for brickmaking.

**American bond:** That bond consisting of five to seven stretcher courses between headers.

**Anchor:** **1.** A piece of material usually metal, used to attach building parts (e.g., plates, joists, trusses, etc.) to masonry. **2.** Masonry materials or a metal tie used to secure stone in place.

**Anchored type ceramic veneer:** Thicker sections of ceramic veneer held in place by grout and wire anchors connected to backing wall.

**Angle brick:** Any brick shaped to an oblique angle to fit a salient (sharp) corner.

**Angle iron:** A structural piece of steel in the form of a 90° angle used, in certain situations, to support brickwork.

**ANSI:** American National Standards Institute.

**Apron:** A plain or molded piece of finish below the stool of the window covering the rough edge of the plastering.

**Apron wall:** That part of a panel wall between window sill and wall support.

**Arcade:** A range of arches, supported either on columns or on piers, and detached or attached to the wall.

**Arch:** A curved compressive structural member, spanning openings or recesses; also built flat. Also a form of construction in which a number of units span an opening by transferring vertical loads laterally to adjacent units and thus to the supports. An arch is normally classified by the curve of its intrados.

**Arch axis:** The median line of the arch ring.

**Arch brick:** **1.** Wedge-shaped brick for special use in an arch. **2.** Extremely hard-burned brick from an arch of a scove kiln.

**Arch buttress:** Sometimes called a flying buttress; an arch springing from a buttress or pier.

**Architectural terra cotta:** Hard-burned, glazed or unglazed clay building units, plain or ornamental, machine extruded or hand-molded, and generally larger in size than brick or facing tile.

**Area wall:** **1.** The masonry surrounding or partly surrounding an area. **2.** The retaining wall around basement windows below grade.

**Arris:** The external edge formed by two surfaces, whether plane or curved, meeting each other.

**Artificial:** Made to resemble a natural product—synthetic.

**Ashlar:** **1.** A squared or cut block of stone, usually of rectangular dimensions. 2. A flat-faced surface generally square or rectangular having sawed or dressed beds and joints.

**Ashlar line:** The main line of the surface of a wall of the superstructure.

**Ashlar masonry:** Masonry composed of rectangular units of burned clay or shale, or stone. Generally larger in size than brick. When properly bonded, it has sawed, dressed or squared beds and joints laid in mortar. Often the unit size varies to provide a random pattern, which is called *random ashlar*.

**ASHRAE:** American Society of Heating, Refrigerating and Air-Conditioning Engineers.

**ASTM:** American Society for Testing and Materials.

**Axis:** The spindle or center of any rotating object. In a sphere, an imaginary line through the center.

## B

**Back arch:** A concealed arch carrying the backing of a wall where the exterior facing is carried by a lintel.

**Back filling:** **1.** Rough masonry built behind a facing or between two faces. **2.** Filling over the extrados of an arch. **3.** Brickwork in spaces between structural timbers, sometimes called *brick nogging*.

**Backboard:** A temporary board on the outside of a scaffold.

**Backing:** That portion of a masonry wall built in the rear of the face and bonded to the face; usually a cheaper class of masonry.

**Backing up:** Laying the inside portion of the wall after the facing of wall has been built header high.

**Backup:** That part of a masonry wall behind the exterior facing.

**Balanced:** Made symmetrical and in correct proportion.

**Basement:** The lower part of a house or building, usually below the ground.

**Bat:** A piece of brick.

**Batter:** **1.** Recessing or sloping masonry back in successive courses; the opposite of corbel. **2.** The slope or inclination of the face or back of a wall from a vertical plane. **3.** The slope backwards of the face of the wall, opposite to *overhand*.

**Batter stick:** A tapering stick used in connection with a plumb rule for building battering surfaces.

**Bay:** Any division or compartment of an arcade, roof, etc. Thus, each space from pillar to pillar in a cathedral, is called a *bay* or *severy*.

**Bay window:** Any window projecting outward from the wall of a building. It can be either square, semicIrcular, or polygonal. If it is carried on projecting corbels, it is called an oriel window.

**Beam:** A piece of timber, iron, steel or other material which is supported at either end by walls, columns, or posts. It is used to support heavy flooring or the weight over an opening.

**Bearing blocks:** Small blocks of stone built in a wall to support the ends of particular beams.

**Bearing wall:** A wall that supports a vertical load in addition to its own weight.

**Bed joint:** **1.** The horizontal layer of mortar on which a masonry unit is laid. **2.** A horizontal joint, or one perpendicular to the line of pressure. **3.** A joint between two horizontal courses of brick.

**Bedford limestone:** A certain formation of limestone rock which is called Bedford, because it was first found at the city of Bedford, Indiana.

**Belt course:** A narrow, continuous horizontal course of masonry, sometimes slightly projected such as window sills. Sometimes called *string course* or *sill course*. Also: A continuous horizontal course, marking a division in the wall plane.

**Beltstones or courses:** Horizontal bands or zones of stone encircling a building or extending through a wall.

**Benches:** Brick in that part of the kiln next to the fire that are generally baked to vitrification.

**Bevel:** The angle that one surface or line makes with another, when they are not at right angles.

**Binders:** Bricks that extend only a part of the distance across the wall.

**Blind bond:** Bond used to tie the front course to the wall in pressed brickwork where it is not desirable that any headers should be seen in the face work.

**Blind header:** A concealed brick header in the interior of a wall, not showing on the faces.

**Block and cross bond:** A combination of the two bonds. The face of the wall is in cross bond and the backing in block bond.

**Blocking:** A method of bonding two adjoining or intersecting walls, not built at the same time, by means of offsets whose vertical dimensions are not less than 8".

**Blocking course:** A course of stone placed on top of a cornice crowning the walls.

**Bond:** **1.** Tying various parts of a masonry wall by lapping units one over another or by connecting with metal ties. **2.** Patterns formed by exposed faces of units. **3.** Adhesion between the mortar or grout and masonry units of reinforcement.

**Body brick:** The best brick in the kiln. The brick that are baked hardest with the least distortion.

**Bond beam:** A horizontal reinforced concrete masonry beam designed to strengthen a masonry wall and reduce the probability of cracking.

**Bond course:** The course consisting of units which over-lap more than one wythe of masonry.

**Bond stone:** Stones projecting laterally (straight back) into the backup wall used to tie the wall together.

**Bonder:** A bonding unit.

**Breaking joint:** Any arrangement of masonry units that prevents continuous vertical joints from occurring in adjoining courses.

**Breast of a window:** The masonry forming the back of the recess and the parapet under the window sill.

**Breast wall:** One built to prevent the falling of a vertical face cut into the natural soil.

**Brick:** A solid masonry unit of clay or shale formed into a rectangular prism while plastic, and burned or fired in a kiln.

**Brick and brick:** A method of laying bricks so that units touch each other with only enough mortar to fill surface irregularities.

**Brick ashlar:** Walls with ashlar facing backed with bricks.

**Brick grade:** Designation for durability of the unit expressed as SW for severe weathering, MW for moderate weathering, or NW for negligible weathering.

**Brick type:** Designation for facing brick that controls tolerance, chippage, and distortion. Expressed as FBS, FBX, and FBA for solid brick, and HBS, HBX, BHA, and HBB for hollow brick.

**Brick veneer:** The outside facing of brickwork used to cover a wall built of other material; usually refers to brick walls covering a wood frame building.

**Brickwork:** Masses of wall built of bricks laid in mortar.

**Broken range:** Masonry construction in which the continuity of the courses are broken at intervals.

**Bugged finish:** A smooth finish produced by grinding with power sanders.

**Building brick:** Brick for building purposes not especially treated for texture or color. Formerly called *common brick*.

**Bull header:** A brick laid on its edge showing only the end of the face of the wall.

**Bull stretcher:** A brick laid on edge so as to show the broad side of the brick on the face of the wall.

**Bullnose:** Convex rounding of a member, such as the front edge of a stair tread, a concrete block, or window sill.

**Buttered joint:** A very thin mortar joint made by scraping a small quantity of mortar with the trowel on all edges of the brick and laying it without the usual mortar bed.

**Buttering:** Placing mortar on a masonry unit with a trowel.

**Buttress:** **1.** Masonry projecting from a wall and intended to strengthen the wall against the thrust of a roof or vault. **2.** A vertical, projecting piece of stone or brick masonry built in front of a wall to strengthen it.

**Buttress, flying:** A detached buttress or pier of masonry at some distance from a wall, and connected to it by an arch or a portion of an arch, so as to discharge the thrust of a roof or vault on some point.

# C

**C/B ratio:** The ratio of the weight of water absorbed by a masonry unit during immersion in cold water to weight absorbed during immersion in boiling water. An indication of the probable resistance of brick to freezing and thawing. Also called *saturation coefficient*.

**Camber:** The relatively small rise of a Jack arch.

**Capacity insulation:** The ability of masonry to store heat as a result of its mass, density, and specific heat.

**Capital:** Column cap.

**Cavity wall:** A wall built of masonry units arranged to provide a continuous air space 2" to 3" thick. Facing and backing wythes are connected with rigid metal ties.

**Cell:** One of the hollow spaces in building tile.

**Cement:** A burned mixture of clay and limestone pulverized (crushed) for making mortar or concrete.

**Centering:** Temporary formwork for the support of masonry arches or lintels during construction.

**Ceramic color glaze:** An opaque colored glaze of satin or gloss finish, obtained by spraying the clay body with a compound of metallic oxides, chemicals, and clays. It is burned at high temperatures, fusing glaze to body, making them inseparable.

**Ceramic veneer:** Architectural terra cotta, characterized by large face dimensions and thin sections.

**Chain bond:** The building into masonry of iron bars, chain, or heavy timbers.

**Chase:** A continuous recess built into a wall to receive pipes, ducts, etc.

**Chat sawed:** Description of a textured stone finish, obtained by using chat sand in the gang sawing process.

**Chimney lining:** Fire clay or terra cotta material made for use inside of a chimney.

**Clay:** A natural, mineral aggregate consisting essentially of hydrous aluminum silicate; it is plastic when sufficiently wetted, rigid when dried, and vitrified when fired to a sufficiently high temperature.

**Clay mortar mix:** Finely ground clay, used as a plasticizer for masonry mortars.

**Cleanout:** An opening in the first course of a reinforced concrete masonry wall for removal of mortar protrusions and droppings. Also, an opening under a fireplace for removing ashes.

**Clear ceramic glaze:** Same as ceramic color glaze except that it is translucent or slightly tinted with a gloss finish.

**Clinker brick:** A hard-burned brick whose shape is distorted or bloated due to nearly complete vitrification.

**Clip:** A portion of a brick cut to length.

**Clip course:** The course of brick resting on a clip joint.

**Clip joint:** A joint of abnormal thickness to bring the course up to the required height. In no case should a clip joint be over 1/2" thick.

**Clipped header:** A bat placed to look like a header for purposes of establishing a pattern. Also called *false header*.

**Closer:** The last masonry unit laid in a course. It may be whole, or a portion of a unit.

**Closure:** A quarter or three-quarter brick to close when required; the end of a course, as distinguished from a half brick.

**Code (building):** A set of laws or regulations governing the location, materials, and workmanship in the construction of buildings.

**Collar joint:** The vertical, longitudinal (often mortared) joint between wythes of masonry.

**Column:** A vertical member whose horizontal dimension, measured at right angles to the thickness, does not exceed three times its thickness.

**Common American bond:** The bond in which from five to seven stretcher courses are laid between headers.

**Common bond:** Several courses of stretchers followed by one course of either Flemish or full headers.

**Common brick:** See *building brick.*

**Common brickwork:** The wall built out of the ordinary and cheaper classes of brick, where appearance is not an important consideration.

**Composite wall:** Any bonded wall with wythes constructed of different masonry units.

**Concrete:** A mixture of Portland cement, aggregates, and water.

**Concrete brick:** A solid concrete masonry unit usually not larger than 4" by 4" by 12".

**Concrete masonry unit:** A hollow or solid unit made of Portland cement and suitable aggregates. Units are often referred to by the type of aggregate used in their manufacture: cinder block, lightweight block, etc.

**Coping:** The material or masonry units forming a cap or finish on top of a wall, pier, pilaster, chimney, etc. It protects masonry below from penetration of water from above.

**Corbel:** A shelf or ledge formed by projecting successive courses of masonry out from the face of the wall.

**Corbel out:** To build out one or more courses of brick, or stone, from the face of a wall to form a support for timbers.

**Corbel table:** A projecting cornice or parapet, supported by a range of corbels a short distance apart, which carry a molding above which is a plain piece of projecting wall forming a parapet, and covered by a coping. Sometimes small arches are thrown across from corbel to corbel to carry the projection.

**Corner block:** A concrete masonry unit with a flat end for construction of the end or corner of a wall.

**Cornice:** The projection at the top of a wall finished by a blocking course, common in classic architecture. Also, a molded projecting stone at the top of an entablature.

**Course:** **1.** One of the continuous horizontal layers of units, bonded with mortar in masonry. **2.** A horizontal row of brick in a wall. **3.** Each separate layer in stone, brick, or other masonry.

**Course bed:** Stone, brick, or other building material in position upon which other material is to be laid.

**Coursed ashlar:** Ashlar set to form continuous horizontal joints.

**Cove:** A concave molding used in the composition of a cornice.

**Cramps:** Bars of iron having their ends turned at right angles to the body or the bar in order to enter holes in the faces of adjacent stones to hold them in place.

**Cross bond:** Bond in which the joints of the second stretcher course come in the middle of the first; a course composed of headers and stretchers intervening.

**Cross joint:** The joint between the two ends of the brick.

**Crowding the line:** Laying the bricks in such a way as to prevent the line or string from being clear of the face of the brickwork. In other words, building with a tendency to make the wall overhang.

**Crown:** The apex of the arch ring. In symmetrical arches, the crown is at mid-span.

**Culling:** Sorting brick for size, color, and quality.

**Culls:** The brick rejected in culling.

**Curtain wall:** A nonbearing wall. Built for the enclosure of a building, it is not supported at each story.

**Cut stone:** Finished, dimensioned stone ready to set in place.

# D

**Damp course:** A course or layer of impervious material which prevents capillary entrance of moisture from the ground or a lower course. Often called *damp check.*

**Dampproofing:** One or more coatings of a compound that is impervious to water. Usually applied to the back of stone or face or back of wall.

**Deformed bar:** A reinforcing bar with irregular surfacing for producing a better bond with grout than can be obtained with a smooth bar.

**Dentil:** The cogged, or tooth-like, members which project under a cornice; they are used for decorative effect.

**Depth:** The depth of any arch is the dimension which is perpendicular to the tangent of the axis. The depth of a jack arch is taken to be its greatest vertical dimension.

**Diagonal bond:** A form of raking bond where the bricks are laid in an oblique direction in the center of a thick wall or in paving.

**Diaper:** Any continuous pattern in brickwork of which the various bonds are examples. It is usually applied, however, to diamond or other diagonal patterns.

**Dimensioned stone:** Stone precut and shaped to specified sizes.

**Dog's tooth:** Bricks laid with their corners projecting from the wall face.

**Dowels:** Straight metal bars used to connect two sections of masonry; also, a two-piece instrument for lifting stones.

**Draft:** A margin on the surface of a stone cut approximately to the width of the chisel.

**Drip:** A projecting piece of material, shaped to throw off water and prevent its running down the face of wall or other surface. Also, a slot cut in the bottom of a projected stone, to interrupt the capillary attraction of rain water.

**Dry stone walls:** Walls that are of any of the classes of masonry with the exception that the mortar is omitted. They should be built according to the principles laid down for the class to which they belong.

**Dry-press brick:** Brick formed in molds under high pressures from relatively dry clay (5% to 7% moisture content).

**Dutch arch:** A flat arch whose voussoirs (wedge-shaped pieces) are laid parallel to the skewback on each side of the center.

**Dutch bond:** The arrangement of bricks forming a modification of Old English bond made by introducing a header as the second brick in every alternate stretching course, with a three-quarter brick beginning the other stretching courses.

**Dwarf wall:** A wall or partition which does not extend to the ceiling.

# E

**EBM:** See *Engineered Brick Masonry.*

**Eccentricity:** The normal distance between the centroidal axis of a member and the parallel resultant load.

**Economy brick:** Brick whose nominal dimensions are 4" by 4" by 8".

**Edgeset:** A brick set on its narrow side instead of on its flat side.

**Effective height:** The height of a member to be assumed for calculating the slenderness ratio.

**Effective thickness:** The thickness of a member to be assumed for calculating the slenderness ratio.

**Efflorescence:** A deposit of white powder or crust on the surface of brickwork which is due to soluble salts in the mortar or brick, drawn out to the surface by moisture.

**Enclosure wall:** An exterior nonbearing wall in skeleton frame construction. It is anchored to columns, piers, or floors, but not necessarily built between columns or piers, nor wholly supported at each story.

**End-construction tile:** Tile designed to be laid with axes of the cells vertical.

**Engineered brick:** Brick whose nominal dimensions are 3.2" by 4" by 8".

**Engineered brick masonry:** Masonry in which the design is based on a rational structural analysis.

**English bond:** **1.** Usually called Old English bond, the bond that is made by alternate courses of stretchers and headers with a 2" piece, or closer, next to the corner header. **2.** Alternate courses of headers and stretchers.

**English cross bond:** A variation of English bond made by putting one header next to the corner stretcher on alternate stretching courses.

**Expansion anchor:** A metal expandable unit inserted into a drilled hole that grips stone by expansion.

**Expansion joint:** A vertical joint or space to allow for expansion due to temperature changes.

**Exterior wall:** Any outside wall or vertical enclosure of a building other than a party wall.

**Extrados:** The convex curve which bounds the upper extremities of an arch.

# F

**Face: 1.** The exposed surface of a wall or masonry unit. **2.** The surface of a unit designed to be exposed in the finished masonry.

**Face brick:** A well-burned brick especially prepared, selected, and handled to provide attractive appearance in the face of a wall.

**Face shell:** The side wall of a hollow concrete masonry unit.

**Faced wall:** One in which facing and backing are bonded to exert common action under load.

**Facing: 1.** Any material, forming a part of a wall, used as a finished surface. **2.** The projecting courses at the base of a wall for the purpose of distributing the weight over an increased area; a footing.

**Facing brick:** Bricks made especially for facing purposes, often treated to produce surface texture. They are made of selected clays, or are treated to produce desired color.

**Facing tile:** Tile for exterior and interior masonry with exposed faces.

**Fat mortar:** Mortar containing a high percentage of cementitious components. It is a sticky mortar which adheres to a trowel.

**Field:** The expanse of wall between openings, corners, etc., principally composed of stretchers.

**Filling in:** The process of building in the center of the wall between the face and back.

**Filter block:** A hollow, vitrified clay masonry unit, sometimes salt-glazed, designed for trickling filter floors in sewage disposal plants.

**Fire brick:** Brick made of refractory ceramic material that will resist high temperatures.

**Fire clay:** A grade of clay that can stand a great amount of heat without softening or burning up; therefore, used for fire bricks.

**Fire division wall:** Any wall that subdivides a building so as to resist the spread of fire. It is not necessarily continuous through all stories to and above the roof.

**Fire resistive material:** See *Noncombustible material.*

**Fire Stop:** A projection of brickwork on the walls between the joists to prevent the spread of fire or vermin.

**Fire wall:** Any wall that subdivides a building to resist the spread of fire and that extends continuously from the foundation through the roof.

**Fireproofing:** Any material or combination of materials protecting structural members to increase their fire resistance.

**Fireproofing tile:** Tile designed for protecting structural members against fire.

**Fixed arch:** An arch whose skewback is fixed in position and inclination. Plain masonry arches are, by nature of their construction, fixed arches.

**Flare header:** A header of darker color than the field of the wall.

**Flashing:** A thin impervious material placed in mortar joints and through air spaces in masonry to prevent water penetration and/or provide water drainage.

**Flat arch:** An arch whose top and bottom are flat or practically flat. It involves the same principles of stress and strain as a segmental arch, but has voussoir (wedge-shaped pieces) extended up and down to reach the level lines.

**Flat stretcher course:** A course of stretchers set on edge and exposing their flat sides on the surface of the wall. Frequently done with brick finished for the purpose of the flat side, such as enameled or glazed brick.

**Flemish bond:** The arrangement of bricks made by alternating headers and stretchers in each course. The position of each header is in the center of the stretcher above and below.

**Flemish bond (double):** The arrangement of the bricks which gives Flemish bond on both sides of the wall.

**Flemish cross bond:** Any bond having alternate courses of Flemish headers and stretcher courses. The Flemish headers being plumb over each other and the alternate stretcher courses being crossed over each other.

**Flemish double cross bond:** A bond with odd numbered courses made up of stretchers divided evenly over each other. The even numbered courses are made up of Flemish headers in various locations with reference to the plumb of each other.

**Flemish garden bond:** Bricks laid so that each course has a header to every three or four stretchers.

**Float:** A flat, broad-bladed wood or metal hand tool.

**Floating concrete:** Leveling a fresh concrete surface with a float. Also, creating sufficient surface paste needed for troweling the concrete. Floating is performed immediately after a concrete surface has been consolidated and struck off.

**Floor brick:** Smooth dense brick, highly resistant to abrasion, used as finished floor surfaces.

**Floor tile:** Structural units for floor and roof slab construction.

**Flue:** A passage in a chimney especially for the exit of smoke and gases. One or more may be enclosed in the same chimney.

**Flue lining:** A smooth one-celled hollow tile for protecting flues.

**Flush:** Having the surface even with the adjoining surface.

**Flushed:** Filled up to the surface.

**Footing:** The broadened base of a foundation wall or other superstructure.

**Foundation wall:** That portion of a load-bearing wall below the level of the adjoining grade, or below first floor beams or joists.

**Frame high:** The height of the top of the window or door frames; the level at which the lintel is to be laid.

**Frog:** A depression in the bed surface of a brick. Sometimes called a *panel.*

**FTI:** Facing Tile Institute.

**Full header:** A course consisting of all headers.

**Furring tile:** Tile designed for lining the inside of exterior walls and carrying no superimposed loads.

**Furring:** A method of finishing the interior face of a masonry wall to provide space for insulation, prevent moisture transmittance, or to provide a level surface for finishing.

# G

**Gang saw:** A machine with multiple blades used to saw rough quarry blocks into slabs.

**Garden wall bond:** A name given to any bond particularly adapted to walls two tiers thick. A bond consisting of one header to three stretchers in every course.

**Gauge brick: 1.** Bricks that have been ground or otherwise produced to accurate dimensions. **2.** A tapered arch brick.

**Gauge:** To measure for a particular purpose. Some tools, for particular measurement can be termed as gauge; for example, a *gauge stock* with courses of brick marked thereon.

**Glazed tile:** Tile that is finished with glasslike surface.

**Gothic arch:** A pointed arch made of two segments whose arcs meet at the crown.

**Green brickwork:** Brickwork in which the mortar has not yet set.

**Grounds:** Nailing strips placed in masonry walls as a means of attaching trim or furring.

**Grout:** A cementitious component of high water-cement ratio, permitting it to be poured into spaces within masonry walls. Grout consists of Portland cement, lime, and aggregate. It is often formed by adding water to mortar.

# H

**Hacking: 1.** The procedure of stacking brick in a kiln or on a kiln car. **2.** Laying brick with the bottom edge set in from the plane surface of the wall.

**Hard-burned:** Nearly vitrified clay products that have been fired at high temperatures. They have relatively low absorptions and high compressive strengths.

**Head joint:** The vertical mortar joint between ends of masonry units. Often called *cross joint.*

**Header: 1.** A masonry unit that overlaps two or more adjoining wythes of masonry to tie them together. Often called *bonder.* **2.** A brick laid on its flat side across the thickness of the wall so as to show the end of the brick on the surface of the wall. **3.** A stone having its greatest dimension at right angles to the face of the wall.

**Header block:** Concrete masonry units made with part of one side of the height removed to provide space for bonding with adjoining units, such as brick.

**Header bond:** Bond showing only headers on the face, each header divided evenly on the header under it.

**Header course:** A course composed entirely of headers.

**Header high:** The height up to the top of the course directly under a header course.

**Header joint:** A joint between the ends of two bricks in the same course. Also called a *vertical joint.*

**Header tile:** Tile containing recesses for brick headers in masonry faced walls.

**Heading course:** A continuous bonding course of header brick. Also called *header course.*

**Headway:** Clear space or height under an arch, or over a stairway, etc.

**Heart bonds:** When two headers meet in the middle of the wall and the joint between them is covered by another header.

**Hearth:** That portion of a fireplace level with the floor, upon which the fire is built. The rear portion extending into the fire opening is known as the back hearth.

**Herringbone bond:** Bricks laid in an angular or zigzag fashion resembling the bone structure of a herring.

**High bond mortar:** Mortar that develops higher bond strengths with masonry units than normally developed with conventional mortar.

**High-lift grouting:** The technique of grouting masonry in lifts up to 12'.

**Hollow masonry unit:** One whose net cross-sectional area in any plane parallel to the bearing surface is less than 75% of the gross.

**Hollow wall:** A wall built of masonry units arranged to provide an air space within the wall. The separated facing and backing are bonded together with masonry units.

**Hydrated lime:** Quicklime treated with sufficient water to convert the oxides to hydroxides. Hydrated lime is the usual material used to add lime to mortar.

# I

**Incise:** To cut inwardly or engrave, as in an inscription.

**Initial rate of absorption:** The weight of water absorbed expressed in grams per 30 sq. in. of contact surface when a brick is partially immersed for one minute. Also called *suction*.

**Initial set:** The first setting action of mortar; the beginning of the set.

**Interlocking:** The binding of particles, one with another.

**Intrados:** The concave curve that bounds the lower side of the arch. See *soffit*. The distinction between soffit and intrados is that the intrados is linear, while the soffit is a surface.

**IRA:** See *Initial Rate of Absorption*.

# J

**Jack arch:** One having horizontal or nearly horizontal upper and lower surfaces. Also called *flat* or *straight arch*.

**Jamb:** The side of an opening, such as a window or door.

**Jamb block:** A concrete block especially formed with a slot for holding the jambs of window or door frames.

**Joint:** The narrow space between adjacent stones, bricks, or other building blocks, usually filled with mortar.

**Jointer:** A tool used for smoothing or indenting the surface of a mortar joint.

**Jointing:** The process of facing or tooling the mortar joints.

**Joists, combination tile and concrete:** A floor or roof system consisting of reinforced concrete and structural clay tile.

**Jumbo brick:** A generic term indicating a brick larger in size than the standard. Some producers use this term to describe oversize brick of specific dimensions manufactured by them.

# K

**Key:** The relative position of the headers of various courses with reference to a vertical line.

**Keystone:** The center masonry unit of the arch.

**Kiln:** A furnace oven or heated enclosure used for burning or firing brick or other clay material.

**Kiln run:** Brick or tile from one kiln that have not been sorted or graded for size or color variation.

**King closer:** A brick cut diagonally to have one 2-in. end and one full-width end.

# L

**Laitance:** An accumulation of fine particles on the surface of fresh concrete due to an upward movement of water.

**Lap:** The distance one brick extends over another.

**Lateral support:** The means whereby walls are braced either vertically or horizontally by columns, pilasters, crosswalls, beams, floors, roofs, etc.

**Lateral thrust:** The pressure of a load which extends to the sides.

**Laying overhand:** Building the further face of a wall from a scaffold on the other side of the wall.

**Laying to bond:** Laying the brick of the entire course without a cut brick.

**Lead:** The section of a wall built up and racked back on successive courses. A line is attached to leads as a guide for constructing a wall between them.

**Lean mortar:** Mortar that is deficient in cementitious components. It is usually harsh and difficult to spread.

**Light hard:** A term applied to red brick that are not the hardest in the kiln. Although suitable for carrying moderate loads, they are not able to withstand alternate freezing and thawing as well as the hard brick.

**Lightweight aggregate:** Aggregate made up of granular, puffy materials such as cinders, pumice, or expanded shale.

**Lime:** The base of mortar, the result of limestone burned in a kiln until the carbon dioxide has been driven off.

**Lime putty:** Slaked lime in a soft puttylike condition before sand or cement is added.

**Lime, hydrated:** See *Hydrated lime*.

**Line:** The string stretched taut from lead to lead as a guide for laying the top edge of a brick course.

**Line pin:** A metal pin used to attach line used for alignment of masonry units.

**Linear foot:** A foot measurement along a straight line.

**Lintel:** A beam placed over an opening in a wall.

**Lintel block:** U-shaped or W-shaped concrete block used in construction of horizontal bond beams and lintels.

**Load-bearing tile:** Tile for use in masonry walls carrying superimposed loads.

**Loadbearing wall:** A wall that supports any vertical load in addition to its own weight.

**Lock:** Any special device or method of construction used to secure a bond in masonry.

**Low-lift grouting:** The technique of grouting as the wall is constructed.

# M

**Major arch:** Arch with spans greater than 6' and equivalent to uniform loads greater than 1000 lb. per ft. Typically a Tudor arch, semicircular arch, Gothic arch, or parabolic arch. Has rise to span ratio greater than 0.15.

**Mantel:** A shelf projecting beyond the chimney breast above the fireplace opening.

**Mason:** A worker skilled in laying brick, block, or stone; such as a brickmason, blockmason, or stonemason.

**Masonry:** 1. Brick, tile, stone, etc., or combination thereof, bonded with mortar. 2. That branch of construction dealing with plaster, concrete construction, and the laying up of stone, brick, tile, and other such units with mortar.

**Masonry cement:** A mill-mixed mortar to which sand and water must be added.

**Masonry unit:** Natural or manufactured building units of burned clay, stone, glass, and gypsum.

**Minor arch:** Arch with maximum span of 6' and loads not exceeding 1000 lb. per ft. Typically Jack arch, segmental arch, or multicentered arch. Has rise to span ratio less than or equal to 0.15.

**Modular masonry unit:** One whose nominal dimensions are based on the 4' module.

**Mortar:** A plastic mixture of cementitious materials, fine aggregate, and water.

**Mortarboard:** A board about 3' square laid on the scaffold to receive the mortar ready for the use of a bricklayer.

**Mortarbox:** The box in which the mortar is mixed and softened by water for use.

# N

**Natural bed:** The surface of a stone parallel to the stratification.

**Nominal dimension:** The dimension equal to the actual masonry dimension plus the thickness of one mortar joint.

**Noncombustible:** Any material that will neither ignite, nor actively support combustion in air at a temperature of 1200°F when exposed to fire.

**Nonloadbearing tile:** Tile designed for use in masonry walls carrying no superimposed loads.

**Nonloadbearing wall:** A wall that supports no vertical load other than its own weight.

**Norman brick:** A brick whose nominal dimensions are 2 2/3" by 4" by 12".

## O

**Offset:** A course that sets in from the course directly under it. Also called *setoff, setback,* etc.; the opposite of corbel.

**Open end block:** A concrete block with an end web removed for placing the block around vertical steel reinforcement.

**Outrigger:** A joist projecting out of a window to support an outside scaffold.

**Overhand work:** An entire wall built with a staging located on only one side of the wall.

**Overhang:** A face of the wall leaning from the vertical away from the wall.

## P

**Panel wall:** A nonloadbearing wall in skeleton form construction, wholly supported at each story.

**Parapet:** A wall or barrier on the edge of an elevated structure for the purpose of protection or ornament.

**Parapet wall:** That part of any wall entirely above the roof line.

**Parging:** The process of applying a coat of cement mortar on masonry.

**Partition:** An interior wall, one story or less in height.

**Partition tile:** Tile for use in interior partitions.

**Party wall:** A wall used for joint service by adjoining buildings.

**Paving:** Regularly placed stones or bricks forming a floor.

**Paving brick:** Vitrified brick especially suitable for use in pavements where resistance to abrasion is important.

**Peach basket:** A template against which the entire head of a tall chimney is built.

**Perforated wall:** One that contains a considerable number of relatively small openings. Often called *pierced wall* or *screen wall.*

**Perpend bond:** Signifies that a header extends through the whole thickness of the wall.

**Pick and dip:** A method of laying brick whereby the bricklayer simultaneously picks up a brick with one hand and, with the other hand, enough mortar on a trowel to lay the brick. Sometimes called the *Eastern or New England method.*

**Pier:** An isolated column of masonry. Also, a block of brickwork usually between two openings which is built to support arches, or to carry beams or girders.

**Pilaster:** **1.** A wall portion projecting from either or both wall faces and serving as a vertical column and/or beam. **2.** A pillar of brickwork, rectangular in form, used as a supplement to a pier, usually projecting one-third of the thickness of the wall.

**Pitch:** To square a stone.

**Pitch stone:** Stone having the arris clearly define by a line beyond which the rock is cut away by the pitching chisel, so as to make approximately true edges.

**Plastic:** In the form of a sticky paste.

**Plinth:** The square block at the base of a column or pedestal. In a wall, the term plinth is applied to the projecting base or water table, applied to the projecting base or water table, generally at the level of the first floor.

**Plumb bob:** The lead weight to make taut the plumb line.

**Plumb rule:** A tool (mason's level) used to aid in building surface in a vertical plane.

**Plyform:** A special plywood used to make forms for concrete.

**Pointing:** Troweling mortar into a joint after the masonry units are laid.

**Pointing trowel:** A small tool used for filling joints on the exposed surface of the wall.

**Portland cement:** A type of cement manufactured by combining, burning, and finely grinding a combination of lime, silica, alumina, and iron oxide. It is capable of hardening through a chemical reaction when mixed with water.

**Prefabricated brick masonry:** Masonry construction fabricated in a location other than its final in service location in the structure. Also known as *preassembled, panelized,* and *sectionalized brick masonry.*

**Pressed bricks:** Those that are pressed in the mold by mechanical power before they are burned or baked.

**Prism:** A small masonry assemblage made with masonry units and mortar. Primarily used to predict the strength of full scale masonry members.

**Pugging:** A coarse kind of mortar laid on the boarding between floor joists to prevent the passage of sound; also called *deafening.*

**Putlog:** The cross support of the scaffold that holds the scaffold planks or platform.

## Q

**Queen closer:** A cut brick having a nominal 2" horizontal face dimension.

**Quoin:** A projecting right angle masonry corner. Also, projecting courses of brick at the corners of building as ornamental features.

## R

**Racking:** Laying the lead or end of the wall with a series of steps so that when work is resumed, the bond can be easily continued. More convenient and structurally better than toothing.

**Raggle:** A groove in a joint or special unit to receive roofing or flashing.

**Rake:** The end of a wall that racks back.

**Raking bond:** Brick laid in an angular or zigzag fashion.

**Random ashlar:** Ashlar set with stones of varying length and height so that neither vertical nor horizontal joints are continuous.

**Range work:** In this construction, a course of any thickness, once started, is continued across the entire face, but all courses need not be of the same thickness.

**RBM:** Reinforced brick masonry. Also, *reinforced clay masonry.*

**Recess:** A depth of some inches in the thickness of a wall such as a niche, etc.

**Reglet:** A recess to receive and secure metal flashing.

**Reinforced concrete:** Concrete that has iron and steel rods and pieces to enable it to withstand greater stress and strain.

**Reinforced masonry:** Masonry units, reinforcing steel, grout and/or mortar combined to act together in resisting forces.

**Relieving arch:** One built over a lintel, flat arch, or smaller arch to divert loads, thus relieving the lower member from excessive loading. Also known as *discharging* or *safety arch.*

**Return:** Any surface turned back from the face of a principal surface.

**Reveal:** That portion of a jamb or recess that is visible from the face of a wall back to the frame placed between jambs.

**Riprap:** Rough stones of various sizes placed irregularly and compactly to prevent scour by water.

**Rise:** The distance at the middle of the arch between the springing line and intrados or soffit.

**Rodding concrete:** An up-and-down action with a tamping rod.

**Rolled:** A brick laid with an overhanging face.

**Roman brick:** Brick whose nominal dimensions are 2" by 4" by 12".

**Rowlock:** A brick laid on its face edge so that the normal bedding area is visible in the wall face. Frequently spelled rolok.

**Rowlock course:** Bricks set on edge.

**Rubble:** Field stone or rough stone as it comes from the quarry.

**Running bond:** Same as *stretcher bond.*

## S

**Sag:** A depression in a horizontal line, meaning that there is a slight fall below the level. Usually refers to the bricklayer's line which in a long distance, will fall below the level because of its own weight, no matter how tightly it is stretched.

**Salmon brick:** Relatively soft, under-burned brick, or named because of color. Sometimes called *chuff* or *place brick.*

**Salt glaze:** A gloss finish obtained by thermochemical reaction between silicates of clay and vapors of salt or chemicals.

**Sand:** Small grain of mineral, largely quartz, which is the result of disintegration of rock.

**Scaffold height:** The height of the unfinished wall which requires another raising of the scaffold to continue the building.

**Scale box:** A derrick box made with an open top and one open end.

**Scant:** A slight slope inwards from the plumb line.

**SCR acoustile:** A side-construction two-celled facing tile, having a perforated face backed with glass wool for acoustical purposes.

**SCR brick:** Brick with a nominal dimension of 2 2/3" by 6" by 12".

**SCR building panel:** Prefabricated, structural ceramic panels, approximately 2 1/2 in. thick.

**SCR insulated cavity wall:** Any cavity wall containing insulation that meets rigid criteria established by the *Structural Clay Products Institute, BIA.*

**SCR masonry process:** A construction aid providing greater efficiency, better workmanship, and increased production in masonry construction. It utilizes story poles, marked lines, and adjustable scaffolding.

**SCR:** Structural Clay Research (trademark of the *Structural Clay Products Institute, BIA*).

**Screeding (strike-off):** The operation of leveling a concrete surface. Performed by moving a straightedge across the top of the side forms.

**Scutch:** A tool resembling a pick on a small scale with flat cutting edges, for trimming bricks for particular uses.

**Segmental arch:** An arch whose intrados and extrados make the line of a half circle.

**Segreation:** The tendency of coarse aggregate (stone) to separate from the mortar (cement paste and sand) as concrete is placed.

**Selects:** The bricks accepted as the best after culling.

**Set:** A change from a plastic to a hard state, also a name given for the chisel used for cutting bricks, also called a *bolster*.

**Set-in:** The amount that the lower edge of a brick on the face tier is back from the line of the top edge of the brick directly below it.

**Sewer brick:** Low absorption, abrasive-resistant brick intended for use in drainage structures.

**Shale:** Clay which has been subjected to high pressures until it has hardened.

**Shank:** That part of the trowel between the blade and the handle or hold.

**Shear wall:** A wall that resists horizontal forces applied in the plane of the wall.

**Shot sawed:** Description of a finish obtained by using steel shot in the gang sawing process to produce random markings for a rough surface texture.

**Shoved joints:** Vertical joints filled by shoving a brick against the next brick when it is being laid in a bed of
mortar.

**Side-construction tile:** Tile intended for placement with axes of cells horizontal.

**Silica:** A mineral contained in the clay used for brickmaking.

**Sill block:** A solid concrete masonry unit used for sills of openings.

**Sill high:** The height for the window sill upon which the window frame rests.

**Single-wythe wall:** A wall containing only one masonry unit in wall thickness.

**Skewback:** The inclined surface on which the arch joins the supporting wall. For jack arches the skewback is indicated by a horizontal dimension.

**Slenderness ratio:** Ratio of the effective height of a member to its effective thickness.

**Slump mold or cone:** A standard metal mold in the form of a truncated cone with a base diameter of 8", a top diameter of 4", and a depth of 12", used to fabricate a concrete specimen for the slump test.

**Slump of concrete:** A measure of consistency or fluidity of concrete equal to the number of inches of subsidence of a truncated cone of concrete released immediately after molding in a standard slump cone.

**Slushed joints:** Vertical joints filled, after units are laid, by "throwing" mortar in with the edge of a trowel. (Generally, not recommended.)

**Smoke chamber:** The space in a fireplace immediately above the throat where the smoke gathers before passing into the flue and nar-

rowed by corbeling to the size of the flue lining above.

**Soap:** A brick or tile of normal face dimensions, having a nominal 2" thickness.

**Soffit:** The underside of a beam, lintel, or arch.

**Soft-burned:** Clay products that have been fired at low temperature ranges, producing relatively high-absorption and low-compressive strengths.

**Soft-mud brick:** Brick produced by molding relatively wet clay (20% to 30% moisture). Often a hand process. When insides of molds are sanded to prevent sticking of clay, the product is sand-struck brick. When molds are wetted to prevent sticking, the product is water-struck brick.

**Solar screen:** A perforated wall used as a sun-shade.

**Solar screen tile:** Tile manufactured for masonry screen construction.

**Soldier:** A stretcher set on end with face showing on the wall surface.

**Solid masonry unit:** One whose net cross-sectional area in every plane parallel to the bearing surface is 75% or more of the gross.

**Solid masonry wall:** A wall built of solid masonry units, laid contiguously, with joints between units completely filled with mortar.

**Spall:** A small fragment removed from the face of a masonry unit by a flow or by action of the elements.

**Span:** The distance to be covered by an arch, lintel, beam, or girder, between two abutments or supports; the width of an opening.

**Spandrel:** The triangular portion of the wall contained between the arches where a horizontal line is drawn from crown to crown.

**Spandrel wall:** That part of a curtain wall above the top of a window in one story and below the sill of the window in the story above.

**Splay:** A slope or bevel, particularly at the sides of a window or door.

**Split block:** Concrete masonry units with one or more faces that have a rough surface from being split during manufacture.

**Springer:** The stone from which an arch springs. In some cases this is a capital or import; in other cases the moldings continue down the pier. The lowest stone of a gable is sometimes called a springer.

**Springing course:** The course from which an arch springs.

**Springing line:** The upper and inner edge of the line of skewbacks on an abutment.

**Stack:** Any structure or part thereof which contains a flue or flues for the discharge of gases.

**Stacked ashlar:** Ashlar set to form continuous vertical joints.

**Stiff-mud brick:** Brick produced by extruding a stiff but plastic clay (12% to 15% moisture) through a die.

**Story high:** The height for the floor joists.

**Story pole:** A marked pole for measuring masonry coursing during construction.

**Straightedge:** A board having an edge trued and straight, used for leveling and plumbing.

**Stretcher:** A masonry unit laid with its greatest dimension horizontal and its face parallel to the wall face.

**Stretcher bond:** The arrangement of bricks in courses consisting entirely of stretchers.

**Strike-off:** 1. Action of removing concrete in excess of that which is required to fill the form

evenly. 2. The name applied to the timber straightedge used to level the concrete with form edges.

**Stringing mortar:** The name of a method where a bricklayer picks up mortar for a large number of bricks and spreads it before laying the bricks.

**Struck joint:** Any mortar joint that has been finished with a trowel.

**Structural steel:** Steel beams, girders, and columns used for building purposes, particularly for high buildings such as skyscrapers.

**Superstructure:** That part of a building which is above ground.

# T ▬▬▬▬▬▬▬

**Temper:** To mix so that the mortar is in the proper condition for use.

**Template:** Any form or pattern, such as centering, over which brickwork can be formed.

**Tender:** A laborer who helps masons. A general name covering hod and pack carriers and wheelbarrow handlers.

**Termite shield:** A barrier, usually of sheet metal, placed in masonry work to prevent the passage of termites.

**Three-quarter:** A brick with one end cut off, usually measures about 6 in. in length.

**Throat:** An opening at the top of a fireplace through which the smoke passes to the smoke chamber and chimney.

**Through bonds:** Bonds that extend clear across from face to back.

**Tie:** Any unit of material which connects masonry to masonry or other materials.

**Tier:** One of the 4" or one-brick layers in the thickness of a wall.

**Tile, structural clay:** Hollow masonry building units composed of burned clay, shale, fire clay, or mixtures thereof.

**Tooling:** Compressing and shaping the face of a mortar joint with a special tool other than a trowel.

**Toothing:** Constructing the temporary end of a wall with the end stretcher of every alternate course projecting. Projecting units are toothers.

**Traditional masonry:** Masonry in which design is based on empirical rules which control minimum thickness, lateral support requirements, and height without a structural analysis.

**Trig:** The bricks laid in the middle of a wall between the two main leads to overcome the sag in the line, and also to keep the center plumb in case there is a wind bearing upon the line.

**Trim:** Stone used as sills, copings, enframements, etc., with the facing of another material.

**Trimmer arch:** An arch, usually a low rise arch of brick, used for supporting a fireplace hearth.

**Trowel:** A flat, broad-bladed steel hand tool used in the final stages of finishing operations to impart a relatively smooth surface to concrete slabs or other unformed concrete surfaces.

**Tuck pointing:** The filling in with fresh mortar of cut out or defective mortar joints in masonry.

**Two-inch piece:** A closer about one-quarter of a brick in length used to start the bond from the corner.

# V

**Veneer:** A single wythe of masonry for facing purposes, not structurally bonded.

**Veneered wall:** A wall having a masonry facing that is attached to the backing but not bonded to exert common action under load.

**Vertical joint:** See *head joint*.

**Virtual eccentricity:** The eccentricity of a resultant axial load required to produce axial and bending stresses equivalent to those produced by applied axial loads and moments. It is normally found by dividing the moment at a section by the summation of axial loads occurring at that section.

**Vitrification:** The condition resulting when kiln temperatures are sufficient to fuse grains and close pores of a clay product, making the mass impervious.

**Voussoir:** One of the wedge-shaped masonry units that forms an arch ring.

# W

**Wall:** A vertical platelike member, enclosing or dividing spaces and often used structurally.

**Wall plate:** A horizontal member anchored to a masonry wall to which other structural elements can be attached. Also called *head plate*.

**Wall tie, cavity:** A rigid, corrosion-resistant metal tie which bonds two wythes of a cavity wall. It is usually steel, 2/16" in diameter, and formed in a *Z* shape or a rectangle.

**Wall tie, veneer:** A strip or piece of metal used to tie a facing veneer to the backing.

**Wall tie:** A bonder or metal piece that connects wythes of masonry to each other or to other materials.

**Walls, bearing:** A wall supporting a vertical load in addition to its own weight.

**Walls, cavity:** A wall in which the inner and outer wythes are separated by an air space, but tied together with metal ties.

**Washing down:** Cleaning the surface of the brick wall with a mild solution of muriatic acid after it is completed and pointed.

**Water retentivity:** That property of a mortar that prevents the rapid loss of water to masonry units of high suction. It prevents bleeding or water gain when mortar is in contact with relatively impervious units.

**Water table:** A projection of lower masonry on the outside of the wall slightly above the ground. Often a damp course is placed at the level of the water table to prevent upward penetration of ground water.

**Waterproofing:** Prevention of moisture flow through masonry due to water pressure.

**Weathering:** The process of decay brought about by the effect of weather conditions.

**Web:** The cross wall connecting the face shells of a hollow concrete masonry unit.

**Weep holes:** Openings in mortar joints of facing material at the level of flashing, to let moisture escape.

**Wire cut brick:** A brick having its surfaces formed by wires cutting the clay before it is baked.

**With inspection:** Masonry designed with the higher stresses allowed under EBM. Requires the establishing of procedures on the job to control mortar mix, workmanship, and protection of masonry materials.

**Without inspection:** Masonry designed with the reduced stresses allowed under EBM.

**Workability:** That property of fresh concrete or mortar that determines the ease with which it can be mixed, placed, and finished.

**Working drawings:** Drawings that show sufficient detailed information including sizes and shapes from which sufficient interpretation can be obtained to properly build the object described by the drawings.

**Wythe:** **1.** Each continuous vertical section of masonry one unit in thickness. **2.** The thickness of masonry separating flues in a chimney. Also called *withe* or *tier*.

# Index